# *Caribbean* Tsunami Hazard

# *Caribbean* Tsunami Hazard

NSF CARIBBEAN TSUNAMI WORKSHOP
MARCH 30 - 31, 2004 SAN JUAN BEACH HOTEL, PUERTO RICO

Editors

**Aurelio Mercado-Irizarry**
University of Puerto Rico, Puerto Rico

**Philip Liu**
Cornell University, USA

**World Scientific**

EW JERSEY · LONDON · SINGAPORE · BEIJING · SHANGHAI · HONG KONG · TAIPEI · CHENNAI

*Published by*

World Scientific Publishing Co. Pte. Ltd.
5 Toh Tuck Link, Singapore 596224
*USA office:* 27 Warren Street, Suite 401-402, Hackensack, NJ 07601
*UK office:* 57 Shelton Street, Covent Garden, London WC2H 9HE

**British Library Cataloguing-in-Publication Data**
A catalogue record for this book is available from the British Library.

**CARIBBEAN TSUNAMI HAZARD**
**Proceedings of the NSF Caribbean Tsunami Workshop**



ISBN-13 978-981-256-535-8
ISBN-10 981-256-535-3

**We would like to dedicate this workshop and its proceedings to**

**Mr. James (Jim) Lander**

**for his contribution to raising tsunami awareness in the Caribbean**

# Sponsors

RECINTO UNIVERSITARIO DE MAYAGUEZ
C.A.A.M.
UNIVERSIDAD DE PUERTO RICO

NSF

AGENCIA ESTATAL PARA EL MANEJO DE EMERGENCIAS
ADMINISTRACION DE DESASTRES
PUERTO RICO

APOYO • PROTECCION • ACCION

Marine Sciences, UPRM

Sea Grant
Universidad de Puerto Rico

## Preface

Because of its climate and natural beauty, the Caribbean region coastal population is growing at a very fast pace. Everyone who lives in this region is fully aware of the hurricane threat and is, to a great extent, prepared for dealing with the hurricane hazard. On the contrary, most of the people living in this region are not informed of what has been accurately described as "the forgotten hazard", which is the tsunami. According to Lander *et al.* (2002)[1], since 1498 there have been twenty-seven verified true tsunamis and an additional nine as "very likely true tsunamis". In the past 136 years there have been three destructive tsunamis in the northeastern Caribbean, the last one being in 1946. In the Caribbean Sea region the potential sources for tsunami generation include earthquakes, sub-aerial and submarine landslides and underwater volcanic explosions. And, as described in these proceedings, it seems as if the tsunami risk is underrated if based on just these historical events.

This workshop follows a very successful 1997 Caribbean Tsunami Workshop, which was sponsored by the University of Puerto Rico Sea Grant Program. The 1997 workshop led to an increase in awareness and research funding for this latent hazard, but mainly in the island of Puerto Rico and, mostly, for earthquake generated tsunamis. Seven years later, it is the main objective of this workshop to inform the international tsunami community of this hazard in the Caribbean Sea and the diversity of tsunami generating sources in the region. At the same time, we hope this workshop will serve as a motivation for the tsunami modeling community to apply their skills and knowledge in addressing the multiple regional tsunami threats which, in combination with the huge increase in coastal development, present a very worrisome scenario. We aim to bring into the limelight these threats so that more islands in the Caribbean will become involved in tsunami hazard and mitigation. The fact that the last destructive tsunami occurred in 1946 is of serious concern for the regional emergency response authorities because of the complacency that this lack of frequent occurrences might have created. Finally, it is another goal to continue discussions for the establishment of a Regional Tsunami Warning Center in the Caribbean.

To achieve the objectives, the workshop was planned as follows:

- Presentations of the nature of the tsunami generation sources in the Caribbean region.
- Presentation of paleotsunami studies showing the geophysical evidence of past tsunamis, earthquakes, landslides, and volcanic explosions.

---

[1] LANDER, J. F., WHITESIDE, L. S. and LOCKRIDGE, P. A., 2002. A brief history of tsunamis in the Caribbean. *Science of Tsunami Hazards*, 20(2), 57-94.

- Presentations of past and on-going efforts in addressing the Caribbean tsunami hazard from the different potential sources.
- Presentations of recent Caribbean-related programs and research capable of addressing the potential impacts of the tsunami hazard in the Caribbean region.
- Presentations on national and international tsunami efforts and their potential role in the Caribbean
- Discussions on future action items. This might include future modeling efforts, data requirements (topography, bathymetry), the establishment of a warehouse for Caribbean geophysical data, and potential funding sources.

The activity started with Maul's presentation of the need for a Caribbean tsunami warning system, and his proposal has been presented to the Intergovernmental Oceanographic Commission of UNESCO. Then it was time to present where the hazard lies, by reviews of the potential tsunami causes in the region. McCann presented an overview of the earthquake and submarine landslide-induced tsunami threat and the potential location of their sources. The submarine landslide threat was also addressed by ten Brink *et al.*, while Pararas-Carayannis presented a very comprehensive discussion of the tsunami threat bought about by the active volcanoes present at the Lesser Antilles islands, due to both subaerial and submarine landslides. Sigurdsson *et al.* discussed recent observations of the Kick'em Jenny submarine volcano, which typically has been associated with the possibility of an underwater explosion-induced tsunami, but the observations also add the possibility of the future threat of a tsunami by flank collapse as the volcano continues to grow. Related with Kick'em Jenny, Gisler *et al.* presented their results of a sophisticated numerical 3-D compressible hydro code (used for asteroid impacts simulations), confirming that an explosion from the volcano would present only a local hazard, not a regional one, as has been the concern for many years. Yalciner *et al.*, compared the tsunami history of the Caribbean and Mediterranean Seas regions.

Moya/Mercado and Kelletat/Scheffers presented their results of paleotsunami studies in the region. The first study was concentrated on the island of Puerto Rico, where sedimentary layers were associated with the 1918 tsunami and two pre-historical ones. The second study was carried out at the islands of Barbados, St. Martin and Anguilla and concluded "that the real tsunami risk in the Intra Americas Seas for the population and in particular for all infrastructure and investments in the local economics is certainly higher than hitherto noticed." Related with the issue of tsunami deposits, Yeh made a presentation of a model used to predict wave forces in the runup zone.

Mercado *et al.*, von Hillebrandt/Huérfano and Mendoza/Huérfano talked about their work in the Puerto Rico Tsunami Warning and Mitigation Program (http://poseidon.uprm.edu), involving flood mapping, education and outreach, waveform analysis procedures for the rapid identification of shallow earthquake source parameters for both local and regional events, and the development of a tsunami warning system for Puerto Rico and the US Virgin Islands. Zahibo *et al.* reviewed the 1867 US Virgin Islands tsunami, including its simulation, while Watlington discussed the implications for these islands of a repetition of a similar event nowadays.

Kong and McCreery discussed the role that both the International Tsunami Information Center and the Pacific Tsunami Warning Center could play in mitigating the tsunami hazard in the Caribbean region. Pancake talked about the National Earthquake Engineering Simulation, an NSF-sponsored program that seeks to revolutionize the way in which earthquake engineering research is carried out through shared facilities supporting collaborative research at unprecedented scales. This includes the new tsunami basin located at the Oregon State University.

The rest of the presentations involved discussions of the results of recent, more basic, tsunami-related research. Fritz, Lynett/Liu and Liu presented results related to laboratory and numerical simulations of landslide tsunamis. Okal/Synolakis described attempts to identify critical properties of tsunamis generated by both earthquake dislocations and landslides. Finally, Borrero/Synolakis described the tsunami flood mapping effort in California.

This workshop was sponsored by the US National Science Foundation under Grant No. CMS-0331704. The co-sponsors were the Department of Marine Sciences at the University of Puerto Rico at Mayagüez, and the Sea Grant Program at the University of Puerto Rico. We would like to give special thanks to the late Dr. Clifford Astill, Program Manager, Geotechnical and Geohazards Systems Program, who enthusiastically supported the workshop. We kindly acknowledge the help of Ms. Maritza Pagan, Ms. Nereida Santiago, and Mr. Harry Justiniano. Support for the printing of these proceedings came from the Puerto Rico State Emergency Management Agency and the Faculty of Arts and Sciences at the Mayagüez Campus of the University of Puerto Rico.

Aurelio Mercado
Philip L.-F. Liu

# Contents



## Articles

### Need for a Caribbean Tsunami Warning System



### Overview of Potential Tsunami Sources in the Caribbean Sea Region

**The Tsunami Evidence in the Caribbean: Paleotsunami Studies**



**Recent Caribbean Programs and Studies**

**Speakers**

| | |
|---|---|
| Borrero, José | Department of Civil Engineering, University of Southern California |
| Fritz, Hermann | Department of Civil Engineering, Georgia Institute of Technology |
| Gisler, Galen | Los Alamos National Laboratory, NM |
| Huerfano, Victor | Puerto Rico Seismic Network, Department of Geology, University of Puerto Rico |
| Kelletat, Dieter | Department of Geography, Universitat Duisburg-Essen, Germany |
| Kong, Laura | Director, International Tsunami Information Center, Hawaii |
| Liu, Philip L.-.F. | Department of Civil and Environmental Engineering, Cornell University, NY) |
| Lynett, Patrick | Coastal and Ocean Engineering Division, Department of Civil, Texas A&M University, TX |
| Maul, George | Departtment of Marine Environmental Systems, Florida Institute of Technology |
| McCann, William | Earth Scientific Consultants, Westminster , CO |
| McCreery, Charles | Director, Pacific Tsunami Warning Center, Hawaii |
| Mendoza, Carlos | Consultant, Golden, CO |
| Mercado, Aurelio | Departtment of Marine Sciences, University of Puerto Rico |
| Moya, Juan C. | Departtment of Geography, Southwest Texas State University, TX |
| Okal, Emile | Departtment of Geological Sciences, Northwestern University |

| | |
|---|---|
| Pancake, Cherri | School of Electrical Engineering and Computer Science Oregon State University |
| Pararas-Carayanis, George | Consultant, Hawaii |
| Sigurdsson, Haraldur | Dept. of Geological Sciences, University of Rhode Island |
| Synolakis, Costas | Department of Civil Engineering, University of Southern California) |
| ten Brink, Uri | USGS-Woods Hole, MA |
| von Hillebrandt, Christa | Director, Puerto Rico Seismic Network, Geology Department, University of Puerto Rico |
| Watlington, Roy | University of the Virgin Islands |
| Yeh, Harry | Department of Civil, Construction and Environmental Engineering, Oregon State University |
| Yalciner, Ahmet | Department of Civil Engineering, Ocean Engineering Research Center, Middle East Technical University, Turkey |
| Zahibo, Narcisse | Laboratory of Tropical and Atmospheric Physics, University of Antilles Guyane, Guadeloupe |

NSF CARIBBEAN TSUNAMI WORKSHOP
San Juan Beach Hotel
March 30-31, 2004
MAR 31 2004

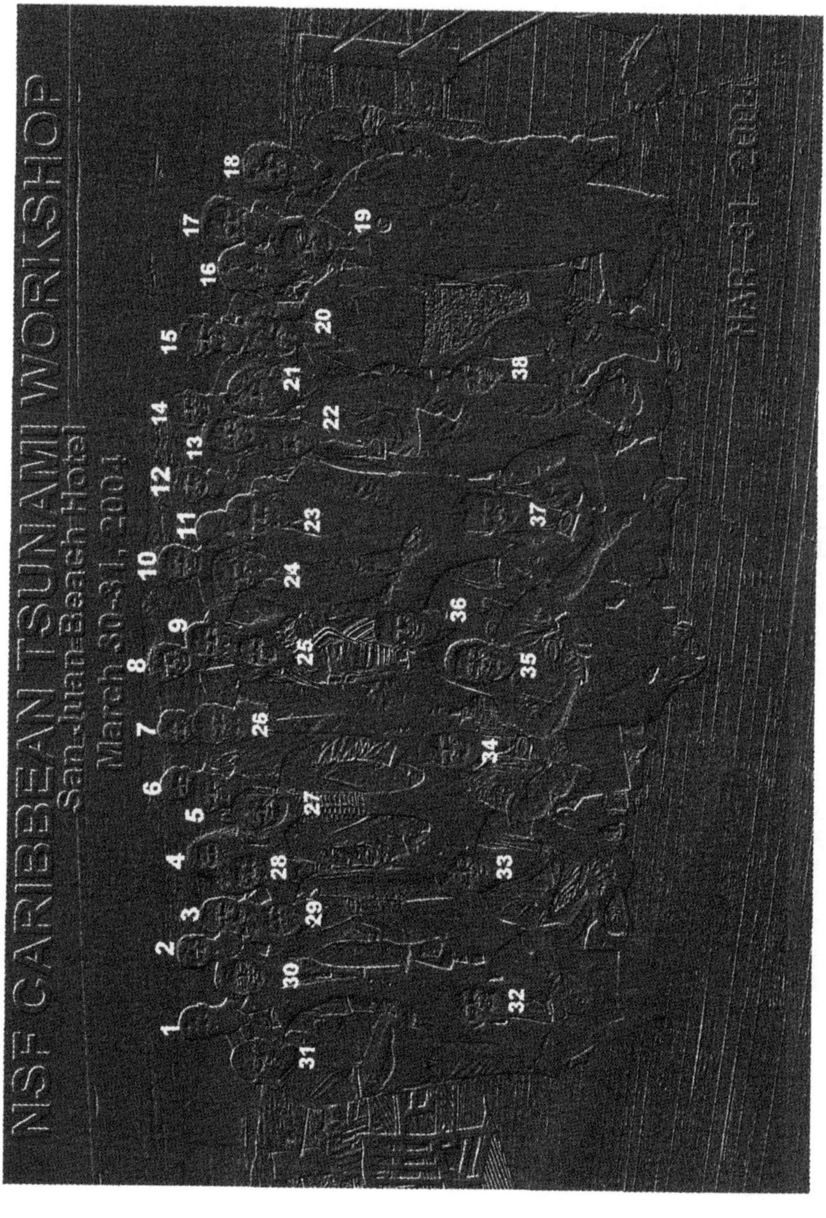
NSF CARIBBEAN TSUNAMI WORKSHOP
San Juan Beach Hotel
March 30-31, 2004
1
2
3
4
5
6
7
8
9
10
11
12
13
14
15
16
17
18
19
20
21
22
23
24
25
26
27
28
29
30
31
32
33
34
35
36
37
38
MAR 31 2004

1. Narcisse Zahibo
2. Kurt Grove – Sea Grant Program, University of Puerto Rico
3. Dieter Kelletat
4. Rachel Gross – US National Weather Service
5. Ahmet Yalciner
6. Charles McCreery
7. José Borrero
8. Costas Synolakis
9. Patrick Lynett
10. Roy Watlington
11. Emile Okal
12. George Maul
13. Galen Gisler
14. Walter Snell – US National Weather Service
15. Haraldur Sigurdsson
16. Judy Zachariasen – Private Consultant
17. Uri ten Brink
18. Martitia Tuttle – Private Consultant
19. Jesus Frau – Federal Emergency Management Agency
20. Christa von Hillebrandt
21. Carol Prentice - US Geological Survey
22. José Lebrón – Federal Emergency Management Agency
23. Alejandro de la Campa - Federal Emergency Management Agency
24. José Bravo - FEMA
25. Marie González – Federal Emergency Management Agency
26. William McCann
27. George Pararas-Carayannis
28. Harry Yeh
29. Cherri Pancake
30. Philip L.-F. Liu
31. Sonny Beauchamp – Puerto Rico Emergency Management Agency
32. Carlos Mendoza
33. Victor Huérfano
34. Juan C. Moya
35. Aurelio Mercado
36. Julio Oms – US Geological Survey
37. Herman Fritz
38. Laura Kong

# ARTICLES

# Need for a Caribbean Tsunami Warning System

# THE CASE FOR AN ATLANTIC TSUNAMI WARNING SYSTEM

GEORGE A. MAUL
*Florida Institute of Technology*
*Melbourne FL 32901*

More than 25% of Earth's tsunami events occur in the North Atlantic Ocean, notably in the Mediterranean Sea and the Caribbean Sea, but basin-wide events are also known. In the Caribbean over the last 150 years, death-related tsunami events are recorded in Venezuela (1853 and 1906), Virgin Islands (1867), Panama (1882), Puerto Rico (1918), Dominican Republic (1946), Costa Rica (1991), amongst others. The century-and-a-half Caribbean death toll is conservatively estimated as more than 2,500 persons, over four times as many as Hawaii, Alaska, and the US West Coast combined. The Subcommission for the Caribbean and Adjacent Regions of the Intergovernmental Oceanographic Commission (IOCARIBE), as a continuance to the 1993 Workshop on Small Island Developing States, began the process of formulating a Tsunami Warning System for the Intra-Americas Sea. The four elements of the TWS in order of importance are: education, warning, management, and research. With the guidance of ITSU, the Tsunami Warning System in the Pacific, an Intra-Americas Sea Tsunami Warning System proposal was approved by the Intergovernmental Oceanographic Commission General Assembly in 2002. Now to find funding and implement its elements before the next ones inundate our shores.

## Introduction

Seismic sea waves, *tsunami* in Japanese, are natural occurrences in all of Earth's oceans and marginal seas. Whenever an energy impulse – earthquake, landslide, meteor, and volcano – causes a rapid vertical water motion, a tsunami can be generated. Even strong meteorological events such as a hurricane have been implicated in causing earthquakes and indirectly a potential tsunami (O'Loughlin and Lander, 2003). With more than one-fourth of all tsunami events occurring in the North Atlantic Ocean (Bryant, 1991), a regional tsunami warning system should form an integral part of any natural-hazards network.

For the non-technical reader, for whom this article is intended, a short list of definitions follows:

*Tsunami* – a series of sea waves, typically caused by a submarine earthquake, submarine or subaerial landslide or volcano.
*Wavelength* – horizontal distance from the crest of one wave to the crest of the next wave.
*Waveheight* – vertical distance from wave crest to wave trough.
*Wave Amplitude* – one-half the waveheight measured ± about the instantaneous mean sea level.

*Wave Period* – time between the passage of one wave crest and the next wave crest.
*Run-Up* – vertical distance from the instantaneous mean sea level to maximum land elevation to which the water rises.
*Wave Celerity* – horizontal speed of advancing wave crest.
*IOC* – Intergovernmental Oceanographic Commission of UNESCO
*Intra-Americas Sea* – geographic region encompassing the oceanic and coastal zones of the Caribbean Sea, Gulf of Mexico, Bahamas, Bermuda, and Guianas.

Linear wave theory describes deep-ocean tsunami waves quite well, until they steepen and break. Tsunami waves are very long (order of L = 100 km) compared to the depth of the ocean (order of Z = 4 km), and so their celerity is given by $c = \sqrt{gZ}$, where g is gravity (9.8 $m \cdot s^{-2}$) and Z is the water depth. For an average ocean depth of Z = 4,000 m, c = 198 $m \cdot s^{-1}$ or approximately 388 knots (nautical miles per hour) – jet aircraft speed. By comparison, the winning Olympic Games 100-meter dash will be run at c = 10 $m \cdot s^{-1}$, which is the celerity of a tsunami in Z = 10 meters water depth! Hence the only safe thing that can be done is to *immediately* evacuate the coastal area and head for high ground if an earthquake is felt or a tsunami warning is issued.
Linear wave theory also gives the energy per unit meter of wave front (E) in a tsunami wave as $E = \rho g H^2 L/8$, where $\rho$ is the density of sea water (~ 1025 $kg \cdot m^{-3}$), g as before is the acceleration of gravity, H is the waveheight, and L is the wavelength. In the open sea the waveheight of a tsunami may be 1 m, and if the wavelength is L = 100 km, the energy per unit meter of the wave front is $1 \times 10^9$ joules, or about $3.3 \times 10^6$ watts; for a typical tsunami wave period of 5 minutes this is enough to power 33,000 100-watt light bulbs – per meter of wavefront! But as the $H_{deep}$ = 1-meter waveheight in deep water encounters shallow water, say $Z_{shallow}$ = 10 m, the waveheight increases according to $\sqrt{\frac{Z_{deep}}{Z_{shallow}}} = \frac{H^2_{shallow}}{H^2_{deep}}$. Again choosing $Z_{deep}$ = 4,000 m, $H_{shallow}$ = 8 m, or about the height of a three-story building. Clearly these energetic sea waves are an ocean phenomenon requiring the greatest respect if life and property are to be protected.

**Atlantic Tsunami Risks**

The sources of a tsunami in the Atlantic are similar to those in any ocean, namely: submarine earthquake, submarine landslide, submarine volcano, and meteor impact. There must be a rapid vertical displacement of water for the wave to form. If there is a subaerial event such as a landslide that slips into the sea or there is a collapse of a caldera whereby a passage to the sea is catastrophically formed, a tsunami may well be created.

In the following appendix the whole text of the Intra-Americas Sea Tsunami Warning System proposal is presented.

**References**

Bryant, E.A. *Natural Hazards*. Cambridge University Press, Cambridge, New York, Melbourne, 294 pp., 1991.

Ingham, M. *The Princess Deliverance*. Hummingbird Works, South Dennis, MA 02660, 43 pp., 2001.

O'Loughlin, K.F., and J.F. Lander. *Caribbean Tsunamis: a 500-year history from 1498-1998*. Kluwer Academic Publishers, Dordrecht, Boston, 263 pp., 2003.

Maul, G.A. On the role of IOCARIBE in a Caribbean Tsunami System: Science, Engineering, Management, and Education. *Marine Geodesy*, 22(1), pp: 53-62, 1999.

Maul, G.A. Atlantic Tsunami Warnings Are Needed. Guest Editorial, *Sea Technology*, 44(3), pg: 7, 2003.

Watlington, R.A., and S.H. Lincoln. *Disaster and Disruption in 1867: Hurricane, Earthquake, and Tsunami in the Danish West Indies*. University of the Virgin Islands, St. Thomas, USVI 00803, 134 pp., 2001.

UNESCO

# AN INTRA-AMERICAS SEA TSUNAMI WARNING SYSTEM PROJECT PROPOSAL

Intergovernmental Oceanographic Commission

# LIST OF CONTENTS

**Page**

## EXECUTIVE SUMMARY

Tsunamis are among the world's most destructive coastal hazards. They occur in all the world's oceans, inland seas, and many other large bodies of waters. Scientists have provided a record of past tsunamis that clearly indicates the potential for future tsunami damage within the Intra-Americas Sea and they have warned governments to take urgent action. The sharp increase in coastal population density, the intense development of harbours and urban infrastructure, and the exploitation of mineral resources in coastal areas, all set up a potential disaster of catastrophic proportions.

This proposal is for a 3-year programme to develop a tsunami warning system for the Intra-Americas Sea with the ultimate goal to save lives and property. It is based on an end-to-end principle – from data collection to the provision of services and issuing warnings. It includes efforts in developing data collection sites, communication arrangements, development of tsunami warning centres (national as well as regional), distribution of information, and raising the level of tsunami education and awareness. The proposal considers the ways of contributing to earthquake and tsunami research, and to capacity building and human resources development through creating educational programmes and organizing workshops and conferences.

The proposal is based on the experience gained by the IOC Member States of the Pacific in operating the Tsunami Warning System in the Pacific and the collective wisdom of scientists, engineers, managers and citizens. It is focussed on preventive measures that is fully in line with the conclusions of the IDNDR and is a part of the International Strategy on Disaster Reduction. Procedures for strengthening the links among the national geophysical services, meteorological services, marine agencies, and counter-disaster and environmental organizations are proposed, as well as ways of establishing co-operation with regional and international organizations dealing with disaster reduction.

## INTRODUCTION

Tsunamis are among the most destructive and complex natural disasters. They have been responsible for great loss of life and extensive destruction of property throughout the world. The significance of this hazard has greatly increased in the last 20 years due to the rapid growth and development of coastal areas in the developing and developed nations of the world. This is the result of a population explosion and of technological and economic developments in the coastal zones. Thus, mitigating of the effects of tsunamis is of considerable and increasing importance to the socio-economic development of Member States.

Tsunami events have been recorded in the Intra-Americas Sea (IAS) since the 16th Century (Lander *et al.*, 1999). Evidence for significant paleotsunamis is also found in the sediments of the Netherlands Antilles at 400-500 ybp, 1500 ybp and 3500 ybp (Scheffers and Kaletat 2001). Tsunami events are both local in origin and from distant sources but occur at the rate of one or more severe occurrences per century (e.g., Venezuela, 1530; Jamaica, 1692; Martinique, 1755; St. Thomas, 1867; Puerto Rico, 1918; Dominican Republic, 1946; etc.). The great Lisbon earthquake of 1755 created a tele-tsunami with 6 and 7 meter-high waves in the Lesser Antilles; its effect on the less populated areas such as the eastern Bahamas, Florida and Bermuda is unknown. So counter to the common perception (Gonzalez, 1999), the Atlantic Ocean, as well as the Pacific Ocean, is subject to these destructive sea waves (Annex I).

Although there have been deadly tsunamis in the Intra-Americas Sea this last century (1918: 42 persons; 1946: 1,790 persons by some recent reports; Lander *et al.*, 1999), it is the event of 1867 in the US Virgin Islands (Watlington and Lincoln, 1997) that is very reminiscent of the 1998 tsunami disaster in Papua New Guinea: juxtaposed earthquake epicentre; large nearly instantaneous tsunami; travel time in minutes; dense coastal population centres; uninformed populace.

If today re-occurred the 6-meter high tsunami wave observed in 1867 entering St. Thomas' Charlotte Amalie and simultaneously the 7 to 9 meter wave entering St. Croix's Christiansted Harbour, the 10-fold increase in population density, the cruise ships, petroleum carriers, harbour infrastructure, hotels and beach goers, would all be at immediate risk. Without preparation and warning it would be a disaster of catastrophic proportions.

In addition to the harbour investments since 1867 there now are nearby power plants, petrochemical complexes, marinas, condominiums, schools and other coastal structures. If several cruise ships are in Charlotte Amalie when the 1867 event re-occurs, direct economic damage of between US$500,000,000 and US$1,000,000,000 are quite possible; indirect damages (post event fires, disease, search and rescue, debris removal, electrical and telecommunication reconstruction, chemical and fuel tank failures, hazardous material cleanup, vegetation loss, salt water intrusion and environmental stress) could significantly raise these estimates.

The Caribbean Plate boundary is marked by active sub-aerial and submarine volcanoes, steep underwater slopes and numerous earthquakes. Certain submarine earthquakes, volcanic eruptions or sub-aerial and submarine landslides can generate tsunamis. In the central Lesser Antilles there have been several major volcanic sector collapses in the last 10-20 thousand years, and these are potentially very efficient tsunami-generating events.

To minimize 'false alarms' it must be quickly determined whether a seismic event creates a wave in the juxtaposed ocean. Seismic sea wave detection therefore requires both seismic and sea-level observations, integrated into a real-time operational telecommunications network. Some local IAS tsunamis if detected by such an operational network could easily provide 15 minutes of forewarning to many coastal site residents in the Caribbean Sea, Bahamas and the eastern USA, including the Gulf of Mexico and the Straits of Florida. Fifteen minutes warning is on a par with tornado warnings and is adequate, with the proper preparation, for most of the population at risk to evacuate to safe locations.

**BACKGROUND**

The Intergovernmental Oceanographic Commission (IOC) established in 1960 is performing a vital role in ocean science, services and international affairs by providing the essential framework for co-ordination and leadership of intergovernmental co-operation in understanding, observing, predicting and ultimately protecting the world's oceans.

The implementation of the IOC programmes through the collective efforts of its Member States has always been a central theme of the IOC. It has been based on the belief that the programmes can be best handled through a regional approach.

Within the IOC are 7 regional bodies, one of which is the Sub-Commission for the Caribbean and Adjacent Regions, or IOCARIBE. The IOCARIBE is responsible for programmatic development for the Caribbean Sea and Adjacent Regions and has its office in Cartagena, Colombia.

The IOCARIBE has several programmes that it attempts to develop through co-operation with Member States including: Ocean Science in Relation to Living Resources; Ocean Science in Relation to Non-Living Resources; Ocean Processes and Climate; Training, Education and Mutual Assistance; and Regional GLOSS, a component of the Global Sea Level Observing System (GLOSS), amongst others. The issue of an IAS tsunami is a concern of the IOCARIBE Group of Experts on Ocean Processes and Climate which is oriented towards physical oceanography and marine meteorology, and which has cross-cutting interests through several of these programmes.

The IOC has gained a worldwide reputation through implementation of the Tsunami Warning System in the Pacific (ITSU). This is a long-standing and successful programme, which helps to warn of and mitigate the impacts of devastating tsunamis. In operation since 1964, the IOC Tsunami Programme has assisted in saving thousands of lives and billions of dollars in the Pacific and contributed effectively to the objectives of the International Decade for Natural Disaster Reduction (IDNDR).

Although the IOC Tsunami Programme has been centered in the Pacific for many years, there is no special regional significance to this beyond the proneness of the region to large earthquakes and hence to tsunamis. Recently, the Member States of the IAS, Indian Ocean and Mediterranean have been requesting more attention to tsunami warning in their respective areas.

In 1993, IOC with the assistance of the United Nations Environment Programme, proposed and conducted a Workshop on Small Islands Oceanography in relation to Sustainable Economic Development and Coastal Area Management in direct response to the needs of Small Island Developing States. The meeting was hosted by the Government of France in Martinique. One of the numerous issues discussed in Martinique was that of tsunami hazards. Many Small Island Developing States are located in the tropical waters of the Pacific and Indian Ocean, and are notably vulnerable to tsunami. It was noted too that the IAS and the Caribbean Sea in particular is a seismically active region and has a history of tsunami-caused by earthquakes and volcanoes.

At SC-IOCARIBE-V, the Fifth Session of the IOC Sub-Commission for the Caribbean and Adjacent Regions, the Recommendation of the IOCARIBE Group of Experts on Ocean Processes and Climate to hold a Caribbean Tsunami Workshop was adopted. The Eastern Caribbean Centre of the University of the Virgin Islands hosted the 2-day workshop in 1996 in St. John, at the request of IOCARIBE, which was attended by some 17 scientists. The attendees made it very clear (IOC, 1996) that the IAS has a record of significant tsunami-caused deaths and is at substantial risk for others (Smith and Shepherd, 1994).

The St. John scientific meeting led to the 1997 workshop at the University of Puerto Rico (Mercado, 1997). At Mayaguez, there were in attendance approximately 150 concerned citizens of the region, civil defence and government officials, scientists and tsunami warning experts, both local and from abroad. The Mayaguez meeting emphasized broader issues including education, warning, management, as well as research (Mercado and McCann, 1998; McCann, 1998; IOC Circular letter No. 1579, 1998, Gusiakov, 1999; Mofjeld *et al.*, 1999; Maul, 1999, Lander *et al.*, 1999).

All attendees at the June 1997 meeting in Puerto Rico were requested to contact their head-of-state with a statement of concern. In July 1997, USA President Clinton was formally informed of the recommendations made at the Mayaguez workshop. It focussed on 4 mitigation measures: education, warning, management and research.

In 1999, the proposal to establish an IAS Tsunami Warning System was encouraged by the officers of the International Co-ordination Group for the Tsunami Warning System in the Pacific (ICG/ITSU) during their intersessional meeting in Hawaii (IOC, 1999). Later, in April 1999, in conjunction with SC-IOCARIBE-VI, the IOCARIBE Tsunami Steering Group of Experts gathered in San Jose, Costa Rica to draft this proposal. The Chairman of the Group wrote the final proposal details with input from interested parties including the IOC secretariat, the Pacific Tsunami Warning Centre and the International Tsunami Information Centre.

In 2000, the Thirty-third Session of the IOC Executive Council recommended to bring the experience and knowledge of the Pacific and Caribbean experts in Tsunami research and mitigation together in order to finalize the Tsunami Warning System (TWS) project proposal for the IAS region. The Mayaguez meeting was organized in December 2000 in compliance with the recommendation, and the draft project proposal was formulated.

In October 2001, the Eighteenth Session of ICG/ITSU was invited to comment on the text of the project proposal and propose modifications for submission of a final version to the IOCARIBE Sub-Commission's next session. After discussions, the ICG/ITSU accepted the proposal in principal and formed a working group to address the remaining outstanding issues, take into account the comments that had been offered, and put the proposal into a final form (IOC, 2001). That work was completed and is the proposal presented here.

## JUSTIFICATION

It is clear from the above that the tsunami threat in the IAS region is real and may cause tremendous damage to life, property, and infrastructure in all countries bordering the IAS. Everything possible must be done to minimize these disastrous effects.

The design of the IAS Tsunami Warning System will be implemented based largely on existing infrastructure and resources, e.g., sea-level measuring systems such as GLOSS, RONMAC and CPACC, co-operation with the WMO and OAS is envisaged; the seismic stations operated by Member States of the region and communication systems like those of WMO(GTS) and EMWIN will be involved. In setting up the TWS, the current institutional structures will be taken into account. The system will involve many organizations with varying expertise and emphasis in different system components. The Member States of IAS bordering both the Pacific and IAS regions have already gained invaluable experience in operating and contributing to the Tsunami Warning System in the Pacific. This experience should be exploited to the largest degree possible. The challenge is to develop an integrated network based on available experience, which will accommodate development of new requirements and capabilities.

Though there is some success in the region with respect to the collection of sea level and seismic data and in communicating them operationally to users, and some experience in tsunami warning systems operations, there are big gaps in data collection and communication systems and in tsunami mitigation in the region. There is a need for considerable improvement so as to bring about a better response and minimize loss of life and property. The gaps include:

- Need for reliable coastal and deep ocean monitoring stations;
- Need to verify and update existing sea-level and seismic stations to acceptable international standards;
- Need for additional TREMORS systems;
- Need for communication and data exchange systems that meet the requirements of the TWS;
- Need for numerical modelling to estimate the hazard, arrival time of tsunami and the inundation zones of the worst cases;
- Need for a special tsunami awareness programme and appropriate training;
- Need for a nationally and regionally co-ordinated approach to the establishment of the TWS;
- Need for links with existing relevant organizations, programmes and projects in the region.

This is not an all-embracing list; it is only a first approximation. It may be extended and adapted to new demands and challenges, when the project is being implemented.

As the TWS in IAS will be developed as a system to support sustainable social and economic development, welfare, and safety, the capacity building and awareness are essential as well as better scientific understanding of the tsunami phenomenon.

The project will develop and apply state-of-the-art tsunami models for inundation map development, tsunami risk assessment and evacuation plan formation. The project will place a strong emphasis on education for appropriate public response. The timely response to tsunami warnings of the public and of the disaster prevention and preparedness organizations is extremely important to minimize the loss of life and property.

## SYSTEM COMPONENTS

To mitigate the tsunami hazard, it is critical to accurately assess the nature of the threat posed by the hazard, to design and implement a warning technique and to prepare at-risk areas for appropriate actions to reduce the impact of the hazard. That is why the Tsunami Warning System structure in the IAS region will be based on the following 3 pillars: hazard assessment, warning and preparedness.

### Hazard Assessment

To find out what type of warning system is required, a background approach is used. As the first step, an assessment of the tsunami hazard for each coastal community will be made to identify populations and assets at risk and the level of that risk. To assist in this effort a historical tsunami database for the Caribbean (Lander et al., 1999) will be improved and made available in a form more easily and rapidly accessible to users. A graphical interface for these data has already been developed by the Tsunami Laboratory of the Siberian Division of the Russian Academy of Sciences (example output is shown on their web site at http://omzg.sscc.ru/tsulab/carib.html). For communities with very limited or no past data, numerical models of tsunami inundation can provide estimates of areas that will be flooded in the event of a local or distant tsunamigenic earthquake. This information can later be used for creating tsunami evacuations maps and procedures.

The experience of the IOC ICG/ITSU in its implementation of the Historical Tsunami Database (HTDB) and Tsunami Inundation and Modeling Exchange (TIME) projects for the Pacific region will be very useful.

### Warning

To alert coastal communities that danger from a tsunami is imminent, an appropriate warning system is required. Tsunami warning systems rely on seismic data and earthquake analysis for the rapid initial warning, and on sea level data for confirming and evaluating the tsunami and for continuing, upgrading, or cancelling the warning. Warning systems also rely upon a variety of communication methods to receive seismic and sea level data and to issue messages to appropriate authorities.

Seismic Subsystem

The IAS has several seismic reporting systems in place. The Seismic Research Unit at the University of the West Indies has been in operation since 1952 and the Puerto Rico Seismic Network has operated the data centre for the Middle-America Seismograph Consortium (MIDAS) since 1998. A protocol has been established for participating seismic networks of the Caribbean, North America, Central America and South America to submit data on significant events in the region in near real time. This issue has already been widely discussed at the two Caribbean Tsunami workshops already held, St. John, USVI (IOC, 1996) and the Mayaguez UPR Workshop in 1997 (Mercado, 1997; Maul, 1999), and at the Peru 1997 meeting of the ICG/ITSU-XVI and the Hawaii 1999 meeting of the Officers of the ICG/ITSU (IOC, 1999). At several MIDAS meetings, the desirability of having seismic data exchanged in real-time has been discussed for tsunami warning applications.

Other seismic systems operated by the French in Martinique and Guadeloupe and by the Seismic Research Unit of the University of the West Indies from Saba to Trinidad have experience in detecting potential tsunamigenic events. The Eastern Caribbean has a very high density of seismographic stations -- each major island has at least one 3-component digital broadband seismograph.

The Caribbean Development Bank has funded the Seismic Research Unit to instrument Kick 'em Jenny volcano and several sites in nearby islands for detection of volcano-generated tsunamis. The Intra-Americas Seas Tsunami Warning System will capitalize on this parallel effort through co-operation and communication.

TREMORS (Tsunami Risk Evaluation through seismic MOment from a Real-time System) is considered the most effective stand-alone warning system for tsunamis generated by earthquakes. It is based on a single 3-component broadband seismometer connected to a personal computer where special software is continuously running. It automatically detects the arrival of seismic waves from any large earthquake, locates the epicentre, and computes the seismic moment. Depending on the results, the system is capable of sending a short message, including all the results, through INMARSAT. It can also send a warning to a telephone and set off an alarm tone on the personal computer. TREMORS is capable of detecting and analyzing potentially tsunamigenic earthquakes at regional distances within the IAS and far distances such as across the Atlantic.

In order to organize effective warnings there should be a minimum of at least 6 TREMORS stations as the Caribbean region is vast and the future warning system should monitor 2 distant regions (the first: the eastern and northern part of the Caribbean from Cuba to Trinidad, and the second: the western part where the Cocos plate subducts under the Caribbean plate). Three stations will be installed in each of the regions running at the same time. However, taking into account possible downtime at any of the stations, it is recommended to have 8 stations for an optimum configuration. It is proposed to have sites in the following states: Colombia, Costa Rica, Cuba, Guadeloupe, Mexico, Puerto Rico, St. Vincent and Venezuela. Ideally, it is desirable to have additional TREMORS (in addition to the proposed 8) in the Central American Zone, in the northern part of the Caribbean and the West Indies islands. In the proposed locations there already exist or are plans to install broadband seismic stations. The characteristics of the existing stations are in Annex II. The institutions responsible for these stations must be consulted to see whether TREMORS can be added on. The only additional cost for existing broadband stations will be the INMARSAT transceiver and the TREMORS software.

With the proposed set of broadband stations and TREMORS systems, large magnitude events can be quickly recorded without the signal going off scale and the seismic parameters can subsequently be determined. The goal is for this network to be capable of notification of a major earthquake in the IAS within 2 minutes of the initial rupture. This notification will be followed within 3 minutes by detailed seismic parameters that provide an understanding of the likelihood of a tsunami. The TREMORS network will also be capable of notification of a major trans-Atlantic earthquake within about 10 minutes of its occurrence with detailed seismic parameters within about 45 minutes.

The seismic warning system must be complimented by first order estimates of travel times and wave heights for all the most likely sources based on numerical simulations.

Sea-Level Subsystem

Water level gauges are an essential element of TWS. When strategically located they are used to quickly confirm the existence or non-existence of tsunami waves following an earthquake, to monitor the tsunami's progress, to help estimate the severity of the hazard and to provide a basis for declaring the hazard over. Water level gauges may also be the only way to detect tsunamis in cases where there is no seismic data or when the tsunami is not earthquake-generated.

The TWS in the IAS region needs a reliable network of operational coastal and deep ocean monitoring stations. The inventory of existing coastal stations is in Annex III.

This list shows there are probably a sufficient number of coastal stations, although verification of some of them needs be made. Several existing stations need upgrading to international standards for tsunamis as defined by ICG/ITSU. Many of the sea-level gauges are mechanical instruments without telemetry capability. The programmes of OAS, such as CPACC (Caribbean: Planning for Adaptation to Climate Change) and RONMAC (Water Level Observation Network for Central America) offer the best opportunity for establishing a regional sea-level network because each instrument transmits its data via GOES and are GPS-located. The position of the CPACC and RONMAC stations and their technical specifications are in Annex IV.

There may be a need for some new stations on certain islands (for example, Isla de Aves, San Andres, Swan Island, St. Martin, La Blanquilla). Some existing sea-level stations require installation of additional hardware in the Data Collection Platform (DCP) and the acoustic controller to be used for tsunami monitoring. Key criteria for selection of sea-level stations should be spatial location.

Warning Centre(s)

Two approaches can be considered: 1) a TWS with one or more central warning centres to receive and intrerpret the TREMORS and sea level data and generate regional or widespread warnings, and 2) a TWS where data are sent directly to each Member State of the IAS for their own local interpretation and warning. Either approach or a combination of both can be implemented. There are trade-offs in cost and capabilities regarding these options. A central centre must have at least 5 full-time professional personnel to enable one-person 24-hour operations, but it can provide expertise in the interpretation of the seismic and sea level data to help reduce false warnings and it may ultimately be able to provide wave height forecasts based on numerical simulations. The warning, however, may be somewhat slower due to this layer of interpretation. By receiving the field data directly, each Member State has the possibility to respond immediately to TREMORS transmissions and later sea level data. However, this also requires a 24-hour response capability and expertise in interpreting the data. The decision regarding the most suitable way to configure the system is left to the Member States of the IAS. However, for this proposal only a system with a central warning centre(s) will be further discussed.

Communication and Data Exchange Subsystem

Tsunami Warning Systems have unique and extensive communication requirements. Seismic and water level signals must be sent from remote sites, often without power or telephone lines, and warning messages must be transmitted quickly and reliably to users having different means of access.

In the IAS region the distances to be covered range from less than a kilometre to hundreds of kilometres.

The communication system should be as independent of the normal communication circuits as possible. This will enable users to get data in and information out regardless of what happens to the local communication infrastructure.

Data from individual TREMORS systems will be transmitted via INMARSAT C to the warning centre(s) and/or Member States using equipment purchased as a part of the seismic component. After the data are evaluated by a warning centre, the appropriate tsunami bulletin (informational, advisory, watch, or warning) is transmitted to the San Juan Weather Forecast Office (WFO) via INMARSAT C. It is then forwarded into the Emergency Managers Weather Information Network (EMWIN) for transmission to individual Member States. This will require an INMARSAT ground station at the San Juan WFO and EMWIN receivers and software at each of the 33 Member States of the IAS. Sea level data that are transmitted via the GOES satellite will be relayed from the downlink at Wallops Command and Data Acquisition facility to the NWS central communications gateway where it will be sent to the San Juan WFO or entered into EMWIN for distribution. Methods to acquire sea level data from across the Atlantic, possibly via METEOSAT, for the evaluation of teletsunamis will be explored.

The public Internet, which has greatly improved and expanded in the IAS region, should be considered as an alternate or supplemental method for communication needs in the TWS.

Dissemination of tsunami watch or tsunami warning messages to the public will be accomplished using the existing meteorological communication network and systems like GTS or the Aeronautical Fixed Telecommunications Network (AFTN). Messages can also be transmitted widely through commercial circuits, such as telex and the Internet. On local or national levels, tsunami messages can also be sent over text or voice circuits designed for national defense or other emergencies. Messages can also be sent by telephone or fax.

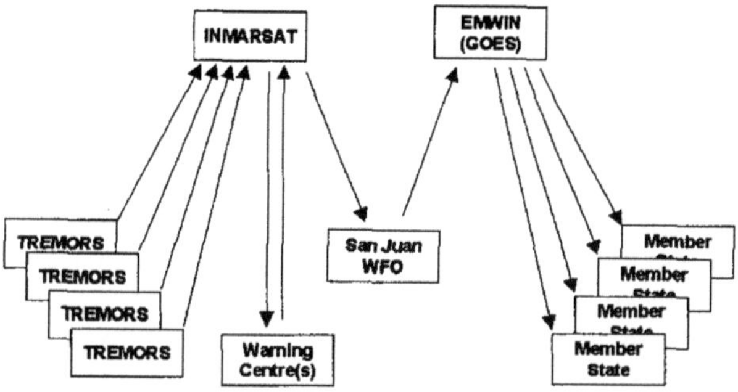


*Figure 1. The proposed flow of seismic data and tsunami bulletins. Multiple TREMORS systems report the detection of large earthquakes and then the earthquake parameters to the warning centre(s) via INMARSAT. The centre(s) subsequently transmits appropriate tsunami bulletins to the San Juan Puerto Rico Weather Forecast Office for relay into the Emergency Managers Weather Information Network (EMWIN). Member States receive the bulletins via EMWIN from the GOES satellite using an inexpensive receiver and software that can be programmed to trigger an audible alarm, commercial page, or telephone call if a tsunami warning is received.*

Within each country consideration will be given to strengthening existing communication networks and implementing new ones.

**Preparedness**

Activities in this category take place in order to achieve an appropriate level of preparedness for a warning of impending danger from the tsunami. They will include creating and disseminating knowledge about potential tsunami inundation, about the warning system, about evacuation preparation and procedures, and regarding land use planning. A community-wide effort of tsunami hazard awareness is essential to educate the residents as to the appropriate action to take in the event of a tsunami. A public information campaign will be mounted to make sure that the information gets through to the general public.

The response of the public and disaster prevention and preparedness organizations to tsunami warnings is most important to minimize loss of life and property. The current reaction by the emergency management community in the area is slow. The lack of a current quick response capability to a warning is another argument in favour of the establishment of a warning system.

One of the major efforts in operation of the warning system will be one of education not only of the public but of those agencies and people who have to respond/react to natural disasters. A concerted effort using sea-level and seismic stations and communication facilities will require additional training through the organization of workshops and training courses.

Regional workshops will be held to develop an appropriate set of pilot projects, e.g., modelling and analysis; data and information management; methods of risk assessment and tsunami warning system operations. It is recommended to use the expertise of the TEMA programme of IOC and the equivalent programmes of WMO, UNEP and other international organizations in implementing these activities.

Awareness education will include the implementation of an education programme for schools to prepare students at all age levels, the co-ordination of periodic public drills to maintain the preparedness level, the development of a search and rescue plan, and the involvement of community organizations to educate all sectors of the population at risk. Communities must be committed to a continuous long-term institutionalized education programme as tsunamis are infrequent events and succeeding generations may forget tsunami safety lessons.

The need for 3 areas of training is identified: seismic station operation (equipment and station maintenance, seismology, TREMORS software, etc.); tsunamis generated by submarine landslides and volcano eruptions (nature, evaluation of tsunami risk, organization of underwater and post-disaster surveys, etc.); and numerical modeling (travel time charts, historical tsunamis, TIME, etc.).

The IOC has long experience in assisting countries in implementing tsunami awareness and education programmes. Written educational material in English, Spanish, French and Russian, educational curriculums, videos and reports from communities with comprehensive awareness programmes are available through ITIC and the IOC Secretariat. New flyers and brochures must be developed using the ICG/ITSU experience to reflect better regional peculiarities and culture. English, French, and Spanish versions will be widely distributed through the IOC Regional Office and other appropriate channels to schools, civic organizations and religious institutions. More generic natural hazard brochures should be identified and updated to include tsunami information. Finally, a professionally produced multi-lingual video should be taped and made available for broadcast and duplication (this should be encouraged by a special copyright freedom).

It is anticipated that a general reluctance by certain business, insurance, and political groups will be encountered. Recruiting these persons and organizations in the context of thoughtful, non-hysterical, planned contact will be challenging and rewarding. It is essential not to overstate the hazard, yet to be firm and apolitical in such dealings. In this regard, the cultural context of the community must be appreciated and respected. Regular and persistent communication by mail and in person will be required to convince reluctant sectors of the community to refocus their perspectives.

## NON-EARTHQUAKE GENERATED TSUNAMIS

Tsunamis generated by submarine landslides and volcanic eruptions cannot be detected automatically at the present time. The time an eruption or landslide will occur in a specific zone generally cannot be predicted with enough accuracy and further research must be carried out. Nevertheless, it is possible to survey coastal and submarine slopes and evaluate the potential tsunami hazard from volcanoes or landslides for harbours and coastal villages situated close to the zone of generation.

Volcanoes do not generate tsunamis without a preliminary period of activity. The volcanoes themselves should be monitored in order to recognize when they might be about to generate a tsunami. The Kick 'em Jenny monitoring project funded by the Caribbean Development Bank is an example of how such monitoring should be organized and implemented.

For on-land volcanoes there will be a need to know more about past tsunamigenic events such as major collapses from the volcanoes of the central Lesser Antilles. Natural collaborators in implementing this study will be from Caribbean proper, primarily from Venezuela whose coasts and islands are exposed to tsunamis from the Lesser Antilles.

## SUMMARY OF BENEFITS AND BENEFICIARIES

Member States of the IAS region are the first countries to benefit from the project by the development of institutional, national and regional infrastructures for the tsunami warning system. In addition, European and African countries will benefit from the IAS TWS in case transatlantic tsunamis occur.

The target beneficiaries of the project will be all coastal communities of the countries concerned. The users of the tsunami information are likely to be governmental agencies, including safety, search and rescue and environmental departments, planners for coastal land use, individual enterprises, insurance companies, scientists and the public.

The benefits of the project can be summarized as: reduction in the loss of life and property; reduction in damage to infrastructure and land; increased stability of local economy and more dependable investment; improved cost-effective coastal engineering including design of local coastal defenses; increased knowledge in seismology and tsunamis; increased scientific and technical capacity and finally, strengthening of existing national and regional institutions, facilities and programmes, as well as of co-operation between Member States and governmental and non-governmental organizations with tsunami-related interests in the region.

To give a quantitative estimate of the potential benefits of the projected TWS in terms of dollars is difficult, as potential products and beneficiaries are diverse and the cost details are not easily available. In documented cases, cost/benefit ratios of weather warning services have been put in the range of 1:55 to 1:217. For the development of a regional SeaWATCH system, it is in the range of 1:20 to 1:50. We may expect that benefits provided by the products and services of TWS to different users will also be in this range. While the system, when established, will not avoid all damage and will depend on the capacity of decision makers to respond effectively in advance to the threat and then to the warning, it will have the potential to save many millions of dollars at a cost of a few million dollars.

## MANAGEMENT AND ADMINISTRATION

If the proposal is adopted by the IOCARIBE Regional Sub-Commission it will then be sent to the IOC Executive Council session of 2002 for approval. At the same time, it should also be distributed to interested regional and international in organizations for information and support. It will be accompanied by an Action Plan based on the proposal developed by the IOCARIBE Tsunami Steering Group of Experts with the assistance of a consultant(s). The proposal, if approved, will be presented to potential donors for funding. The OAS will co-operate with IOC in looking for funds, especially within the region. There will be a need to bring together all the stakeholders so as to achieve an integrated approach in designing and implementing the TWS.

Individual nations should put their faith and resources in a collective tsunami warning system. At the national level, responsibilities must be assigned to individuals to ensure that someone is looking after a particular element of the system. Each country will create a national co-ordination committee to regularly evaluate the progress of the project and recommend appropriate actions for the better development of its national system to the governmental authorities concerned. The ICG/ITSU will help the countries of the region by developing guidelines for national participation in the TWS.

The implementation of the project will be the responsibility of the national authorities as far as possible within the existing institutions along the national action plans, with clear objectives and milestones for the development of each element of the system.

There is a recognition that Small Island nations in the region may find it difficult to participate in, contribute to the TWS and make long-term commitments to the system's operations. External resources from regional/international funding agencies will be essential and sought.

The management of the project will be carried out in accordance with the agreements among participating governments and funding institutions. The implementation will be undertaken under the general direction of the participating member countries with the technical assistance of international agencies and appointed experts/consultants.

The project will have a Chief Technical Adviser(s) from inside or outside the region whose main function will be to advise national authorities concerned on technical and scientific matters related to the project, as well as on plans for the mitigation of tsunami hazards; and to provide overall supervision of the project to ensure satisfactory implementation of its various components.

If the region decides to take a centralized warning approach, then a Regional Tsunami Warning Centre for the IAS might be established, for example, at the Mayaguez Campus of the University of Puerto Rico, which has the necessary experience in tsunami research and mitigation and the required facilities. It is recommended to also create a Regional Tsunami Information Centre in the same place under the direct supervision of the Chief Technical Adviser. There will also be arrangements for regular monitoring and reviews of progress in achieving specific objectives of the project.

It is recommended that an IAS TWS Project Steering Group (PSG) be formed with the Chairman of the existing IOCARIBE Tsunami Steering Group of Experts, the Chief Technical Adviser and representatives of project funding agencies and system members. Among the responsibilities of the PSG, will be to review the Plan of Operations and Implementation schedule prepared by the Executing Agency and advise on matters related to the implementation phase.

The IOCARIBE Tsunami Steering Group of Experts, consisting of scientific leaders appointed by countries with the assistance of the ICG/ITSU and IUGG Tsunami Commission, will provide scientific backing to the project and will help build bridges between the scientific community and policy makers.

Taking into account the existing successful practice of implementation of many international projects and programmes, Working Groups may be established to deal with specific problems of the region related to the TWS project and their causes.

Once funding is obtained, the OAS or another regional organization will be contracted as the Executing Agency (EA) with the responsibility for the detailed design and implementation of the programme. The Project Co-ordinator will be nominated in consultation with the IOC, and with national and international agencies sponsoring the project.

The EA will be responsible for the technical supervision and administrative co-ordination of its implementation. It will also design and implement a public awareness campaign on the tsunami hazard and the need for tsunami preparedness. This campaign will make extensive use of already available material and will engage the national and regional disaster management offices in its implementation.

It will establish, as required, co-operative agreements or memoranda of understanding with the regional institutions and national governments or their specialized agencies. It will be accountable to the agency(ies) providing the funding for the implementation of the system and will be responsible for the timely production of the necessary technical and financial reports.

The following are among the principal responsibilities of the EA:

- Develop a detailed Plan of Operations for the installation of a fully operational IAS Tsunami Warning System (IAS-TWS);
- Contract a full-time Project Co-ordinator and provide the Co-ordinator with the necessary technical and administrative support for the duration of the project;
- Establish and maintain regular and effective communications with the IAS-TWS Steering Group, participating countries, and regional/sub-regional institutions;
- Develop and implement a public education and awareness campaign aimed at increasing the awareness of the tsunami hazard and generating the necessary support and participation in the IAS-TWS among vulnerable communities and key institutions;
- Develop and implement the necessary institutional arrangements and co-operation agreements with participating institutions and governments;
- Contract the necessary short-term technical expertise for the detailed design of the IAS-TWS system components and for their installation in the field;
- Procure all equipment and technical services necessary for the establishment and full operation of the IAS-TWS;
- Develop technical and administrative management procedures for the functioning of the IAS-TWS following its establishment;
- Produce timely progress and financial reports to the PSG and for the agency(ies) funding the implementation phase;

- Participate in negotiations with donors for funding of the continued operation of the IAS-TWS;
- Ensure the necessary co-ordination with other related programmes and projects;
- Launch the system.

The outcome to be delivered by the EA will be a fully operational IAS Tsunami Warning System co-ordinated by a Technical Office located in one of the participating countries/territories and with full participation of specialized regional/sub-regional institutions and national agencies (the system members).

**IMPLEMENTATION**

Full implementation will require 3 years. The 3-year (funded) implementation phase is as follows: Year 1 is dedicated to naming an interim Director, establishing an office and expanding the educational effort beyond that. Included in this first year will be an educational video and establishing regular annual meetings of government and non-government activists. The office location will be a political decision but one based on the 4 technical requirements (elements) of the *Tsunami Caribbean Workshop:* 1. Education. 2. Warning. 3. Management, and 4. Research. By the end of Year 2, an operational staff and office will be established. Much of the second year's efforts will have been to install the additional field instrumentation and communications links necessary for broadcast warnings. The research component will have been defined at this time and the management effort in place. Year 3 is the first full operational period and will have activities in all 4 elements. As with the other start-up years, the governmental parties will review the effort and instruct the TWS Project Co-ordinator as appropriate.

The Project Co-ordinator of the IAS-TWS will make quarterly reports to the PSG. The annual meeting of the Member States will be the normal means of communication to government and non-governmental organizations. These meetings shall be highlighted with press releases and other activities to promote public awareness.

**BUDGET**

A three-year implementation and budget is proposed based on the recurring theme from the 1997 Caribbean Tsunami Workshop: education, warning, management, and research.

The project will organize a training program on the concepts and techniques of hazard assessment and mapping, vulnerability analysis, and the assessment and mitigation for key planning and sectoral agencies, including the national disaster offices. The core elements of this program will be offered at the regional level, accompanied by specific follow-up activities at the national level, including the formulation of mitigation strategies and measures, and the linkage of these measures to existing national and sectoral development plans. Training, Education, and Mutual Assistance (TEMA) Programmes will focus on the CPACC methodology of in-country training of the observers and technicians. An annual TEMA within-region technical meeting for involved personnel and supervisors is required. Education, capacity-building, awareness, and TEMA are estimated at $192,000, $75,000 of which is directed to broad public outreach, signage, in-service teacher training, and teaching materials.

There is a requirement for reliably operating coastal real-time reporting tide gages, as well as monitoring stations using isolated small islands. From the inventory of existing coastal stations (Annex III), there seem to be sufficient coastal stations, but many are not real-time reporting systems. Several of the existing stations need upgrading to the ITSU standard for tsunamis. New island stations needed as a minimum include Isla de Aves, San Andres, Swan Island, St. Martin, and La Blanquilla. Summary of costs using a judicious mix of ITSU and CPACC standards is $340,000 (Annex IV). Upgrading existing CPACC-type stations with tsunami monitoring capability requires installation of additional hardware in the DCP equipment and the acoustic controller. The DCP for tsunami monitoring purposes requires that all sea level stations use GOES satellite links to transmit information. A backup GOES ground station is recommended, at an additional cost of $35,000. Not all existing tide stations are needed for tsunami warning, but they will greatly enhance the research component (where spatial location is a key selection criterion), and add to the other benefits stemming from modernization.

To provide emergency managers with timely earthquake magnitude, location, and tsunami probability information, TREMORS systems are recommended for the seismic analysis. This seismic component recommendation includes upgrading three existing ultra-broadband seismic stations, and the installation and start-up of five new TREMORS stations, for a total cost of $265,000. Included in this budget are the software installation, and the training necessary for competent operation. New TREMORS stations will be located at existing seismic centres (Annex II) and will include training on-site plus one in-region workshop for all participants. This seismic educational component is in addition to the activities described in the second paragraph of this section of the proposal. As with the sea-level component and the communications network, the seismic budget includes broader goals for capacity building and technology transfer.

Warning centre operations require a minimum of 5 full-time employees to have one person on duty continuously. The Puerto Rico Seismic Network, for example, would require two-person shifts for this work, with one professional analyst and one electronic technician on duty each shift. They estimate the additional cost to their current operations at $275,000 per year, with up to 10% per year additional for salary adjustments. Centre operations are not anticipated until the second and third year of the project. Thus, the total cost for warning centre operations in this proposal is estimated at $625,000.

The Unit for Sustainable Development and Environment of the Organization of American States (OAS/USDE) has expressed interest in acting as Executing Agency (EA) for the IAS Tsunami Warning System. In that capacity, the OAS/USDE would be responsible for the detailed design and implementation of the Program, expanding on their experience and success with CPACC. The outcome to be delivered by the EA is a fully operational IAS Tsunami Warning System, including components in education, warning, management, and research. The implementation will be coordinated by a technical office located in one of the participating countries, and with full participation of specialized regional/sub-regional institutions and national agencies (the program members). A three-year effort is required including procuring funding, equipment purchase and installation, training, operations start-up, and turnover of a fully operational system. The OAS/USDE estimate for the three-year EA activity is $469,000.

Research needs include improved bottom topography data; tsunami wave arrival amplitude estimation; potential for tsunamigenic volcanic eruptions; potential for landslides and submarine slumping; detailed fault structure; inundation mapping; paleo-tsunami dating and description; and improvements to the tsunami and earthquake historical databases, amongst others. Research is fundamental to the graduate education of the next generation of oceanographers, coastal engineers, seismologists, volcanologists, coastal zone managers, geochemists, psychologists, emergency managers, environmental economists, sociologists, infrastructure planners, and so forth in the region. Accordingly, a substantial research element is included, and is budgeted at $250,000.

Budget Summary:

| | |
|---|---|
| Education, Capacity Building, and Awareness US$192,000 | |
| Sea-level Infrastructure Modernization US$375,000 | |
| Seismic Infrastructure Modernization US$265,000 | |
| Warning Centre Operations US$625,000 | |
| Communications Infrastructure Modernization US$250,000 | |
| Executing Agency US$469,000 | |
| Tsunami Research US$250,000 | |
| TOTAL | US$2,426,000 |

**REFERENCES**

Gonzalez, F.I., 1999. Tsunami! *Scientific American*, 280(5), pp: 56-65.

Gusiakov, V., 1999. Tsunami Travel-Time Software Package for the Caribbean Region, Puerto Rico Civil Defense (in progress).

IOC, 1996. *IOCARIBE Tsunami Warning System Workshop*, St. John, US Virgin Islands, 23-24 May 1996. IOC Workshop Report No., Intergovernmental Oceanographic Commission of UNESCO, Paris, 9 pp., + IV Annexes.

IOC, 1999. *International Co-ordination Group for the Tsunami Warning System in the Pacific* (ICG/ITSU) Officers Meeting, 26-29 January 1999, Honolulu, Hawaii. Intergovernmental Oceanographic Commission of UNESCO, Paris, IOC/INF-1115, Summary Report, 7 pp. + VIII Annexes.

IOC, 2001. *International Co-ordination Group for the Tsunami Warning System in the Pacific, Eighteenth Session, Cartagena, Colombia 8-11*

*October 2001*. Intergovernmental Oceanographic Commission of UNESCO, Paris, Summary Report.

Lander, J.F., K.F. O'Loughlin, and L.S. Whiteside, 1999. Caribbean Tsunamis: A 500-Year History, 1498 to 1998. *Natural Hazards* (in press).

Maul, G.A., 1999. On the role of IOCARIBE in a Caribbean Tsunami System: Science, Engineering, Management, and Education. *Mar. Geod. J.*, 22(1), pp: 53-62.

McCann, W., 1998. Tsunami Hazard of Western Puerto Rico from Local Sources: Characteristics of Tsunamigenic Faults. Sea Grant College Program, University of Puerto Rico (submitted).

Mercado, A., 1997. Overlooked threat of tsunamis to the Caribbean discussed. *EOS, Transactions American Geophysical Union*, 78(50): December 16, 1997.

Mercado, A., and W. McCann, 1998. Numerical Simulation of the 1918 Puerto Rico Tsunami. *Natural Hazards*, 18, pp: 57-76.

Mofjeld, H.O., F.I. Gonzalez, and J.C. Newman, 1999. Tsunami Prediction in Coastal Regions. Chapter 14 in: *Coastal Ocean Prediction*, ©American Geophysical Union, Coastal and Estuarine Series No. 56, Washington DC, pp: 353-375.

Smith M.S., and J.B. Shepherd, 1994. Explosive Submarine Eruptions of Kick'em Jenny Volcano: Preliminary Investigations of the Potential Tsunami Hazard in the Eastern Caribbean Region. In: Ambeh, W.B. (ed.), *Proceedings, Caribbean Conference on Natural Hazards: Volcanoes, Earthquakes, Windstorms, Floods.* University of the West Indies, Mona Campus, pp: 249-260.

Watlington, R.A., and S.H. Lincoln, 1997. *Disaster and Disruption in 1867: Hurricane, Earthquake, and Tsunami in the Danish West Indies.* Eastern Caribbean Center, 131 pp.

## Annex I. Preliminary List of Historical Caribbean Tsunamis

| Date | Area | Location of Effects | Runup (m) | Comments |
|---|---|---|---|---|
| 1530 09 01 | Venezuela | Paria<br>Cumana<br>Cubaqua | 7.3 | Ground opened emitting black salt water and asphalt. Mountain at the side of the Gulf of Cariaco was cleft (earthquake). A fort and many houses destroyed, but not clear whether due to the wave, the earthquake, or both. |
| 1543 | Venezuela | Venezuela | | Waves noted. City of Cumana destroyed by earthquake? |
| 1688 03 01 | Jamaica | Port Royal, Jamaica | | Shocks felt throughout the island and waves damaged ships in Port Royal. A ship at sea was damaged by a hurricane. |
| 1690 04 16 | Leeward Is. | Charlotte Amalie, Virgin Is.<br>Charleston, Nevis | | The sea withdrew from Charlotte Amalie, St. Thomas, (16.5 to 18.5 m). Earthquake of intensity IX caused landslides on volcanic Nevis Peak which caused the sea to withdraw 201m from Charleston before returning in 2 minutes. |
| 1692 06 07 | Jamaica | Port Royal, Jamaica | | Earthquake and subsidence destroyed the city. Ships overturned, frigate washed over tops of buildings. Along the coast of Ligance (possibly Liguanea Plain) the sea withdrew 18.3 or 274m, exposing the bottom; upon returning the water overflowed the greater part of the shore. At Yallhouse (possibly Yallahs) the sea is said to have retired about 1.6km. At Saint Anns Bay a large wave was report. 2000 people killed by the earthquake and tsunami. |
| 1755 11 01 | Lisbon, Portugal | Saba<br>St. Martin<br>Antigua & Dominica<br>Barbados<br>Martinique<br>Santiago de Cuba | 7.0<br>4.5<br>3.6<br>1.5-1.8 | At St. Martin, a sloop anchored in 4.6 m of water was left lying broadside on the dry bottom. At Barbados, the wave had a period of 5 minutes and the water was black as ink. This could be a local landslide tsunami or seiche triggered by the Lisbon wave. At Martinique, at some places the water was reported to have withdrawn for 1.6km and at other places it flowed into the upper level rooms of the houses. The lowlands on most of the other French Islands were inundated. There is a report of Santiago de Cuba being nearly inundated in 1755, but the month and day were not given. This is probably from the Lisbon tsunami. |
| 1761 03 31 | Lisbon, Portugal | Barbados | | An earthquake near Lisbon, Portugal caused an extraordinary flux and reflux of the sea at Barbados. |
| 1766 06 11 | Cuba | Jamaica | | An earthquake lasting 1-1/2 to 7 minutes hit Cuba. Ships at sea 7.2km from the coast of Jamaica rolled so much that their gunwales were immersed in the water. Ships in deep water would not experience a tsunami. Either the ships were near the coast or in shoaling water or the wave was a storm wave but no storm was reported. |
| 1766 08 21 | Venezuela | Cumana, Venezuela | | Very violent shocks raised Cumana and caused the island of Orinoco to sink and disappear. In many placed the water surface was disturbed. This is a possible tsunami report. |
| 1767 04 24 | Martinique & Barbados | Martinique<br>Barbados | | The sea was much agitated and ebbed and flowed in an unusual way. |
| 1770 06 03 | Haiti | Golfe de la Gonave, Haiti | | La Saline Mountain foot partly submerged. The sea inundated 7.2km inland. |
| 1775 | Hispaniola & Cuba | Hispaniola<br>Cuba | | Three earthquakes reported and waves did extensive damage. |
| 1780 10 02 | Jamaica | Savanna la Mar, Jamaica | 3.0 | An earthquake occurred during a hurricane. The sea rose to a height of 3m at 0.8km from the beach and swept away a number of houses. Ten people were killed by the wave and at least 40 more by the storm. All vessels in the bay were dashed to pieces or drive onshore. |
| 1781 09 01 | Jamaica | Jamaica | | In 1781 a series of waves and disastrous earthquakes nearly ruined the Island. |
| 1787 10 27 | Jamaica | Montego Bay, Jamaica | | A small local shock was felt at Montego Bay and the vessels in the harbor were agitated. Mallet reports earthquakes in Jamaica on Oct. 1 and 21 at Kingston and Port Royal. This would be a low validity report as no wave was cited and the agitation may have been a report of a seaquake effect. |
| 1802 03 19 | Leeward Is. | Antigua<br>St. Christopher | | Earthquakes were reported in February and March with the largest on this date. It was accompanied by great agitation of the sea. Intensity IV |

| | | | | |
|---|---|---|---|---|
| 1802 05 05 | Venezuela | Orinoco River, Venezuela | | Earthquakes at Cumana caused the water of the Orinoco River to rise so high as to leave part of the bed dry. This could describe wave action near the mouth of the river, or bore action. |
| 1812 11 11 or 12 | Jamaica | Jamaica | | The sea was much agitated following an earthquake. This could describe wave action or seaquake action. |
| 1823 11 30 | Martinique | Saint-Pierre Harbor, Martinique | | At 3:10pm, a strong undulation (earthquake) was followed by a tidal wave which caused some damage in Saint-Pierre Harbor. |
| 1824 09 13 | Guadeloupe | Plymouth, Montserrat | | Earthquakes were felt at Basse Terre on the 9th and on the 13th; there was remarkable rise and fall of the tide at Plymouth, Montserrat. There had been a terrible storm and heavy rain on September 7-9. |
| 1825 09 20 | British Guiana | Demerara County, British Guiana | | Local earthquake and oscillations of the sea were noted. An earthquake was also noted at Trinidad, Tobago, St. Vincent, and Barbados. |
| 1831 12 03 | Trinidad and St. Christopher | Trinidad<br>St. Christopher | | An earthquake occurred. The sea was in a state of violent agitation. Note the large distance between reporting areas. An earthquake was also reported in Grenada, St. Vincent, British Guiana. |
| 1837 07 26 | Martinique | Martinique | | Several shocks accompanied by a large wave occurring during a hurricane. Source of wave uncertain. |
| 1842 05 07 | Guadeloupe | Guadeloupe; Basse Terre<br>Deshaies, & Sainte Rose<br>Bequia Is.<br>St. Johns, Virgin Is.<br>Charlotte Town, Grenada<br>Haiti; Cap Haitien, Port-de-Paix, Fort Liberte, Mole St. Nicolas, & Santiago de los Caballeros | 0.9<br>8.3<br>1.8<br>3.1 | A strong earthquake produced waves with heights reported; a wave carried away all floatable objects at Deshaies and Sainte Rose; at Gouyave, Grenada (Charlotte Town), there was some damage; at Haiti, a destructive tsunami struck the north coast; at Mole Saint-Nicholas, Ca-pHaitien, there was extensive destruction caused by the earthquake and tsunami; at Port-de-Paix the sea receded 60 and the returning wave covered the city with 5m of water. About 200 of the city's 3,000 inhabitants were killed by the earthquake and tsunami. It was observed at Fort Liberte, Mole Saint-Nicholas, and Santiago de los Caballeros. At Hispaniola, there was destruction on north coast. Note the large area of this event which suggests a teletsunami, but the earthquake was felt at Haiti, Jamaica, Puerto Rico, and other islands. Note also the missing locations such as Puerto Rico for which no tsunami report is available although there are reports from Haiti and the Virgin Islands. |
| 1843 02 08 | Antigua | | | An earthquake was felt at Point-a-Pitre, Guadeloupe, St. Lucia, St. Kitts, Montserrat, Martinique, and other islands. The sea rose 1.2m but sank again immediately. |
| 1853 07 15 | Venezuela | Cumana, Venezuela | | A violent earthquake in Cumana followed by a tsunami. |
| 1860 03 08 | Hispaniola | Hispaniola; Golfe de la Gonaves, Cayes, Acquin, & Anse-a-Veau | | An earthquake was reported from Port-au-Prince and Anse-a-Veau. Waves were reported from Golfe de la Gonaves, Cayes, and Acquin. At Anse-a-Veau, the sea withdrew and broke with a crash on the shore. |

**Annex I. Preliminary List of Historical Caribbean Tsunamis (continued)**

| Date | Area | Location of Effects | Runup (m) | Comments |
|---|---|---|---|---|
| 1867 11 18 | St. Thomas, Virgin Is. | St. Thomas, Charlotte<br>Amalie &<br>Altona<br>St. Croix, Fredricksted,<br>Christensted & Gallows Bay<br>Puerto Rico, San Juan,<br>Fajardo, Yabucoa &<br>Vieques<br>British Virgin Is.:<br>Tortola, Road Town,<br>Peter is., Saba, St.<br>Christopher, St. Martin & St.<br>Barthelemy<br>St. Johns, Artigua<br>Guadeloupe; Basse-Terre,<br>Deshaies, Isles des Saintes,<br>Fond du Cure, Pointe-a-Pitre<br>& Sainte-Rose<br>Martinique<br>St. Vincent<br>Grenada, Charlotte Town | <br>6.0<br><br>7.6<br><br>1.5<br><br><br><br>1.5<br><br><br><br>3.0<br>1.0<br><br><br>10.0<br><br><br>3.0 | At Charlotte Amalie the height was 2.4m above the sea level at the wharf, and the lower part of the city was flooded. The water receded nearly 100 and returned as a wave 4.5 to 6m high, swamping small boats in the harbor. The wave penetrated 76m inland. The USS De Soto was damaged. 11-12 people were killed. At Altona, houses were washed far inland and there was some damage at Hassel Is. At Christensted, St. Croix, waves swept inland 91m, and at Gallows Bay, 20 houses were damaged. At Fredericksted, the sea withdrew and returned as a wall of water 7.6m high leaving the USS Morongahela stranded. Five were killed, 3-4 injured, and 20 houses were damaged. At Puerto Rico, at San Juan, the river water rose 0.9-1.5m and at Vieques, high waves were observed. At Fajardo, a very small wave was reported, and at Yabucoa the sea retreated and inundated 137m on its return. In the British Virgin Islands, at Peter Is., a wave was noted and people fled to Tortola. At Roadtown, Tortola, a 1.5m wave swept some houses away. At Saba, there was some damage. At St. Christopher, the wave was also observed. At St. Martin and St. Barthelemy, there was some damage. At St. Johns, Artigua, the wave had a height of 3.0. At Basse-Terre, Guadeloupe, the height was 1.0m with the sea retreated far from coast. At Deshaies, houses in village were destroyed. At Isles des Saintes, there was a slight swell, and at Fond du Cure, houses inundated to a depth of 1m. At Point-a-Pitre, there was a slight swell, and at Sainte-Rose, a 10m wave. The sea withdrew 100m and flooded and damaged houses on return. It was observed at Martinique and St. Vincent had unusually high water. At Grenada, Gouyave (Charlotte Town) the height was 3m and at St. George, 1.5m. |
| 1868 03 17 | Puerto Rico | St. Thomas, Charlotte Amalie | 0.6 | An earthquake and tsunami were observed at Arroyo and Naguabo. At St. Thomas, Charlotte Amalie, there was a small recession and flooding. |
| 1874 03 11 | Lesser Artilles | Dominica<br>Thomas, Virgin Is. | | A submarine shock to the southeast of St. Thomas shook the island and ships in the harbor. Simultaneously, the water in the bay, then perfectly still, appeared turbid as though clouded by sand and mud. A little later strong ripples from the south agitated the water surface lasting some time. This probably was the tsunami and the earlier effects from the seismic waves agitating the bottom. At Dominica, the steamer Corsica reported a series of heavy rollers in the harbor lasting half an hour and rendering communication with the shore impossible. They did not feel the earthquake. The reduced effects at Charlotte Amalie may indicate a source on the eastern side of the island. |
| 1881 09 12 | Jamaica | Kingston, Jamaica | 0.46 | An earthquake was felt on the island and a wave was reported from the north coast. At Kingston Harbor, the water rose about 46cm. Berninghausen felt that this wave was not caused by the earthquake, but does not give any reason for his conclusion. |
| 1882 09 07? | Panama | Northeastern Panama, San Bias Archipelago | | Mine reports an earthquake for this date observed in Colombia, Panama, Nicaragua, and Ecuador but does not mention a tsunami. Camacho reported the tsunami but did not give details or a date. |
| 1883 08 27 | Indonesia | St. Thomas, Virgin Is. | | A tidal wave occurred on August 27. The water receded from the shore three times. A sharp shock of an earthquake was felt on the following evening. This would have been an effect of Krakatoa Volcano eruption which created air waves widely recorded in Hawaii, Alaska, South Sandwich Islands, Great Britain, and elsewhere. The location of the Caribbean at near antipodal distances may have resulted in larger effects and these effects should have been observable through the islands. While these effects have been considered to not be true tsunamis elsewhere, this description sounds like a true tsunami was generated. |

**Annex I. Preliminary List of Historical Caribbean Tsunamis (continued)**

| Date | Area | Location of Effects | Runup(m) | Comments |
|---|---|---|---|---|
| 1887 | Haiti | Haiti; Mole Saint Nicholas, Anse-d'Hainault, & Pointe Tiburon | | The epicenter was apparently near the Barlett Trough a short distance southwest from Mole Saint Nicholas. At Jeremie the sea withdrew 20m and returned with a rush. Waves were noted at Mole Saint Nicholas, Anse-d'Hainault, Pointe Tiburon, and other ports. Heck mistakenly identified the area as in the Philippines. Milne reports the earthquake felt at Port-de-Paix, Haiti and Inagua Island, Bahama Islands. |
| 1907 01 14 | Jamaica | Jamaica:<br>Annotto Bay<br>Hope Bay,<br>Orange Bay, Sheerness Bay,<br>Saint Anns Bay, Buff Bay<br>Ocho Rios & Port Antonia | <br>7.1 | Earthquake damage at Kingston and surrounding territory. Buff Bay was destroyed. Waves noted at Hope Bay, Orange Bay, Sheerness Bay, and Saint Anns Bay. At Annotto Bay, an observer reported the sea receded 73 to 93m, dropping 3 to 3.7m below normal sea level. The returning wave raised the water level 1.8 to 2.4m above normal, sweeping into the lower parts of town destroying house. On higher land it came up 7.6 to 9.1m. At Buff Bay the sea receded some distance from the land. At Port Maria the sea withdrew 25.6m. At Ocho Rios near St. Anna Bay the sea withdrew 69m. At Port Antonia the wave moved a small building near the beach. Waves were also reported from the south coast of Jamaica and seiches were set up in Kingston Harbor. |
| 1911 11 03 | Trinidad | Trinidad | | Some extraordinary waves were noticed on the coast following an explosion of a mud volcano island. This is a volcanic-related tsunami. |
| 1916 04 25 | Panama | Bocas del Toro, Panama | | An earthquake was reported from Bocas del Toro and Alirante, and waves at Bocas del Toro carried debris and canoes 198m inland. |
| 1918 10 11 | Puerto Rico | Puerto Rico:<br>Aguadilla<br>Isabela<br>Cayo Cardona<br>El Boqueron<br>Punta Borinquen<br>Isla Caja de uertos<br>Gaunica<br>Isla Mona<br>Mayaguez<br>Puerto Arecido<br>Punta Agujereada<br>Punta Higuero<br>Rio Culebrinas<br>St. Thomas:<br>Charlotte Amalie &<br>Krum Bay<br>Santo Domingo, Hispaniola<br>Tortola | <br>2.4-3.3<br>1.8<br>0.75<br>2.4<br>4.3<br>1.5<br>0.6<br>3.6<br>1.5<br>0.6<br>6.0<br>4.9<br>3.7<br><br>0.45<br>1.2<br>0.6 | A magnitude 7.5 earthquake caused a wave of 2.4-3.3m above sea level at Aguadilla which destroyed 300 huts and drowned 34 people. At Cayo Cardona water rose 75cm on the west side of the island. At El Boqueron the wave dropped 1.5m and rose 90cm above mean sea level. About 800m southeast near the entrance to the bay the water rose only 45cm. At Punta Borinquen Lighthouse the wave was 4.5m above sea level. In a low area just southwest of the lighthouse the wave penetrated 91m inland. Submarine cables were cut in several places. At Gaunica, 45cm waves observed. At Isla Caja de Muertos water rose 1.5m covering 15m of the beach. At Isla Mona the receding water bared the reef and the returning wave was 3.6m above sea level washing a pier away and flooding a cistern. At Mayaguez, a wave entered the first floors of buildings near the waterfront and destroyed a few native huts and a brick wall was overturned. Water levels reached 40 to 150cm above sea level. At Playa Ponce slight water movements were observed. At Puerto Arecido, waves 30 to 60cm high were observed and a bore about 10c went up the Rio Grande. At Punta Agujereada, waves estimated at 5.5 to 6m uprooted several hundred palm trees and destroyed several small houses. Eight people drowned. At Punta Higuero Lighthouse waves uprooted coconut palms and crossed railroad tracks 4.9m above sea level while 800m southeast of the lighthouse the water rose 2.6 to 2.7m. At Rio Culbrinas, 1000kg blocks of limestone were moved 46 to 76m slightly downhill. Waves were at least 3.7 high. At Rio Grande de Licoza, water receded and rose about 90cm. At St. Thomas, Virgin Islands, Charlotte Amalie the water rose 45cm and at Krum Bay, 1.2m. At Santo Domingo, Hispaniola water of the Rio Ozama fell and rose 60cm with a period of 40 minutes. Waves were noted at Tortola. |
| 1918 10 25 | Puerto Rico | Puerto Rico, Mona Passage | | Submarine cables were cut again and a steamer rolled heavily. Waves were recorded on the tide gage at Galveston, Texas. |
| 1922 05 02 | Puerto Rico | | | A wave was recorded on the Galveston gage which has been associated with a small earthquake in Viegues, but the small earthquake does not see likely to have produced a recordable tsunami. |
| 1929 01 17 | Venezuela | Cumana, Venezuela | | City was destroyed by an earthquake and a steamer offshore was endangered by a huge wave. The tidal wave following the earthquake caused much damage. Many sailboats were wrecked. |

## Annex I. Preliminary List of Historical Caribbean Tsunamis (continued)

| Date | Area | Location of Effects | Runup(m) | Comments |
|---|---|---|---|---|
| 1932 02 03 | Cuba | Santiago de Cuba | | Small waves were reported at the time of an earthquake at Santiago de Cuba. |
| 1946 09 04 | Dominican Republic | Dominican Republic: Martanzas<br>Julia Molina & Samana<br>San Juan, Puerto Rico<br>Bermuda<br>Daytona Beach, Florida<br>Atlantic City, New Jersey | 2.4 | The town was severely damaged and 100 people killed although the wave probably was only 2.4m. At Villa Julia Molina the wave was estimated to be 3.6 to 4.6m high but caused little damage. At Cabo Samana several ebbs and flows were observed. It was recorded at San Juan, Puerto Rico, 36 minutes after the earthquake. It was also recorded at Bermuda at 2:07 after the earthquake and at Daytona Beach, 3:59 and Atlantic City, 4:49. |
| 1946 09 08 | Puerto Rico | Puerto Rico: Aquadilla Mayaguez & San Juan | | The sea retreated 24m and returned. At Mayagues the sea retreated 76m and returned. At San Juan it was recorded on tide gauge 35 minutes after the earthquake. It was an aftershock of the August 4 event. The wave was also recorded with travel times of: Bermuda-2:02, Daytona Beach-4:02, and Atlantic City-4:42. |
| 1953 05 31 | Dominican Republic | Puerto Plata, Dominican Republic | 0.6 | Recorded on the Puerto Plata tide gage at 6cm height. |
| 1955 01 18 | Venezuela | Venezuela, La Vela | | A wave was reported and four ships were wrecked and four waterfront buildings damaged. No earthquake is listed for this time. |
| 1968 09 20 | Venezuela | | | A report of a tsunami has not been verified. |
| 1969 12 25 | Leeward Is. | Barbados<br>Artigua<br>Dominica | 0.14 | Recorded at Barbados, Artigua, and Dominica with a maximum amplitude of 14cm at Barbados. |
| 1985 03 16 | Leeward Is. | Basse-Terre, Guadeloupe | | A magnitude 6.3 earthquake at 14:54 GMT, caused damage and injuries to 6 people at Guadeloupe and minor damage at Montserrat. It was also felt at Artigua, St. Kitts, and Puerto Rico. A several cm tsunami was recorded at Basse-Terre, Guadeloupe. |
| 1989 11 01 | Puerto Rico | Cabo Rojo, Puerto Rico | | A small tsunami was reported. |
| 1991 04 22 | Costa Rica | Bocas del Toro, Panama | | At Bocas del Toro, Panama, people reported that Las Delicias sand bank normally covered by 60 to 90cm of water emerged as the sea receded less than ten minutes after the earthquake and remained above water for five to seven minutes. Afterwards several waves entered the bay with great force flooding 50 to 100m in the flat northern part of the town. At Carenero Island violent waves destroyed dwellings. At San Cristobal Island the sea receded several meters for about 45 minutes. People went on the beach to catch trapped fish. It was also observed at Bastimento, Cristobal-10cm, Puertobelo, W. Panama-60cm, and recorded at Colon. |

**ANNEX II. List of Seismic Stations in the Countries of the IAS Region***

| CODE | NAME | COUNTRY | LAT | LONG | ELEV (m) | INSTRUMENT / TELEMETRY | INSTITUTION |
|---|---|---|---|---|---|---|---|
| OTAV | Otavalo | Imbaibura, Ecuador | 0.238 N | 78.451 W | 3492 | | USGS/IRIS/Escuela Politecnica Nacional |
| PAYG | Puerto Ayora | Galapagos, Ecuador | 0.6741 S | 90.2863 W | 196 | Geotech KS-54000, CMG3-T, FBA all in borehole | USGS/IRIS/Escuela Politecnica Nacional |
| PAYV | | Venezuela | 5.5285 N | 67.5468 W | 91 | CMG-40T, Satellite (Nanometrics) | FUNVISIS |
| BAUV | | Venezuela | 8.9433 N | 68.0415 W | 106 | CMG-40T, Satellite (Nanometrics) | FUNVISIS |
| GUNV | | Venezuela | 10.1617 N | 62.9427 W | 60 | CMG-40T, Satellite (Nanometrics) | FUNVISIS |
| CRUV | | Venezuela | 10.6167 N | 63.1833 W | 608 | CMG-40T, Satellite (Nanometrics) | FUNVISIS |
| TEST | | Venezuela | 10.4700 N | 66.8100 W | 875 | CMG40T, Satellite (Nanometrics) | FUNVISIS |
| GUIV | | Venezuela | 10.647 N | 62.223 W | 50 | CMG-40T, Satellite (Nanometrics) | FUNVISIS |
| BIRV | | Venezuela | 10.4757 N | 66.2893 W | 200 | CMG-40T, Satellite (Nanometrics) | FUNVISIS |
| PTGA | Pitinga | Brazil | 0.731 N | 59.997 W | 137 | | IRIS/USGS/Universidad de Brasilia |
| HDC2 | Heredia 2 | Costa Rica | 10.0237 N | 84.1167 W | 1220 | STS 1 | Observatorio Vulcanologico y Sismologico de Costa Rica, Universidad Nacional, Campus Omar Dengo, Heredia, Costa Rica |
| JTS | Juntas de Abangares | Costa Rica | 10.2908 N | 84.9525 W | 340 | | IRIS-IDA; Observatorio Vulcanologico y Sismologico de Costa Rica, Universidad Nacional, Campus Omar Dengo, Heredia, Costa Rica |
| UNAH | | Honduras | 14.0750 N | 87.1750 W | 1020 | | Universidad Nacional Autonoma de Honduras |
| FDFZ | | Martinique | 14.7333 N | 61.150 W | 510 | | Observatoire de La Montagne Pelee, Institut de Physique du Globe |
| CUIG | Ciudad Universitaria | Mexico | 19.329 N | 99.178 W | 2200 | | Instituto de Geophsica, Universidad Nacional Autonoma de Mexico |
| TEIG | Tepich | Yucatan, Mexico | 20.2263 N | 88.274 W | 69 | | Instituto de Geophsica, Universidad Nacional Autonoma de Mexico |
| UPA | Universidad de Panama | Panama | 8.9810 N | 79.5338 W | 41 | | Instituto de Geociencias, Universidad de Panama |
| GOGA | Godfrey | Georgia, USA | 33.4112 N | 83.4666 W | 150 | | National Earthquake Information Center |
| OXF | Oxford | Mississippi, USA | 34.5118 N | 89.4092 W | 101 | | National Earthquake Information Center |
| TUC | Tucson | Arizona, USA | 32.3097 N | 110.7842 W | 906 | | National Earthquake Information Center |
| ANMO | Albuquerque | New Mexico USA | 34.946 N | 106.457 W | 1740 | | USGS/IRIS |
| DWPF | Disney Preserve | Florida, USA | 28.1110 N | 81.433 W | -142 | Deep Borehole | USGS/IRIS |
| HKT | Hockley | Texas, USA | 29.950 W | 95.833 W | -122 | Salt Dome | USGS/University of Texas at Austin |
| JCT | Junction | Texas, USA | 30.479 N | 99.802 W | 835 | | USGS/Texas Tech University |
| LRAL | Lakeview Retreat | Alabama, USA | 32.8 N | 86.9 W | | | USGS/Alabama Geological Survey |
| LTX | Lajitas | Texas, USA | 29.334 N | 103.667 W | 1013 | | USGS |
| MIAR | Mt. Ida | Arkansas, USA | 34.546 N | 93.573 W | 207 | | USGS |
| NHSC | New Hope | South Carolina, USA | 33.107 N | 80.178 W | 12 | | USGS |
| PLAL | Pickwick Lake | Alabama, USA | 34.982 N | 88.076 W | 165 | | USGS/St. Louis University |
| WMOK | Wichita Mountain | Oklahoma, USA | 34.738 N | 98.781 W | 486 | | USGS |
| SDV | Santo Domingo | Venezuela | 8.8861 N | 70.6333 W | 1580 | | CAR (FUNVISIS), IRIS-USGS |

| | | | | | | | |
|---|---|---|---|---|---|---|---|
| IMO | Isla Mona | Isla Mona, Puerto Rico | 18.111 N | 67.908 W | 50 | CMG-40T, UHF and DDS comm to Central | Puerto Rico Seismic Network, Univ. of Puerto Rico at Mayaguez |
| AGP | Aguadilla | Puerto Rico | 18.4075 N | 67.141 W | 220 | CMG-3ESP, FBA, UHF and DDS comm to Central | Puerto Rico Seismic Network, Univ. of Puerto Rico at Mayaguez |
| CORN | Cornellia Hill | Cabo Rojo, Puerto Rico | 18.163 N | 67.179 W | 100 | CMG-40T, FBA, on site recording | Puerto Rico Seismic Network, Univ. of Puerto Rico at Mayaguez |
| MGP | Maguayo | Lajas, Puerto Rico | 18.007 N | 67.089 W | 60 | CMG-40T, UHF and DDS comm to Central | Puerto Rico Seismic Network, Univ. of Puerto Rico at Mayaguez |
| CDP | Cerro de Punta | Jayuya, Puerto Rico | 18.175 N | 66.591 W | 1300 | CMG-40T, FBA, Microwave and DDS comm to Central | Puerto Rico Seismic Network, Univ. of Puerto Rico at Mayaguez |
| SJG | Cayey | Puerto Rico | 18.1117 N | 66.1500 W | 457 | STS-1/VBB | IRIS-USGS |
| ICMB | Isla Caja De Muertos | Puerto Rico | | | | CMG-40, UHF and DDS comm to Central | Puerto Rico Seismic Network, Univ. of Puerto Rico at Mayaguez |
| CPD | Cerro Pandura | Yabucoa, Puerto Rico | 18.037 N | 65.9147 W | 370 | CMG-40T, UHF and DDS comm to Central | Puerto Rico Seismic Network, Univ. of Puerto Rico at Mayaguez |
| CBYP | Cubuy | Canovanas, Puerto Rico | 18.275 N | 65.8605 W | 140 | CMG-40T, FBA, UHF and DDS comm to Central | Puerto Rico Seismic Network, Univ. of Puerto Rico at Mayaguez |
| MTP | Monte Pirata | Vieques, Puerto Rico | 18.0847 N | 65.5525 W | 300 | CMG-40T, UHF and DDS comm to Central | Puerto Rico Seismic Network, Univ. of Puerto Rico at Mayaguez |
| CULB | Monte Resaca | Culebra, Puerto Rico | | | | CMG-40T, FBA, UHF and DDS comm to Central | Puerto Rico Seismic Network, Univ. of Puerto Rico at Mayaguez |

* Only 3 stations, SJG, HDC2, and PAYG, have a frequency response compatible with TREMORS. In addition, the TREMORS software requires the data to be in mini-SEED or GSE2.1 format.

Annex III. IOCARIBE Regional GLOSS: Sea-Level / Weather Coastal Stations

| Country / Site | Latitude (north) | Longitude (west) | Sponsor | Gauge Type | Trans-mission | GPS | Ancillary Sensors |
|---|---|---|---|---|---|---|---|
| Antigua & Barbuda: Parham | 17° 09' 30" | 61° 47' 20" | GEF/OAS | Acoustic | GOES | Yes | Met., SST |
| Aruba: Sint Nicolas | 12° 26' | 69° 54' | IOC/UNEP | Pressure | None | No | None* |
| Bahamas: Lee Stocking Island | 23° 46' 24" | 76° 06' 20" | GEF/OAS | Acoustic | GOES | Yes | Met., SST |
| Great Inagua | 21° 03' 07" | 73° 38' 47" | GEF/OAS | Acoustic | GOES | Yes | Met., SST |
| Nassau | 25° 05' 10" | 77° 22' 06" | GEF/OAS | Acoustic | GOES | Yes | Met., SST |
| Settlement Point | 26° 30' | 78° 46' | NOAA | Acoustic | GOES | Yes | Met., SST |
| Barbados: Bridgetown | 13° 06' 06" | 59° 37' 42" | GEF/OAS | Acoustic | GOES | Yes | Met., SST |
| Belize: Belize City | 17° 28' 51" | 88° 12' 08" | GEF/OAS | Acoustic | GOES | Yes | Met., SST |
| Bermuda: St. Georges | 32° 23' | 64° 41' | UK | Pressure | None | No | None |
| Cayman Islands: Georgetown | 19° 18' | 81° 26' | UK | Float | None | No | None |
| Colombia: Cartagena | 10° 19' | 75° 35' | NOAA | Float | None | No | None |
| Costa Rica: Puerto Limon | 10° 00' | 83° 01' | Finland | Pressure | GOES | Yes | Met., SST |
| Cuba: Cabo San Antonio | 21° 52' | 84° 57' | National | Float | None | No | None |
| Gibara | 21° 07' | 76° 07' | National | Float | None | No | None |
| Guantanamo Bay | 19° 54.4' | 75° 08.9' | IOC/UNEP | Acoustic | GOES | Yes | None* |
| Siboney | 23° 09' | 82° 21' | National | Float | None | No | None |
| Dominica: Roseau | 15° 18' 20" | 61° 23' 42" | GEF/OAS | Acoustic | GOES | Yes | Met., SST |
| Dominican Republic: Puerto Plata | 19° 49' | 70° 41' | National | Bubbler | None | No | None |
| Barahona | 18° 11' | 71° 07' | National | Bubbler | None | No | None |
| France: Cayene | 4° 56' | 52° 20' | National | Float | None | No | None |
| Kourou | 5° 12' | 52° 39' | National | Pressure | None | No | None |
| Fort-de-France | 14° 36 | 61° 05' | National | Float | None | No | None |
| Basse Terre | 16° 00' | 61° 44' | National | Float | None | No | None |
| Grenada: Prickly Bay | 12° 00' 20" | 61° 45' 56" | GEF/OAS | Acoustic | GOES | Yes | Met., SST |
| Guatemala: Santo Tomas de Castilla | 15° 41.7' | 88° 37.2' | RONMAC | Acoustic | GOES | Yes | Met., SST |
| Guyana: Georgetown | 6° 48.5' | 58° 10.5' | IOC/UNEP | Acoustic | GOES | Yes | None |
| Rosignol | 6° 18' 15" | 57° 30' 45" | GEF/OAS | Acoustic | GOES | Yes | Met., SST |
| Parika | 6° 50' 48" | 58° 23' 06" | GEF/OAS | Acoustic | GOES | Yes | Met., SST |
| Honduras: Cochino Pequeño | 15° 57' 09" | 86° 29' 56" | Smithsonian | Acoustic | GOES | Yes | Met., SST |
| Puerto Cortes | 15° 50.1' | 87° 57.2' | RONMAC | Acoustic | GOES | Yes | Met., SST |
| Jamaica: Discovery Bay | 18° 28' 06" | 77° 25' 00" | GEF/OAS | Acoustic | GOES | Yes | Met., SST |
| Kingston | 17° 56' 54" | 76° 50' 42" | GEF/OAS | Acoustic | GOES | Yes | Met., SST |
| Mexico: Progresso | 21° 17' | 89° 40' | National | Float | None | No | None |
| Puerto Morelos | 20° 50' | 86° 52' | NOAA | Float | None | No | None |
| Tampico | 22° 13' | 97° 51' | National | Float | None | No | None |
| Veracruz | 19° 12' | 96° 08' | National | Float | None | No | None |
| Netherlands Antilles: Curaçao | 12° 07' | 68° 56' | National | Float | None | No | None |
| Nicaragua: Puerto Cabezas | 14° 01.2' | 83° 22.9' | RONMAC | Acoustic | GOES | Yes | Met., SST |
| Panama: Coco Solo (Limon Bay) | 9° 22' | 79° 54' | Canal Zone | Pressure | None | No | None |
| St. Kitts & Nevis: Basse Terre | 17° 17' 24" | 62° 42' 36" | GEF/OAS | Acoustic | GOES | Yes | Met., SST |
| St. Lucia: Castries | 14° 01' 20" | 61° 00' 06" | GEF/OAS | Acoustic | GOES | Yes | Met., SST |
| St. Vincent: Kingstown | 13° 07' 50" | 61° 11' 55" | GEF/OAS | Acoustic | GOES | Yes | Met., SST |
| Trinidad & Tobago: Charlotteville | 11° 19' 25" | 60° 32' 55" | GEF/OAS | Acoustic | GOES | Yes | Met., SST |
| Guayaguayre | 10° 08' 20" | 61° 00' 06" | GEF/OAS | Acoustic | GOES | Yes | Met., SST |
| Port-of-Spain | 10° 38' 56" | 61° 30' 51" | GEF/OAS | Acoustic | GOES | Yes | Met., SST |
| Turks&Caicos: South Caicos | 21° 30' | 71° 31' | IOC/UNEP | Acoustic | GOES | No | Met., SST* |
| USA: Fernandina Beach | 30° 40' | 81° 27' | National | Acoustic | GOES | Yes | Met., SST |
| Miami (Virginia Key) | 25° 47' | 80° 11' | National | Acoustic | GOES | Yes | Met., SST |
| Key West | 24° 33' | 81° 49' | National | Acoustic | GOES | Yes | Met., SST |
| Naples | 26° 10' | 81° 48' | National | Acoustic | GOES | Yes | Met., SST |
| Clearwater Beach | 27° 57' | 82° 48' | National | Acoustic | GOES | Yes | Met., SST |
| Cedar Key | 29° 08' | 83° 06' | National | Acoustic | GOES | Yes | Met., SST |
| Pensacola | 30° 24' | 87° 13' | National | Acoustic | GOES | Yes | Met., SST |
| Grand Isle | 29° 14' | 89° 59' | National | Acoustic | GOES | Yes | Met., SST |
| Galveston Pier 21 | 29° 19' | 94° 47' | National | Acoustic | GOES | Yes | Met., SST |
| Port Isabel | 26° 05' | 97° 16' | National | Acoustic | GOES | Yes | Met., SST |
| San Juan PR | 18° 28' | 66° 07' | National | Acoustic | GOES | Yes | Met., SST |
| La Parguera PR | 17° 59' | 67° 03' | National | Acoustic | GOES | Yes | Met., SST |
| Lime Tree Bay USVI | 17° 42' | 64° 45' | National | Acoustic | GOES | Yes | Met., SST |
| Charlotte Amalie USVI | 18° 21' | 64° 54' | National | Acoustic | GOES | Yes | Met., SST |
| Venezuela: Cumana | 10° 25' | 64° 17' | National | Float | None | No | None |
| La Guaira | 10° 37' | 66° 56' | National | Float | None | No | None |

*Out of Order

Annex IV. Recommended IAS TWS Sea Level Stations

| Country / Site | Latitude (North) | Longitude (West) | Sponsor | Gauge Type | Tsunami Upgrade |
|---|---|---|---|---|---|
| Bahamas: Lee Stocking Island | 23° 46' 24" | 76° 06' 20" | GEF/OAS | Acoustic | $5000 |
| Settlement Point | 26° 30' | 78° 46' | NOAA | Acoustic | $5000 |
| Barbados: Bridgetown | 13° 06' 06" | 59° 37' 42" | GEF/OAS | Acoustic | $5000 |
| Belize: Belize City | 17° 28' 51" | 88° 12' 08" | GEF/OAS | Acoustic | $5000 |
| Bermuda: St. Georges | 32° 23' | 64° 41' | UK | Pressure | $5000 |
| Cayman Islands: Georgetown | 19° 18' | 81° 26' | UK | Float | $8000 |
| Colombia: Cartagena | 10° 19' | 75° 35' | | Acoustic | $35000 |
| Isla de San Andres | 12.5° | 81.8° | | Pressure | $8000 |
| Costa Rica: Puerto Limon | 10° 00' | 83° 01' | Finland | Pressure | $5000 |
| Cuba: Cabo San Antonio | 21° 52' | 84° 57' | | Acoustic | $35000 |
| Guantanamo Bay | 19° 54.4' | 75° 08.9' | | Acoustic | $35000 |
| Dominican Republic: Barahona | 18° 11' | 71° 07' | National | Bubbler | $8000 |
| France: Fort-de-France | 14° 36' | 65° 05' | National | Float | $8000 |
| Honduras: Puerto Cortes | 15.9° | 88.0° | RONMAC | Acoustic | $0 |
| Swan Island | 17.4° | 83.8° | | Pressure | $8000 |
| Jamaica: Kingston | 17° 56' 54" | 76° 50' 42" | GEF/OAS | Acoustic | $5000 |
| Mexico: Puerto Morelos | 20° 50' | 86° 52' | | Acoustic | $35000 |
| Veracruz | 19° 12' | 96° 08' | National | Float | $8000 |
| Netherlands Antilles: Curaçao | 12° 07' | 68° 56' | National | Float | $8000 |
| St. Martin | 18.1° | 63.2° | | Acoustic | $35000 |
| Nicaragua: Puerto Cabezas | 14.1° | 83.6° | RONMAC | Acoustic | $0 |
| Panama: Coco Solo | 9° 22' | 79° 54' | | Acoustic | $35000 |
| Trinidad & Tobago: Charlotteville | 11° 19' 25" | 60° 32' 55" | GEF/OAS | Acoustic | $5000 |
| USA: Key West | 24° 33' | 81° 49' | National | Acoustic | $3000 |
| Clearwater Beach | 27° 57' | 82° 48' | National | Acoustic | $3000 |
| Grand Isle | 29° 14' | 89° 59' | National | Acoustic | $3000 |
| Galveston Pier 21 | 29° 19' | 94° 47' | National | Acoustic | $3000 |
| Port Isabel | 26° 05' | 97° 16' | National | Acoustic | $3000 |
| San Juan PR | 18° 28' | 66° 07' | National | Acoustic | $3000 |
| Venezuela: Isla de Aves | 15.7° | 63.6° | | Pressure | $8000 |
| La Blanquilla | 11.8° | 64.7° | | Pressure | $8000 |

# OVERVIEW OF POTENTIAL TSUNAMI SOURCES IN THE CARIBBEAN SEA REGION

# ESTIMATING THE THREAT OF TSUNAMIGENIC EARTHQUAKES AND EARTHQUAKE INDUCED-LANDSLIDE TSUNAMI IN THE CARIBBEAN

WILLIAM R. MCCANN

*Earth Scientific Consultants*
*Westminster, CO 80021, USA*

Deformation along the margin of the Caribbean Plate is the principal cause of the tsunami threat in the Caribbean. That margin parallels the northern coast of South America, the Lesser Antilles, and extends along the Greater Antilles from Puerto Rico through Jamaica. The eastern boundary of the Caribbean plate near the Lesser Antilles is the locus of subduction of Atlantic seafloor. At least three distinct, shallow tectonic regimes parallel the margin. They are: an outer tectonic belt where the North America Plate bends to enter the subduction zone, the main interface or zone of contact between the plates, and an inner zone of intraplate activity in the overriding Caribbean Plate. The level of seismic activity and tsunami potential in each of these zones is influenced by the presence of aseismic ridges on the downgoing plate. Ridges may increase the probability of tsunami or slow earthquakes, by reactivating thrust faults in the accretionary prism. The northeastern corner of the Caribbean Plate margin has a smooth transition from the relatively simple subduction zone in the Northern Lesser Antilles into a region of oblique convergence. It is a complex margin dominated by microplate tectonics from near Puerto Rico through Hispaniola. Here too the same three tectonic zones can be defined, but the third zone, "intraplate activity in the Caribbean Plate", is more clearly delineated as microplate deformation in a wide plate boundary zone. Strike-slip tectonics dominates the region from Haiti westward to the northern coast of Honduras. Local bends in the transcurrent fault systems lead to vertical tectonics in the form of push-ups and pull-aparts. Some convergence is absorbed along the southernmost margin of Cuba. Either the tsunamigenic events in this region had source mechanisms with significant vertical components, or they triggered submarine landslides that generated the tsunami. Other potential sources lie along the northern coast of South America ands off the east coast of Central America. Manipulation of gridded bathymetric data to produce maps of seafloor slope, slope direction, and roughness as defined by the curvature of contours, can aid the identification of potential slides/slumps, or reveal portions of subduction zones more likely to produce tsunami earthquakes. A survey of the Caribbean Basin reveals several regions of high slope. Many of these are found where great earthquakes have or are likely to occur, possibly triggering submarine landslides. Regions such as the southern Lesser Antilles have steeper seafloor slopes in the backarc region than in the accretionary prism of the subduction zone. It has been suggested that subduction of rough seafloor highs such as seamounts and fracture zones leads to dewatering and stiffening of the accretionary prism and, as the high enters the deeper portions of the subduction zone, activation of thrust faults within the accretionary prism i.e. above the main decollement. If this were the case, it would enhance the possibility of tsunami or slow earthquakes. The subduction of rough bathymetric highs and its corresponding increasing roughness of the accretionary prism would then be another parameter to determine the likelihood of tsunami earthquakes. Using these criteria, several regions of the NE Caribbean stand out as more likely source for tsunami earthquakes.

## 1. Introduction

The Caribbean region is well known for its numerous natural hazards such as Hurricanes, Volcanoes, Earthquakes, and Tsunami. Understanding the nature and scope of each of these threats is essential if one is to attempt to mitigate the negative affects of those hazards. This paper will attempt to layout the general framework of the tsunami hazard in the Caribbean Basin. As tsunami have various general generating mechanisms, the threat will be examined in those terms. I will explore tsunami caused by earthquakes, and submarine landslides/slumps or a combination of the two, but only in passing way mention tsunami as generated by volcanic eruptions. Most of the tsunami threat lies along the heavily populated island chains of the Greater and Lesser Antilles or the coasts of Central and South America rimming the Caribbean Basin. This means that in most cases those most affected by a tsunami will have the least warning of a potential or actual tsunami.

Large shallow earthquakes are principal cause of the tsunami threat in the Caribbean. Five Hundred years of damaging earthquakes form a long, nearly continuous belt of seismic activity surrounding the Caribbean Basin (Figures 1, 2, 3.). Most of those quakes occurred in submarine areas and their source mechanisms included a significant component of vertical motion, two necessary conditions for earthquake-induced tsunami. As will be discussed later, the aforementioned region of enhanced tectonic activity is nearly coincident with the provinces most liable for submarine landslides, at least from a seafloor slope point of view.

For the purposes of the analysis presented herein, I will assume that earthquake-induced tsunami are most efficiently generated by thrust, reverse, or normal faulting earthquakes of magnitude, $M \geq 6.5$, and that they either cause significant deformation of the seafloor or rupture the seafloor. Deformation or rupture caused by strike-slip events are considered unlikely to cause tsunami because of the lack of significant vertical motion in their source mechanisms; events $M \leq 6.5$ may cause tsunami, but would pose no significant threat to populated areas. I will also assume that the primary feature causing tsunami generated by landslides is the slope of the seafloor. While other factors such as nature of the material on the seafloor (volcanic, sedimentary, etc.) and bedding and dip also influence tsunami potential, consideration of those factors is beyond the scope of this work. The reader is referred to the tsunami catalog of O'Loughlin and Lander (2003) as a source for Caribbean tsunami, and the IPGH/MIDAS catalog as well as to historic catalogs as a source for large earthquakes (Shepherd and Lynch, 1992; Tomblin and Robson, 1977; Utrera 1927; McCann, 2004) and to Mann et al., (1995) for tectonics.

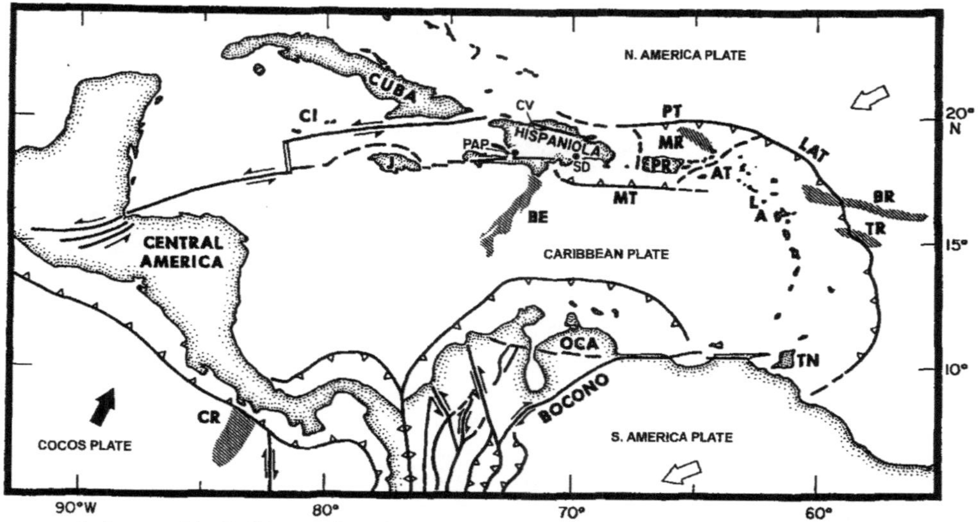


Figure 1. Generalized neotectonic features of the Caribbean region (after Mann and Burke, 1984). Northwestern, western, and eastern margins are comparatively simple; along the other margins, microplates form a buffer zone between the large plates. Convergent margins are characterized by normal faulting in the region near and seaward of the trench axis, thrust events in the zone of interplate contact. Strike-slip event and interspersed reverse and normal events are common along the NW portion of the plate margin. Motion between microplates, are identified in the NE region are complex but predominately extensional in nature. See text for details. Features identified by abbreviations on this map are: the Cayman islands (CI), Port-au-Prince (PAP), Santo Domingo (SD), Cibao valley (CV), Puerto Rico trench (PT), Main ridge (MR), Puerto Rico (PR), Muertos trough (MT), Beata ridge (BE), Anegada trough (AT), Lesser Antilles trench (LAT), Lesser Antilles (LA), Barracuda ridge (BR), Tiburon rise JR), Trinidad (TN), and the Cocos ridge (CR). The white arrows indicate convergence directions of plates relative to the Caribbean according to Jansma et al. (2000).

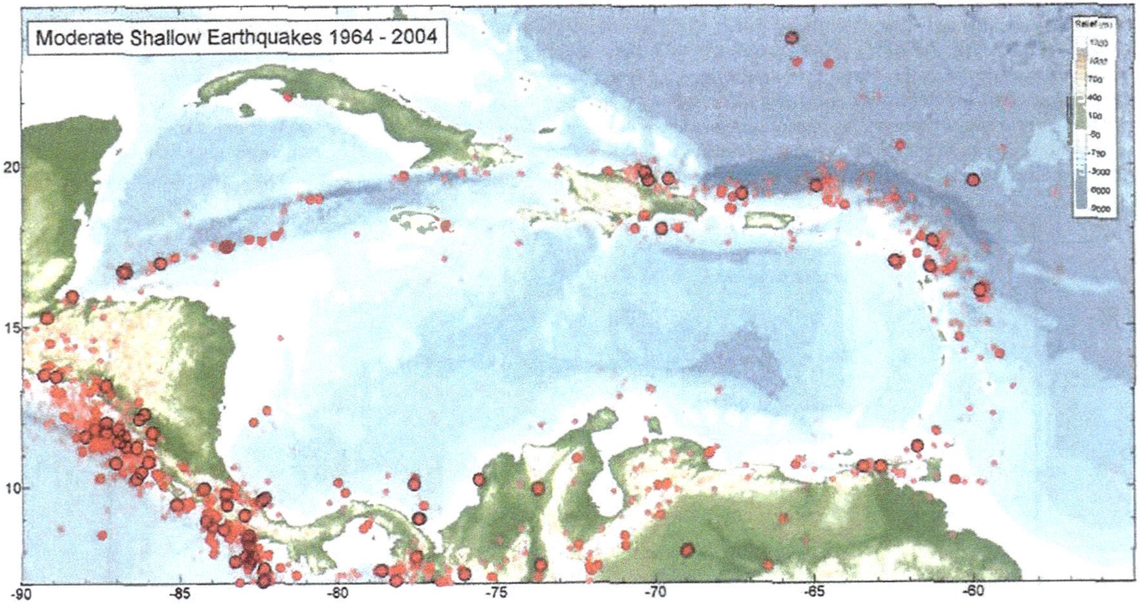


Figure 2. Moderate and smaller events (4.5 ≥M > 6.5) for the period 1964- 2004. Only events with depths of 50 km or less are shown. Events were taken from the IPGH and MIDAS/NEIC catalogs. Spatial distribution of these and stronger/historic events, along with other tectonic considerations are used to estimate earthquake tsunami sources for the Caribbean Basin.

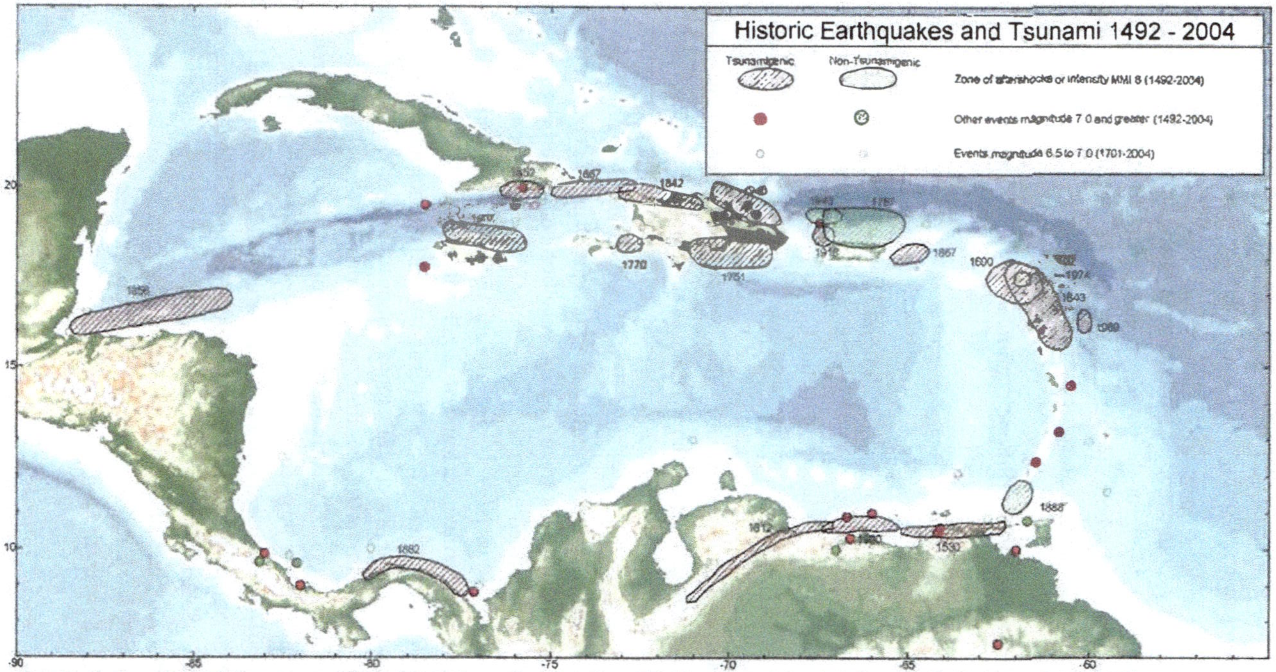


Figure 3. Historic, large (M ≥7.0), and strong (7.0 ≥M ≥6.5) recent earthquakes of shallow focus in the Caribbean Basin for the period 1492-2004. Ruptures zones were estimated using region of Modified Mercalli Intensity of VIII and higher, or from 30 days aftershocks. Events along the Pacific margin of Central America and the interior of South America are not shown. Large earthquakes of the last few centuries cover much of the margin of the Caribbean Plate. Major exceptions are the southern Lesser Antilles and the region west of Jamaica. Historic reporting in the latter region is probably incomplete. Two distinct bands of earthquakes are found from Hispaniola through Cuba and Jamaica. Tsunamigenic events are red others are green.

As noted above, the Caribbean has two primary conditions for tsunami generation, a long belt of strong and large earthquakes of shallow focus, and steep slopes that generally flank the islands of the Greater and Lesser Antilles chains. Unfortunately, both conditions are nearly spatially coincident, with a few notable exceptions, and lie near the populated islands and coasts of the region. I will delve into the basis for both sources of the tsunami threat noting the tectonic and geomorphic framework, and historic events as well.

## 2. Tectonic Framework of Caribbean region

### *2.1. Caribbean Plate*

The Caribbean Plate (CA) is one of a dozen rigid parts of the earth's surface. It is bounded to the east, north and south by the North America (NA) and South America plates (SA), and to the west by the Cocos plate and extends from the east near the Lesser Antilles, to the north near Puerto Rio, Hispaniola, and Jamaica of the Greater Antilles, and to the west along the pacific margin of Central America, and to the south along the northern margin of South America (Figure 1.). If we consider the Caribbean Plate as stationary, then the NA and SA move slowly to the west at about 20 mm/yr (Mann et al., 2002, Dixon et al., 1997). The NA and SA plates subduct beneath the eastern margin of the CA plate descending into the upper mantle and forming a chain of active volcanoes along the Lesser Antilles. The trace of the subduction zone, the Lesser Antilles Trench, is only well defined along the north part of the subduction zone. In the south it is obscured by a large wedge of sediments that the Caribbean plate encountered off the northern margin of South America. Along the northern margin of the CA plate, motion is transpressional, with oblique subduction along the margin near Puerto Rico, collisional subduction near Hispaniola, and dominated by strike-slip tectonics from near Jamaica, westward to Central America. Several microplates distribute in a complex fashion slip between the NA and CA plates along this section of the plate margin (Byrne, et al., 1985; Mann et al, 1995). The western margin of the Caribbean Plate lies in the Pacific basin, and so is out of the scope of this work. However, along the western margin of the Caribbean Sea, there are clear active "intraplate" tectonics, both extensional and compressional, and will be discussed in moredetail later. The southern margin of the Caribbean plate lies along a complex convergent margin off Venezuela and strike-slip faults on land.

### *2.2 Eastern Subduction Zone*

Nearly east-west convergence of the NA and SA plates against the Caribbean plate occurs along the 1000 km long eastern margin of the Caribbean plate. Somewhere to the east of this convergent margin lies the enigmatic boundary between the NA and SA plates. Motion along that boundary is very slow, only a few mm/yr (Muller et al., 1999) but likely to produce occasional earthquakes with significant vertical components of motion. Relatively old Atlantic Ocean seafloor, containing several 1-2 km high aseismic ridges, slides nearly aseismically beneath the overriding CA plate. However, not all of the interplate motions occur aseismically. Great earthquakes in 1843, and probably 1690 ruptured the main interface between the converging plates. Each of these event produced significant tsunami. The downgoing plate bends and, because of its age, is capable of producing large ($M \geq 7$) plate bending (intraplate) earthquakes seaward as well as just landward of the trench axis. The 1969 (Ms 7.2) earthquake is probably a good example of such an event. Active fractures causing NNW-SSE extension (Feuillet et al., 2001) cut the overriding plate from about the latitude of Guadeloupe and northward along the arc. Events such as that in 1974 and 1986 are good examples of quakes occurring on these intraplate fractures. The 1974 earthquake, while large enough to have caused a significant tsunami, did not. In a detailed study of that event, McCann et al. (1984) determined that the rupture plane, while cutting through most of the crust, did not reach the seafloor.

The southern portion of the eastern convergent margin is notably different for three reasons. First the island arc does not exhibit the intraplate fractures as in the north. Second, there is an extremely wide accretionary prism, and third at least in the last 500 years large interplate thrust earthquakes have been relatively sparse. This region may not have the same potential for tsunami as in the north.

### *2.3 Oblique Subduction and transition to strike-slip regime*

The northernmost portion of the eastern subduction zone smoothly changes strike from north to west near the end of the Lesser Antilles and the beginning of the Greater Antilles. The trench axis, forearc, and the islands all participate in this change. Also, the trench axis and forearc both deepen and there is a significant break in the island chain where the Anegada Passage intersects the island chain. Although some convergence occurs along the westerly trending portion of the plate edge, most motion is highly oblique, occurring on a thrust plane dipping gently to the south. In this region, in contrast to the distributed arc deformation in the Lesser Antilles, relatively rigid microplates clearly buffer the motion between the major plates, forming numerous interconnected systems of faults, most of which are submarine. Most faulting between the microplates is high angle normal faulting, thus likely to produce tsunami if the events are sufficiently large and rupture of the seafloor occurs. A major earthquake in 1787 probably ruptured the plate interface north of Puerto Rico. A more recent rupture on that interface occurred in 1943. Neither was associated with tsunami. Large tsunamigenic earthquakes ruptured faults along the microplate margins in 1867, and 1918.

Near Hispaniola, the tectonic environment becomes more complicated and a significant portion of interplate and microplate motion occurs onland. Nevertheless, to the north the western extension of the Puerto Rico Trench, here filled by the shallow Bahama Bank, and the Muertos Trough to the south have both been sources of major thrust earthquakes. Ones in 1946 in the north, and 1751 in the south, produced significant tsunami. As mentioned before, Hispaniola is the transition from oblique subduction in the east, through collisional tectonics, to a strike-slip regime with limited thrust belts in the west. Although there is a transition of tectonic style, there is still a record of tsunamigenic earthquakes, or at least tsunami accompanying earthquakes. Events in 1692 off Kingston Jamaica and 1812 are good examples. Significant thrust belts lie off the south coast of Cuba, and perhaps off the north coast of Jamaica. Most plate motion occurs along well-developed strike-slip faults systems and a few intervening transtensional or compressional segments, and except for those cutting the island of Jamaica, all are submarine.

### *2.4 Strike-Slip Faults, Transtensional and other Regimes*

Extending west from Jamaica and Cuba are two sub-parallel strike-slip fault systems that merge at the Cayman spreading center. West from there is a single fault system that runs north of the North coast of Honduras, and continues onland in Guatemala. Although no vertical deformation has yet been observed on the seafloor in this region, a major earthquake in 1856 was accompanied by a damaging tsunami.

Holcombe et al., (1990) noted a wide transtensional regime off the east coast off Central America. It extends from Jamaica south to panama along the Nicaragua Rise. This region could produce significant earthquakes of shallow focus, and thus represent a significant tsunami risk for the region. A large earthquake in 1941 is a good example of the earthquake threat. That extensional zone nearly abuts a thrust belt along the east coast of Costa Rica and Panama. In the last several hundred years that region has been the source of large earthquake and tsunami. Events such as 1882 and 1991 are good examples of the threat this region poses.

A wide, arcuate accretionary prism along the northern margin of South America, while not very seismically active in the last 500 years, probably poses an earthquake and tsunami threat. The crust in the vicinity of the Beata Ridge, believed to be experiencing slow compression (Mauffret and Leroy, 1999) and that along the old transform fault along the eastern margin of the Yucatan peninsula, and then easterly along the northern margin of Cuba following the Cuba- Bahama Bank suture, probably pose a tsunami threat, albeit poorly defined.

## 3. Tsunami Hazard Based on Potential Seismic Sources

The Caribbean Basin's tsunami threat from seismic sources can be classified based on the tectonic mechanism generating the earthquake source. In Figure 4, the various predominate styles of deformation are noted, and the likely tsunami threat associated with each earthquake source noted. The tsunami threat is consider high, medium or low, based on likely angle of faulting, slip direction (varying from vertical to horizontal), simplicity of margin and history of moderate and large earthquakes, and tsunami.

The eastern subduction zone extending from Trinidad northward along the Lesser Antilles and then westward along Puerto Rico and Hispaniola is composed of three parallel belts of earthquake sources. Seaward at, and just landward of the trench axis from Guadeloupe and to the north and west lies a region of plate bending where normal faulting earthquakes predominate. While these plate-bending events might be less frequent than earthquakes on the plate interface, the nature of their motion, high angle normal faulting, makes them efficient producers of tsunami. Events as large as magnitude 8 could occur in this seismic belt.

Immediately landward of that belt lies the main plate interface along which large interplate thrusting occurs. This main plate boundary is characterized by along strike variations in interplate coupling, and probably a corresponding change in tsunami threat. Roughly the southern half of the convergent margin of the Lesser Antilles lies landward of an extremely wide accretionary prism, a feature at coincides with that segment of the margin producing few large interplate earthquake in the last several hundred years. If this pattern persists into the future reflecting a real change in interplate coupling, then both large earthquakes and tsunami would be fewer in number than in the northern region. The third inner zone is a belt of high angle crustal faults within the arc massif (Feuillet et al., 2002). Such faults, while generally short, could produce

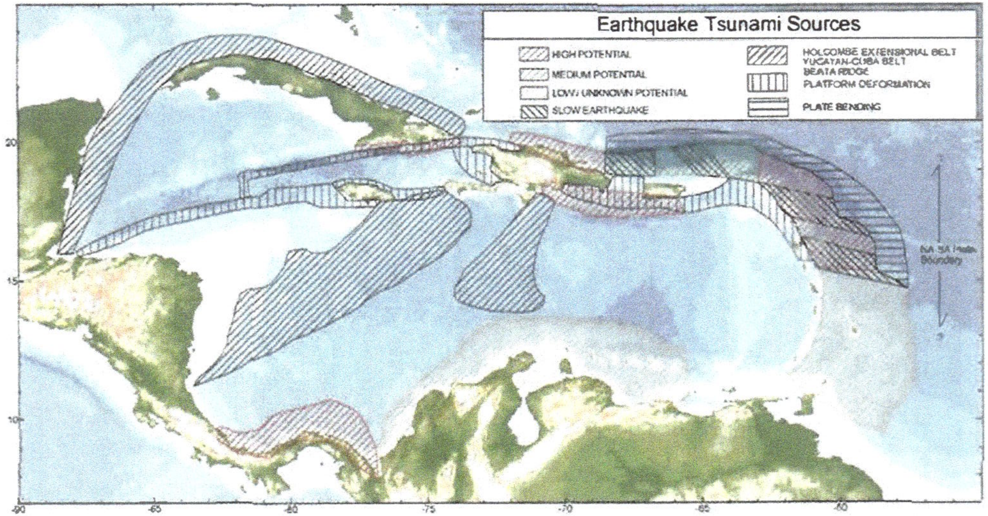


**Figure 4. Earthquake Tsunami Sources for the Caribbean Basin. Convergent margins are subdivided into high medium and low potential for interplate earthquake to generate a tsunami based on historic record, obliquity of convergence, and width of accretionary prism. Seaward of the NE portion of the subduction zone intraplate earthquake associated with the bending of the downgoing plate are probable. Segments of subduction zones near the trailing edges of rough seafloor are noted as possible locations for slow or tsunami earthquakes. Active faults in other portions of the main plate boundary zone, some of which could generate tsunamigenic earthquakes, are also shown. Regions with low rates of deformation but still potentially sources for tsunamigenic earthquakes are shown, these include the transtensional belt identified by Holcombe, deformation near the Beata Ridge identified by Mauffret and Leroy, and the old transform/convergent margin near the Yucatan peninsula and Cuba.**

large earthquakes capable of damaging tsunami.

Great, tsunamigenic earthquakes characterize the hazard of the northeast corner of the Caribbean plate. The presence of aseismic ridges probably plays a role in influencing the occurrence, and limiting the extent of rupture, during these large events. A short segment of the margin near Puerto Rico and the Virgin Islands, and perhaps the northernmost portion of the Lesser Antilles is characterized by a higher level of aseismic slip. Nevertheless, a great earthquake in 1787 probably ruptured the plate interface north of Puerto Rico. As plate motion along this segment of the convergent zone is nearly parallel to the trench axis, little vertical deformation accompanies these large interplate events. No tsunami was observed with the great earthquake in 1787.

The margin north of Hispaniola, while still being classified as convergent in nature has many aspects of a collisional zones, as the buoyant Bahama Bank fills the trench here. Seismic coupling is estimated to be much stronger here, large earthquakes more frequent, and accompanied by tsunami.

Short segments of thrusting are found in association of predominately strike-slip regimes south of Cuba, possibly along the northern coast of Jamaica, and along very short segment of the transform faults that lay between Jamaica and Central America. Another zone of thrusting is found off the Coast of Panama and Costa Rica and south of Hispaniola and Puerto Rico. Each of these regions might produce reverse or thrust faulting earthquakes, possibly producing tsunami. However, the complex nature of the features near Cuba and Jamaica, and their restricted extent, limits the probable size of maximum earthquake (and tsunami). The better developed thrust belts near Hispaniola, and in the SW corner of the Caribbean basin have demonstrated historically, and probably still have the potential for more extensive earthquake ruptures and tsunami. The margin south of Puerto Rico is characterized by strain partitioning into thrusting normal to the margin and strike-slip faulting (with a normal component?) along the investigator fault within the overriding margin (La Forge and McCann, 2004). A similar tectonic environment was found along the margin south of Cuba (Calais et al., 1992). Analogous complexities probably exist in each of the aforementioned thrust belts suggesting an intricate mix of tsunamigenic sources.

Zones of transtension, with normal faulting dominating the local style of deformation are found in the Anegada Passage, the Mona Passage, east and west of Puerto Rico respectively (McCann, 1998, McCann et al., 1994). Other transtensional zones are found in the Morant basin, Yallas Basin, the Cayman Ridge, east, southeast and northwest of Jamaica, and a wide zone extending southwest of Jamaica (Mann et al., 1995; Burke, 1967; Holcombe et al., 1990). The high angle nature of faulting characterizing these regions makes them prime sources for tsunamigenic earthquakes, although the rates of motion of the causative faults is generally relatively low.

## 4. Slow Earthquakes, Subduction of Rough Seafloor, and their Relation Tsunami Hazard

There is a special class of earthquakes that should be noted here because they have a special bearing on tsunami hazard. Most major or great earthquakes rupturing the plate interface at a subduction zones are tsunamigenic. The accompanying tsunami is primarily generated by elastic rebound of the accretionary prism, *i.e. its attendant vertical and horizontal displacement of the* leading edge of the overriding plate, coupled with coseismic deformation along the primary thrust fault or a secondary fault within the stiffer parts of the accretionary prism. In some cases, however, earthquake rupture proceeds more slowly than normal, and is believed to occur within the softer sediments of the accretionary prism. Seismologists call this a "slow" or "tsunami" earthquake (Kanamori, 1972), because of their ability to efficiently generate large tsunami. Bilek and Lay (2002) have noted the generally common nature of small and moderate sized slow earthquakes, suggesting that the conditions for this phenomenon may be widespread in the world's accretionary prisms. If this is true, then these devastating slow or tsunami earthquakes could occur in virtually any subduction zone. While this might be the case, the spatial distribution of large tsunami earthquakes suggest that the subduction of seafloor relief, which demonstrably influences the stability, dynamics and development of accretionary prisms, interplate coupling as well as the occurrences of large earthquakes, may enhance the likelihood of slow earthquakes along certain portions of the world's subduction zones. While a detailed demonstration of the data relating slow earthquakes and subduction of seafloor roughness is beyond the scope of this note, a brief summary outlining the salient points is given and more complete discussion of this issue is reserved for elsewhere (McCann, in prep.).

Under "normal' circumstances, accretionary prism is thought be to in a state of quasi-equilibrium. That equilibrium is related to the mechanical properties of the prism, and the slope of the top and bottom of the prism. That is, the seaward slope of the prism toe is stable just as long as the mechanical conditions of the material in the prism do not change and the slope of the subducting seafloor is also unchanged (Davis, 1983). Subduction of seafloor relief obviously changes this equilibrium state. Positive seafloor relief such as a seamount or a scarp increases the basal slope of the oceanic crust supporting the prism causing the prism to deform, thereby adjusting to a new equilibrium. Dominguez et al. (2000) studied the influence of subducting seamounts and other relief and they found, that as the relief passes deeper into the subduction zone it causes a shadow zone to develop in its wake. That is, the main interface for displacement between the overriding and downgoing material is deflected up along the leading edge of the relief, but does not follow down its trailing edge to return to the interface of the downgoing crust and overriding sediments, but rather cuts through the material of the accretionary prism. This permits the lower portion of the accretionary prism to be carried down into the subduction zone along the trailing edge of the subducting relief. It also means that interplate rupture in this shadow zone occurs within the softer accretionary prism, not at its base. Seismic rupture in softer material is slower than normal due to the low rigidity of the material being broken, i.e. slow earthquakes.

Obviously these observations presume the presence of an accretionary prism, with slow or no erosion, and possible net accretion. These observations also suppose that the relief is limited in extent both vertically and horizontally. If the relief is too high, then it will more likely "collide" with the prism causing substantial deformation, compaction and other radical changes to the condition of the prism; and if the feature is not limited horizontally, it will impinge on the prism never to have a trailing edge. This case would occur if the feature is quasi linear, such as an aseismic ridge, and whose strike is parallel to the local convergence direction.

The observations of Dominguez (2000) provide a mechanism by which subduction of seafloor relief of limited extent may cause a profound, yet transient change, in the character of the plate interface. One sufficiently important to change the character of interplate earthquakes occurring near the trailing edge of the subducting relief from normal rupture velocities to slow or tsunami type. I suggest that this profound change in the subduction process make it more likely that certain margins will suffer tsunami earthquakes. Therefore, regions along the Eastern subduction zone that lay at the trailing edge of a submarine feature of limited relief on the downgoing seafloor have been noted as likely sites for tsunami earthquakes (Figure 4).

## 5. Steep Seafloor, Submarine Slides, and Tsunami Hazard

Tsunamigenic earthquakes are not the only source of tsunami. Some tsunami appear to also be associated with submarine slumps or slides (Watts and Grilli, 2002). However, submarine slides/slumps are primarily triggered by earthquake shaking, and thus in some cases the cause of the tsunami will be uncertain, being possibly generated by an earthquake source, or a submarine slide/slump. The evidence for a submarine slide/slump associated with historic earthquakes is usually found by a reported break in a submarine cable. More ancient slide/slump events are found on images of debris on the seafloor. The record of tsunami in the Caribbean is interesting in that many of them have occurred in regions where tectonic movements are understood to be primarily horizontal in nature (strike-slip source mechanisms). This suggests that submarine slides or slumps, albeit triggered by earthquake shaking, are common in the region. From here on I will assume that earthquake triggered mass movements of the seafloor are slides. I have no evidence to support this assumption, but for the scope of this present work will relate seafloor topography to the hazard of earthquake triggered submarine slides, saving slumps for another time.

It is generally thought that submarine slides occur in regions of unstable seafloor, i.e. region of relatively high seafloor slopes. Perhaps tilted bedding in submarine outcrops also plays a role in this process. In this section I will discuss the spatial distribution of seafloor relief in the Caribbean Basin region and suggest the role that various submarine features may play in the regional tsunami threat. I will presume that the potential for a submarine slide can be simply mapped, at least to the first order, by seafloor slope as tsunami generation is influenced by water depth at generation, and by depth of seafloor where the steep slope is found (Watts and Grilli, 2002). Slopes of seafloor were calculated using the ETOPO-2 database (NGDC, 2001) and then a "reference" tsunami amplitude was calculated using the formulation of (Watts and Grilli, 2002) for submarine slides:

$$A = 0.224\ T\ (b/d)^{1.25}[(\sin(\theta)^{1.29}-0.75(\sin(\theta)^{2.29}+0.17(\sin(\theta)^{3.29}] \quad (1)$$

where A is the tsunami amplitude in meters, T is the maximum slide thickness in meters, b is the initial slide length in meters, d is the mean slide depth in meters, and $\theta$ is the mean incline angle in degrees, (assumed to be seafloor slope). I arbitrarily assume the slide has thickness of 20 meters, length of 1000 meters. At a mean depth (d) of 100 meters, and an incline ($\theta$) of 6°, the tsunami height would be about 4 meters. Figure 5 presents these "reference tsunami amplitudes" for the Caribbean Basin. A similar map could be produced for slides of other dimensions or for slumps.

While the map is self-explanatory, some salient points deserve to be noted. 1- High reference tsunami amplitudes are found near every island, 2- Large amplitudes are found along strike-slip margin in west, and may explain the origin of several of the tsunami in this region; 3- Steep slopes found northeast of the Northern Lesser Antilles, but west of the margin of Southern Lesser Antilles; 4- Numerous other regions near strong seismic sources (of all types) have high amplitudes, and could, therefore, be sources for earthquake-induced submarine slides and tsunami 5- The low slide hazard along the *Pacific* coast of Nicaragua and El Salvador, suggests that shelf/slope submarine landslides may be more of a fundamental problem in the Caribbean basin than around the Pacific Ocean.

## 6. Conclusions

The Caribbean Basin suffers from the presence of several tsunami sources. They include numerous earthquake sources exhibited by zones of major faulting surrounding and within the basin, slide/slump sources as exhibited by steep slopes in relatively shallow water near many of the islands. Using the historic record of earthquakes and the tectonic framework of the basin numerous earthquakes sources have been identified and classified.

The primary source of shallow, potentially tsunamigenic earthquakes is the eastern subduction zone extending from Trinidad north then west through Hispaniola. It is composed of three bands of earthquakes sources, and outer belt representing a zone of plate bending earthquakes, the main plate interface, source of numerous great earthquakes (M > 7 ¾), and finally a band of high angle active faults on the island platform from Guadeloupe to the north and west along the island arc. The main plate interface subdivided into three categories of high, medium and low/indeterminate tsunami potential reflecting the historic earthquake/tsunami record, direction of plate convergence and width of the accretionary prism. A separate supplementary category, the potential for slow or tsunami earthquakes, is added to those regions along the eastern subduction zone recently affected by the subduction of significant seafloor relief.

Other convergent margins are found north of South America indeterminate potential), south of Cuba and along the Caribbean margin of southern Central America (both high potential). Several systems of active faults (primarily strike-slip) extending west of Jamaica to the east margin of Central America also represent a significant earthquake threat. Events in this region may trigger submarine landslides, and, therefore, be accompanied by tsunami. Several zones of active faulting, but with lower rates of deformation, also represent a non-negligible tsunami threat, these include the Holcombe transtensional zone, the Yucatan-Cuba Belt and the Beata Ridge.

Regions of potential tsunami induced by submarine landslides were identified by manipulating ETOPO2 seafloor bathymetry and seafloor slope to develop a map of relative tsunami amplitude for a submarine slide measuring20 m by 1 km. Regions with seafloor slope of 10° or more were superimposed on the tsunami amplitude map to indicate regions of particularly high potential. The Caribbean, because of the presence of several limited island platforms,

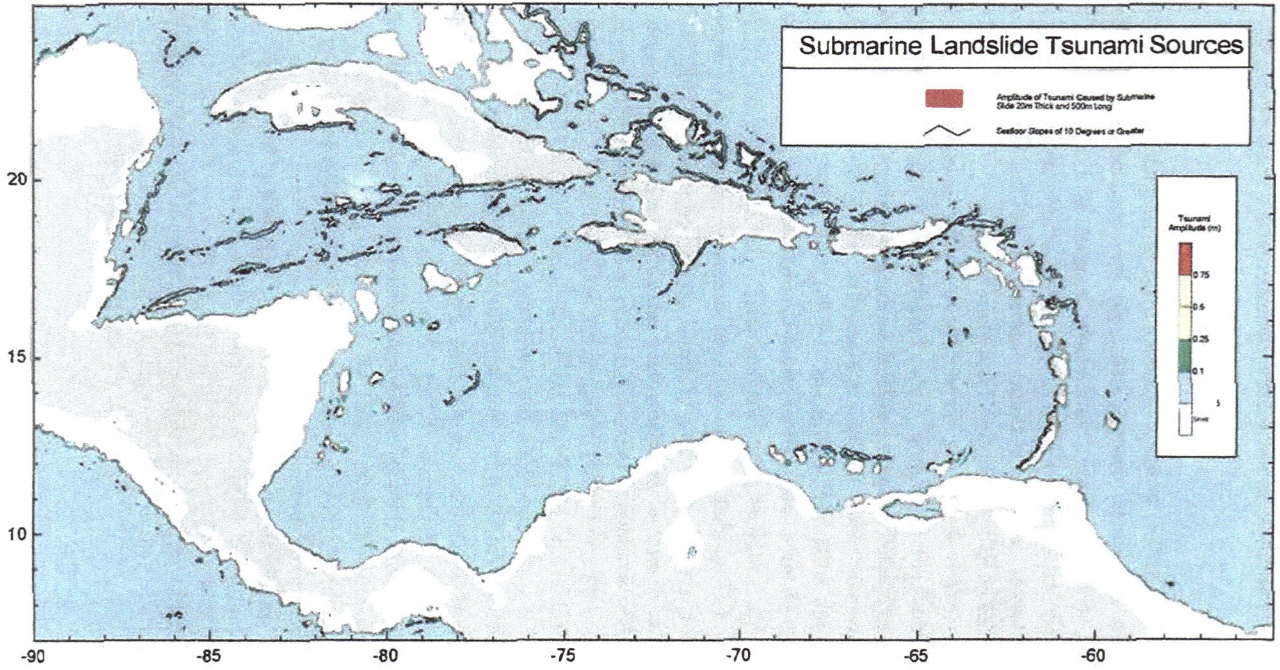


**Figure 5** Landslide tsunami sources in the Caribbean Basin. Regions are color coded by the amplitude of tsunami generated by a 20-meter thick 1 km long submarine landslide. Slides are most tsunamigenic in shallower water, so steeper region near island platforms are typically regions of high tsunami potential. Regions with slopes of 10° or greater are also shown. I consider these regions, in combination with regions color-coded in red, to be particularly dangerous.

seems particularly prone to the tsunami-slide hazard. Similar maps using different slide dimensions could provide a more clear insight to the total slide-tsunami threat and a similar family of maps could provide similar information about slump related tsunami.

Much work needs to be done before we have a good idea of the tsunami threat in the Caribbean. Clearly, the density of nearby sources is high, perhaps higher than in many other parts of the world. Mitigation efforts would benefit from development of:

- An updated, cross-referenced earthquake/tsunami catalog developed by closing the obvious gaps in the IPGH historic catalog, and using improved earthquake locations such as are found in the Engdahl, Centennial and other catalogs (Engdahl. 1998),
- A catalog with the location and character of slides and slumps in the Caribbean region found by executing a search of the seafloor using existing single and multichannel seismic reflection data.
- A digital atlas/database of available and useful geologic and geophysical data for the study of tsunami in the Caribbean Basin (7° to 25° N, 55° -90° W), including but not limited to parameters such as earthquakes and tsunami and location of existing slumps/slides as mentioned above, focal mechanisms, bathymetry, gravity, magnetics, sediment thickness, seismic velocity of sediments, active faults, regional stress, rates of vertical and horizontal deformation, expected ground accelerations from earthquakes, seafloor geology, and bibliography of important tsunami manuscripts.

**7. References**


Bilek S. L. And T. Lay, 2002, Tsunami earthquakes possibly widespread manifestations of frictional conditional stability, Geophysical Res. Letters, 29, GL015215.

Burke, K., 1967, The Yallahs Basin: A Sedimentary Basin Southeast of Kingston, Jamaica, *Marine Geology*, 5, 45-60.

Calais, E., N. Bethoux, B. Mercier de Lepinay, (1992): From Transcurrent Faulting to Frontal Subduction: A Seismotectonic Study of the Northern Caribbean Plate Boundary from Cuba to Puerto Rico, Tectonics,11, 114-123.

Chuy T., and M. Rodriguez (1980) La actividad sísmica de Cuba basada en datos historicos. Investigaciones Sismologicas en Cuba. No 1. La Habana

Davis, D., J. Suppe, F. A. Dahlen, 1983, Mechanics of fold-and-thrust belts and accretionary wedges, J. Geophys. Res, 88, 1153-1172.

Dixon, T. H., and A. Mao, A GPS estimate of relative motion between North and South America, Geophys. Res. Lett., 24, 535 – 538, 1997.

Dominguez, S., J. Malaviella, and S. E. Lallemand, 2000,Deformation of accretionary wedges in response to seamount subductionz:Insights from sandbox experiments, Tectonics, 19, 182-196.

E.R. Engdahl, Van der Hilst, R.D., and Buland, R.P., 1998, Global teleseismic earthquake relocation with improved travel times and procedures for depth determination, Bull. Seism. Soc. Amer. , v. 88, pp. 722-743

Feuillet, N., I. Manighetti, P. Tapponier. 2002, Arc parallel extension and localization of volcanic complexes in Guadeloupe, Lesser Antilles, JGR, 107, b2, 2331.

Feldman, L., Mountains of Fire, Lands that shake: Earthquakes and Volcanic Eruptions in the historic past of Central America (1505-1899), Labyrinthos, 1993, 295p.

Hall, M., 1907, The great earthquake of January 14, 1907 and the after-shocks, *Jamaica Weather Report No*. 337

Holcombe, T., J. Ladd, G. Westbrook, T. Edgar, C. Bowland, 1990, Caribbean marine geology; Ridges and basins of the plate interior, in Dengo, G., Case, J. eds., The Caribbean Region: Boulder, Colorado, Geological Society of America, The Geology of North America, V. H.

Jansma, P. and 6 other authors, 2000, Neotectonics of Puerto Rico and the Virgin Islands, northeastern Caribbean, from GPS Geodesy, Tectonics, 19 (6), 1021-1037.

Jany, I., A. Mauffret, P.Bouysse, A. Mascle, B. Mercier de Lpinay, V. Renard, and F. Stphan. Relev bathymtrique Seabeam et tectonique en dcrochements au sud des Iles Vierges (Nord-Est Caraibes). C.R. Acad. Sc. Paris, t. 304, Srie II, No. 10, 1987

Kanamori, H. (1972), Mechanism of tsunami earthquakes, Phys. Earth Planet. Inter., 6, 346–359.

Kelleher, J., L. Sykes and J. Oliver (1973) Possible Criteria for Predicting Earthquake Locations and Their Application to Major Plate Boundaries of the Pacific and the Caribbean, J. Geophys. Res., 78, 2547-2585.

Mauffret, A., S. Leroy, 1999, Neogene Intraplate Deformation of the Caribbean Plate at the Beata Ridge, in Caribbean Basins. Sedimentary Basins of the World, 4. Edited by P. Mann (Series Editor: K. J. Hsu) 627-669.

Mann, P., E. Calais, J.-C. Ruegg, C. DeMets, P. E. Jansma, and G. S. Mattioli, Oblique collision in the northeastern Caribbean from GPS measurements and geological observations, Tectonics, 21(6), 1057, doi:10.1029/2001TC001304, 2002.

Mann, P, Taylor, F.W. Lawrence Edwards, R., and Teh-Lung Hu, 1995. Actively evolving microplate formation by oblique collision and sideways motion along strike-slip faults: An example from the northeastern Caribbean plate margin. Tectonophysics, 246:1-69.

McCann, W. 2001, Amenaza de Terremoto en La Hispaniola, Conferencia Internacional Sobre Reduccion de Riesgo Sismico en la Region del Caribe, Segundo Seminario Dominicano de Ingenieria Sismica, July, 2001, Santiago, Dominican Republic.

McCann, W., L. Feldman, Maribel, McCann, Catalog of Felt Earthquakes for Puerto Rico ans neighboring Islands 1492-1899, with some additional information for some 20th Century earthquakes, unpubl manuscript.

McCann, William R; Pennington, Wayne D., 1990. Seismicity, Large Earthquakes, and the Margin of the Caribbean Plate. The Geology of North America ed. Vol. H, The Caribbean Region. Geological Society of America, Boulder, CO. 528 pages.

McCann, W. J. Dewey, A. Murphy and S. Harding, 1982, A Large Normal-Fault Earthquake in the overriding wedge of the Lesser Antilles subduction zone: The earthquake of 8 October 1974, Bull. Seism. Soc. Amer., 72, 2267-2283.

McCann, W., 1998, Tsunami Hazard of Western Puerto Rico from Local Sources: Characteristics of Tsunamigenic Faults, Seagrant internal report, Univ. of PR, Mayaguez, Dept of Marine Sciences, 82p.

McCann, W., R. Millan, and J.C. Moya, 1994; Seismic Hazard Map for Puerto Rico, 1994, Department of Natural Resources, San Juan, PR, 60p.

Mendoza, C. And S. Nishenko. (1989). The North Panama Earthquake of 7 september 1882: evidence for active underthrusting. Bull. Seism. Soc. Am., Vol. 79, No 4, 1264-1269.

Muller, R. Ditmar, J-Y Royer, S. C. Cande, Walter R. Rost and S. Maschenkov, 1999, New Constraints on the Late Cretaceuos/ Tertiary Plate Tectonic Evolution of the Caribbean, in Caribbean Basins. Sedimentary Basins of the World, 4. Edited by P. Mann (Series Editor: K. J. Hsu) p33-59.

O'Loughlin, K. F., J.F. Lander, Caribbean Tsunamis: A 500 year History 1498-1998, 2003. Kluwer Academic Publishers, Norwell MA, 263p.

Pacheco, J. F.; Sykes, L. R., 1992. Seismic moment catalog of large shallow Earthquakes, 1900 To 1989. BSSA 82, 3, 1306-1349.

Reid H. and S. Taber, 1920, The Virgin Islands Earthquakes of 1867-1868, Bul. Seism. Soc. Amer, 10 (1), p 9-30.

Shepherd, J. and L. Lynch, 1992. An earthquake catalogue for the Caribbean Part I, the pre-intrumental period 1502-1900, in Report of the Second Technical Workshop Sesimic Hazard Project- Latin America and the Caribbean, p95-158, Melbourne, FL.

Tomblin, J. M. and Robson, G. R., 1977. A catalogue of felt earthquakes for Jamaica with references to other islands in the Greater Antilles. Ministry of Mining and Natural Resources, Mines and Geology Division, Special publication # 2, Hope Gardens, Kingston, Jamaica, 243 pp.

Utrera, Fray C., 1927. Noticias Históricas de Santo Domingo, Vol. 5 y 6, Editora Taller, Santo Domingo, Republicana Dominicana.

Watts P. And Grilli, S., Tsunami generation by submarine mass failure: Part I wavemaker models, Submitted, J. Waterway, Port, Coastal, and Ocean Engineering, 2004

Table 1. Caribbean Basin Earthquakes, 1492-2004, of Magnitude 7.0 or larger, depths 0 to 50 km, as found in IPGH, MIDAS, NEIC or supplementary historic catalogs.

| Agency | IPGH | Year | Month | Day | OT | Long. | Lat. | Depth | Mw | MMI | Zone | T |
|---|---|---|---|---|---|---|---|---|---|---|---|---|
| SISRA | 2 | 1530 | 9 | 1 | 14:30:00 | -64.10 | 10.70 | 10 | 8 | 10 | M4 | yes |
| SAL | 8 | 1564 | 4 | 20 | | -70.20 | 19.25 | 33 | 7 | 9 | | no |
| SISRA | 26 | 1641 | 6 | 11 | 12:45:00 | -66.70 | 10.90 | | | 9 | | no |
| SAL | 44 | 1678 | 2 | 11 | 9:00:00 | -75.82 | 20.00 | 10 | 7 | 8 | | no |
| SISRA | 45 | 1684 | 5 | 5 | 0:30:00 | -64.17 | 10.47 | | | 7 | | no |
| SAL | 49 | 1690 | 4 | 5 | | -62.00 | 17.00 | 33 | 7.5 | 8 | K | yes |
| SAL | 50 | 1692 | 6 | 7 | | -76.50 | 18.00 | 10 | 7.5 | 10 | | yes |
| SAL | 109 | 1751 | 10 | 18 | | -70.00 | 18.50 | 33 | 8 | | M3 | yes |
| SAL | 110 | 1751 | 11 | 21 | | -72.00 | 18.50 | 33 | 7.5 | | M3 | yes |
| SAL | 125 | 1766 | 6 | 11 | 0:30:00 | -76.00 | 19.50 | 10 | 7 | 9 | | no |
| NOAA | | 1766 | 10 | 21 | 9:00:00 | -62.50 | 7.40 | | 7.5 | 9 | | no |
| SAL | 134 | 1770 | 6 | 3 | 19:50:00 | -72.50 | 18.50 | 10 | 7.5 | 10 | K | yes |
| NOAA | | 1787 | 5 | 2 | 15:55:00 | -66.00 | 19.00 | 33 | 8 | | MF | yes |
| SISRA | 188 | 1794 | 9 | 10 | | -64.17 | 10.47 | | | 7 | | no |
| SISRA | 193 | 1797 | 12 | 14 | 23:30:00 | -64.10 | 10.60 | | | 9 | | no |
| NOAA | | 1812 | 3 | 26 | 21:07:00 | -67.00 | 10.00 | | 9.6 | 10 | | yes |
| ROJ | 256 | 1822 | 5 | 7 | 6:30:00 | -83.15 | 9.63 | 20 | 7.5 | 9 | | yes |
| SISRA | 263 | 1825 | 9 | 21 | | -61.70 | 10.80 | | | 8 | | yes |
| SAL | 275 | 1826 | 9 | 18 | 9:08:00 | -76.00 | 19.50 | 33 | 7 | 8 | | no |
| MONT | 276 | 1827 | 4 | 3 | | -83.00 | 9.87 | | 7.5 | 7 | | no |
| SISRA | 367 | 1831 | 12 | 3 | 23:40:00 | -61.45 | 12.40 | | | 7 | | no |
| SISRA | 400 | 1837 | 9 | 10 | 18:30:00 | -66.60 | 10.30 | | | 7 | | no |
| SISRA | 409 | 1839 | 4 | 12 | | -64.17 | 10.47 | | | 7 | | no |
| SAL | 439 | 1842 | 5 | 7 | 22:30:00 | -72.50 | 19.80 | 33 | 8 | 9 | K | yes |
| SAL | 440 | 1843 | 2 | 8 | | -62.00 | 17.00 | 33 | 8 | 10 | K | yes |
| SISRA | 446 | 1844 | 8 | 30 | 7:10:00 | -60.83 | 13.23 | | | 8 | | no |
| SAL | 478 | 1852 | 8 | 20 | 13:36:00 | -75.80 | 20.00 | 10 | 7.5 | 9 | CH | yes |
| SAL | 479 | 1852 | 11 | 26 | | -76.00 | 20.00 | 10 | 7 | 8 | CH | yes |
| | | 1856 | 8 | 9 | 22:47:00 | -86.50 | 17.00 | | 8.3 | | FE | yes |
| SAL | 566 | 1867 | 11 | 18 | 20:00:00 | -65.00 | 18.50 | 33 | 7.5 | 8 | RE | yes |
| VIQU | 680 | 1882 | 9 | 7 | 8:18:00 | -79.00 | 10.00 | | 7.5 | 9 | ME | yes |
| SAL | 734 | 1887 | 9 | 23 | | -74.00 | 20.00 | 33 | 7.75 | 9 | M3 | yes |
| SISRA | 735 | 1888 | 1 | 10 | 13:00:00 | -62.20 | 11.30 | | 7 | 8 | K | no |
| GUT | 768 | 1899 | 6 | 14 | | -77.00 | 18.00 | 0 | 7.8 | | | no |

| | | | | | | | | | | | | |
|---|---|---|---|---|---|---|---|---|---|---|---|---|
| SISRA | 777 | 1900 | 10 | 29 | 9:11:00 | -66.00 | 11.00 | 25 | 7.59 | | M4 | yes |
| AMB | 819 | 1904 | 12 | 20 | 5:42:00 | -82.00 | 9.00 | 35 | 7.44 | 6 | | no |
| MACRO | 832 | 1906 | 2 | 16 | 17:25:00 | -60.50 | 14.50 | 50 | 7 | 8 | | no |
| MACRO | 849 | 1907 | 1 | 14 | 8:00:00 | -76.00 | 18.00 | 10 | 7 | | H | yes |
| TOJ | 977 | 1916 | 4 | 26 | 2:21:14 | -82.10 | 9.60 | 25 | 7.3 | 9 | | yes |
| A-N | 991 | 1917 | 2 | 20 | 19:29:00 | -78.50 | 19.50 | | 7.09 | 6 | | no |
| GUTE | 1006 | 1917 | 7 | 27 | 1:01:18 | -67.50 | 19.00 | 50 | 7 | | | no |
| GUTE | 1056 | 1918 | 10 | 11 | 14:14:30 | -67.50 | 18.50 | | 7.3 | 8 | M2 | yes |
| GUTE | 2172 | 1941 | 4 | 7 | 23:29:17 | -78.50 | 17.75 | | 7 | | | no |
| SISRA | 2263 | 1942 | 5 | 2 | | -66.00 | 11.00 | | | 7 | | no |
| GUTE | 2410 | 1943 | 7 | 29 | 3:02:16 | -67.50 | 19.25 | | 7.5 | | K | no |
| SISRA | 2460 | 1944 | 2 | 6 | 3:50:30 | -62.00 | 10.00 | | | 7 | | no |
| GUTE | 2684 | 1946 | 8 | 4 | 17:51:05 | -69.00 | 19.25 | | 7.8 | | K | yes |
| GUTE | 2698 | 1946 | 8 | 8 | 13:28:28 | -69.50 | 19.50 | | 7.4 | | | yes |
| GUTE | 2850 | 1948 | 4 | 21 | 20:22:02 | -69.25 | 19.25 | 40 | 7.09 | | | no |
| ISC | 12995 | 1969 | 12 | 25 | 21:32:27 | -59.64 | 15.79 | 1 | 7 | | M4 | yes |
| ISC | 17469 | 1974 | 10 | 8 | 9:50:58 | -61.99 | 17.37 | 41 | 7.3 | | M | no |
| RSN | 52787 | 1991 | 4 | 22 | 21:56:49 | -83.15 | 9.63 | 23 | 7.59 | 10 | | yes |
| OSS-SA | 54064 | 1991 | 9 | 24 | 4:56:31 | -77.17 | 8.82 | 3 | 7.69 | 0 | | no |

Table Notes: Agency, denotes and IPGH agency code for the source of the earthquake information, IPGH is IPGH sequence number, when absent event, was not in IPGH catalog, Mw is moment magnitude, MMI is Modified Mercalli Intensity, Zone gives source for inferred rupture zone, if absent, no rupture zone is shown in figures, T is whether or a not a tsunami was observed. Table is complete up to August 13, 2004. Key for Zone source information is as follows: CH, Chuy (1980); H, Hall (1907) ; FE, Feldman (1993) ; K, Kelleher et al. (1973) ; M, McCann et al, (1984) ; M2, McCann (1998) ; ME, Mendoza and Nishenko (1989) ; M4, McCann and Pennington (1990) ; M3, McCann (2001) ; MF, McCann, et al., (2004) ; RE, Reid and Taber, (1920) .

Table 2. Caribbean Basin Earthquakes, 1700-2004, of Magnitude 6.5 to 7.0, depths 0 to 50 km, as found in IPGH, MIDAS or NEIC catalogs.

| Agency | IPGH | Yr | Mo | Dy | OT | Lon | Lat | Dep | Mw | Agc | Mag | Sc | Agc | MMI | Tsunami |
|---|---|---|---|---|---|---|---|---|---|---|---|---|---|---|---|
| SAL | 55 | 1701 | 11 | 9 | | -72.50 | 18.50 | 10 | 6.6 | SAT | 6.5 | Ms | SAL | 7 | No |
| SAL | 117 | 1760 | 7 | 11 | | -75.83 | 20.00 | 10 | 6.6 | SAT | 6.5 | Ms | SAL | | No |
| SAL | 161 | 1784 | 7 | 29 | | -72.28 | 19.78 | 33 | 6.8 | SAT | 6.8 | Ms | SAL | 8 | No |
| SAL | 225 | 1812 | 11 | 11 | 10:50:00 | -76.50 | 18.00 | 20 | 6.8 | SAT | 6.8 | Ms | SAL | 8 | Yes |
| SAL | 240 | 1816 | 12 | 22 | | -60.00 | 13.00 | 50 | 6.6 | SAT | 6.5 | Ms | SAL | 6 | No |
| SAL | 291 | 1827 | 12 | 21 | | -60.75 | 14.50 | 50 | 6.6 | SAT | 6.5 | Ms | DOR | 6 | No |
| SAL | 354 | 1830 | 4 | 14 | 11:30:00 | -72.30 | 18.50 | 10 | 6.6 | SAT | 6.5 | Ms | SAL | 7 | No |
| SAL | 442 | 1844 | 4 | 16 | | -66.00 | 18.75 | 33 | 6.6 | SAT | 6.5 | Ms | SAL | 7 | No |
| SAL | 477 | 1852 | 7 | 7 | 12:25:00 | -76.50 | 19.00 | 33 | 6.6 | SAT | 6.5 | Ms | SAL | 5 | No |
| SISRA | 486 | 1853 | 7 | 15 | | -63.60 | 12.10 | 14 | 6.7 | SAT | 6.7 | Ms | FIE | 9 | Yes |
| SAL | 522 | 1858 | 1 | 28 | 21:45:00 | -76.00 | 19.00 | 10 | 6.6 | SAT | 6.5 | Ms | SAL | 6 | No |
| SAL | 543 | 1860 | 10 | 23 | | -67.50 | 18.50 | 33 | 6.6 | SAT | 6.5 | Ms | SAL | 7 | No |
| SAL | 664 | 1880 | 1 | 23 | | -82.40 | 23.00 | 33 | 6.6 | SAT | 6.5 | Ms | SAL | 8 | No |
| SISRA | 679 | 1882 | 9 | 7 | 7:50:00 | -77.00 | 9.00 | | 6.6 | SAT | 6.5 | Ms | IGE | 9 | No |
| SAL | 753 | 1895 | 5 | 20 | 21:44:00 | -61.00 | 18.00 | 33 | 6.6 | SAT | 6.5 | Ms | SAL | 6 | No |
| SAL | 758 | 1897 | 4 | 29 | 14:15:00 | -61.50 | 16.20 | 10 | 6.6 | SAT | 6.5 | Ms | BAL | 8 | No |
| SAL | 759 | 1897 | 12 | 29 | 11:32:00 | -70.70 | 19.50 | 33 | 6.6 | SAT | 6.5 | Ms | SAL | 8 | No |
| MACRO | 840 | 1906 | 6 | 22 | | -76.00 | 19.50 | 10 | 6.6 | SAT | 6.5 | Ms | JBS | 7 | No |
| AMB | 876 | 1910 | 1 | 1 | 11:02:00 | -85.00 | 17.00 | 10 | 7.0 | SAT | 7.0 | Ms | AMB | | No |
| VIQU | 891 | 1910 | 12 | 22 | 3:24:00 | -82.30 | 9.80 | | 6.6 | SAT | 6.5 | Ms | C&V | 5 | No |
| GUTE | 900 | 1911 | 10 | 6 | 10:16:12 | -70.50 | 19.00 | 0 | 6.9 | SAT | 6.9 | Ms | ABE | | No |
| AMB | 910 | 1912 | 6 | 12 | 12:44:42 | -88.00 | 16.50 | 35 | 6.8 | SAT | 6.8 | Ms | AMB | | No |
| GUTE | 956 | 1915 | 10 | 11 | 19:33:12 | -67.00 | 19.00 | 0 | 6.8 | SAT | 6.8 | Ms | PAS | | No |
| GUTE | 990 | 1916 | 11 | 30 | 13:18:00 | -70.00 | 20.00 | | 6.8 | SAT | 6.8 | Ms | PAS | | No |
| GUTE | 1120 | 1920 | 2 | 10 | 22:07:15 | -67.50 | 18.00 | | 6.6 | SAT | 6.5 | Ms | PAS | | No |
| GUTE | 1294 | 1925 | 9 | 29 | 17:33:50 | -62.00 | 18.50 | | 6.6 | SAT | 6.5 | Ms | PAS | | No |
| ROR | 1306 | 1926 | 2 | 1 | 1:17:47 | -62.40 | 10.88 | 5 | 6.6 | SAT | 6.5 | Ms | PAS | 5 | No |
| AMB | 1311 | 1926 | 3 | 17 | 11:53:36 | -82.50 | 12.50 | 35 | 6.8 | SAT | 6.8 | Ms | AMB | | No |
| GUTE | 1354 | 1927 | 8 | 2 | 0:51:46 | -64.50 | 19.00 | | 6.6 | SAT | 6.5 | Ms | PAS | | No |
| ROR | 1414 | 1928 | 9 | 27 | 0:44:09 | -59.49 | 11.58 | 30 | 6.6 | SAT | 6.5 | Ms | PAS | 7 | No |
| ROR | 1428 | 1929 | 1 | 17 | 11:45:42 | -63.98 | 10.35 | | 6.9 | SAT | 6.9 | Ms | PAS | 9 | Yes |
| GUTE | 1559 | 1932 | 2 | 3 | 6:15:55 | -75.50 | 19.50 | | 6.8 | SAT | 6.8 | Ms | PAS | 8 | yes |
| GUTE | 2413 | 1943 | 7 | 30 | 1:02:30 | -67.75 | 19.25 | | 6.6 | SAT | 6.5 | Ms | PAS | | no |
| SISRA | 2441 | 1943 | 12 | 21 | 13:46:21 | -71.00 | 13.00 | | 6.6 | SAT | 6.5 | Ms | R-I | 7 | no |

| | | | | | | | | | | | | | | | |
|---|---|---|---|---|---|---|---|---|---|---|---|---|---|---|---|
| SISRA | 2443 | 1943 | 12 | 22 | 12:53:00 | -71.00 | 13.00 | | 6.6 | SAT | 6.5 | Ms | R-I | 7 | no |
| SISRA | 2444 | 1943 | 12 | 23 | 15:56:05 | -71.00 | 13.00 | | 6.6 | SAT | 6.5 | Ms | R-I | 7 | no |
| GUTE | 2750 | 1946 | 10 | 4 | 14:45:26 | -68.50 | 18.75 | 50 | 6.5 | SAT | 6.4 | Ms | ABE | | no |
| GUTE | 2796 | 1947 | 8 | 7 | 0:40:20 | -75.25 | 19.75 | 50 | 6.8 | SAT | 6.8 | Ms | PAS | | no |
| SAE | 3521 | 1953 | 5 | 31 | 19:58:35 | -70.40 | 19.68 | | 6.9 | SAT | 6.9 | Ms | SAE | 5 | yes |
| SAE | 4220 | 1957 | 3 | 2 | 0:27:33 | -78.11 | 18.35 | | 6.6 | SAT | 6.5 | Ms | SAE | 8 | no |
| CGS | 4308 | 1957 | 8 | 15 | 8:32:56 | -80.00 | 10.00 | | 6.7 | SAT | 6.7 | Ms | ROT | | no |
| ROR | 4335 | 1957 | 10 | 4 | 5:26:03 | -62.77 | 10.86 | 6 | 6.7 | SAT | 6.7 | Ms | ROR | 9 | no |
| SAE | 5655 | 1962 | 1 | 8 | 1:00:23 | -70.46 | 18.40 | 33 | 6.5 | SAT | 6.4 | Ms | SAE | | no |
| SAE | 5733 | 1962 | 4 | 20 | 5:47:51 | -72.13 | 20.50 | | 6.8 | SAT | 6.8 | Ms | SAE | | no |
| ISC | 7153 | 1964 | 10 | 23 | 1:56:05 | -56.11 | 19.80 | 43 | 6.8 | SAT | 6.2 | mb | ISC | | no |
| SISRA | 10226 | 1967 | 7 | 30 | 0:00:03 | -67.40 | 10.68 | 26 | 6.6 | SAT | 6.6 | Ms | MOS | 8 | no |
| ISC | 10634 | 1967 | 12 | 24 | 20:03:13 | -61.19 | 17.42 | 42 | 6.7 | SAT | 6.1 | mb | ISC | | no |
| 1 | 26226 | 1980 | 8 | 9 | 5:45:07 | -88.52 | 15.93 | 9 | 6.5 | CMT | 6.5 | Mw | CMT | | no |
| ISC | 34819 | 1984 | 6 | 24 | 11:17:11 | -69.30 | 18.02 | 20 | 6.7 | CMT | 6.7 | Mw | CMT | | no |
| SBAC | 44604 | 1988 | 4 | 22 | 4:03:32 | -61.37 | 17.22 | 32 | 6.5 | CMT | 6.5 | Mw | CMT | | no |
| GSM | 55214 | 1992 | 5 | 25 | 16:55:04 | -77.87 | 19.61 | 23 | 6.9 | SAT | 6.9 | Ms | GS | C | no |
| MIDAS | 4676 | 1999 | 7 | 11 | 14:14:15 | -88.261 | 15.702 | 10 | 6.6 | /Ms | 6.1 | mb | USGS | | no |
| PDE | | 2003 | 5 | 14 | 6:03:35 | -58.63 | 18.27 | 41 | 6.7 | GS | | | | | no |
| PDE | | 2003 | 9 | 22 | 4:45:36 | -70.67 | 19.78 | 10 | 6.6 | GS/Ms | | | | | no |

Table Notes: Agency, and Agc denotes IPGH agency code for the source of the respective information, IPGH is IPGH/MIDAS sequence number, when absent event was not in IPGH or MIDAS catalog, Yr, Mo, Dy, Ot, is year month day and origin time (UT) of earthquake, Lon, Lat, Dep are Longitude, Latitude and Depth estimates for hypocenter, Mw is moment magnitude, Mag and Sc is other magnitude and its scale, MMI is Modified Mercalli Intensity, Tsunami is whether or a not a tsunami was observed. Table is complete up to August 13, 2004.

# DEBRIS AVALANCHE FORMATION AT KICK'EM JENNY SUBMARINE VOLCANO

HARALDUR SIGURDSSON
STEVEN CAREY
*Graduate School of Oceanography, University of Rhode Island*
*Narragansett RI 02882*

DOUG WILSON
*NOAA OAR/Chesapeake Bay Office, Annapolis, MD 21403*

Kick'em Jenny submarine volcano near Grenada is the most active volcanic center in the Lesser Antilles arc. Multibeam surveys of the volcano by NOAA in 2002 revealed an arcuate fault scarp east of the active cone, suggesting flank collapse. More extensive NOAA surveys in 2003 demonstrated the presence of an associated debris avalanche deposit, judging from their surface morphologic expression on the sea floor, extending at least 15 km and possibly as much as 30 km from the volcano, into the Grenada Basin to the west. Seismic air-gun profiles of the region show that these are lobate deposits, that range in thickness from tens to hundreds of meters. The debris avalanche deposit is contained within two marginal levees, which extend symmetrically from the volcano to the west. A conservative estimate of the volume of the smaller debris avalanche deposit is about 10 km3. Age dating of the deposits and the flank failure events is in progress, by analysis of gravity cores collected during the 2003 survey. Reconstruction of the pre-collapse volcanic edifice suggests that the ancestral Kick'em Jenny volcano might have been at or above sea level. Kick'em Jenny is dominantly supplied by basalt to basaltic andesite magmas, that are extruded now as submarine pillow lavas and domes or ejected as tephra in relatively minor phreatomagmatic explosions. Geochemical evolution of this volcano has not, however, reached the stage of generation of volatile-rich silicic magmas that might form highly explosive eruptions.

# SUBMARINE SLIDES NORTH OF PUERTO RICO AND THEIR TSUNAMI POTENTIAL

URI S. TEN BRINK

*U.S. Geological Survey, Woods Hole, MA 02543; utenbrink@usgs.gov*

ERIC L. GEIST

*U.S. Geological Survey, Menlo Park, CA 94025; egeist@usgs.gov*

PATRICK J. LYNETT

*Texas A&M University, College Station, TX 77843; plynett@civil.tamu.edu*

BRIAN D. ANDREWS

*U.S. Geological Survey, Woods Hole, MA 02543; bandrews@usgs.gov*

New multibeam bathymetry of the entire Puerto Rico trench reveals numerous retrograde slope failures at various scales at the edge of the carbonate platform north of Puerto Rico and the Virgin Islands. The slumped material comprises carbonate blocks, which fail, at least in initial stages, as a coherent rock mass. This, combined with the fact that the edge of the carbonate platform is steeper than most continental slopes, indicates a higher potential for tsunami run-up than along many other U.S. coasts. The style of slope failure appears to be rock falls, slide blocks and debris avalanches. Secondary failure of the failed carbonate products and of the underlying forearc sediments and rocks may lead to debris flows and turbidity flows. Fissures, discovered in the ocean floor near the edge of the platform, indicate that the process is expected to continue in the future. One of the slope failures, the Arecibo amphitheater, previously thought to represent a single giant slide with a volume of 900-1500 cu. km, appears to comprise smaller failures. The expected maximum tsunami run-up on the northern coast of Puerto Rico from one of these slope failures is <20 m, much lower than previously estimated. A recurrence rate of 100s kyr was calculated for the largest observed failures along the edge of the carbonate platform north of Puerto Rico using simple assumptions. Smaller landslides would presumably occur at a higher recurrence rate. Elsewhere around the island, a 20-km wide failure scarp was discovered in the Upper Mona rift and could be associated with the 1918 tsunami and earthquake that hit northwestern Puerto Rico. Large slope failures were also discovered for the first time on the northern side of the Puerto Rico trench. Because Puerto Rico trench slides occur at large water depths (~6000 m), have large horizontal and vertical dimensions, and the directivity from tsunamis emanating from these slides is toward Puerto Rico, they may be of particular concern and may necessitate further study.

## 1. Introduction

Evidence for tsunamis generated by submarine slope failures has been accumulating during the past decade (e.g., [*Ward*, 2001]). The highly destructive 1998 Papua New Guinea tsunami, for example, was likely amplified in the near field by a landslide-generated tsunami, in addition to the seismogenic tsunami [*Satake and Tanioka*, 2003]. Tsunamis generated by submarine slope failures have only a local effect, because relatively high-frequency waves are being excited by the slides and these high frequencies attenuate quickly with distance from the source [*Ward*, 2001]. However, they can strike with little warning, because of the proximity of the tsunami source to the affected coast and because ground shaking that triggers offshore landslides may be small or lacking. Hence, it may not be possible to provide civil defense authorities with adequate tsunami warning from submarine slides. A better strategy in this case would be to mitigate the hazard by identifying vulnerable coastal regions and modifying their usage, and by educating the public. Numerical simulations of landslides tsunamis and their run-up on shore can help identify these vulnerable coastal regions. Realistic simulations however must rely on a good knowledge of potential sources, a good knowledge of the bathymetry, and an understanding of the style of slope failure.

The region north of Puerto Rico may pose significant tsunami hazards to Puerto Rico, the Virgin Islands, and the Dominican Republic. A tsunami killed 40 people in NW Puerto Rico following a magnitude 7.3 earthquake in 1918 [*Mercado and McCann,* 1998]. Large landslide escarpments have been mapped on the seafloor north of Puerto Rico [*Grindlay*, 1998; *Schwab et al.*, 1991], although their ages are unknown. One of the escarpments, the Arecibo Amphitheater, was simulated as a single failure with run-up on the north shore of Puerto Rico possibly reaching 55 m [*Mercado et al.*, 2002]. In this paper we use multibeam bathymetry [*ten Brink et al.,* 2004a] and re-processed seismic reflection profiles to describe and quantify slope failures and to identify sites of potential future failures. The data indicate that slope failure is likely an incremental process even within the amphitheater. A numerical simulation for landslide-generated tsunami is presented for the north coast of Puerto Rico from one such failure in the Arecibo Amphitheater and indicates a run-up height only 1/3 of the previous estimate.

## 2. Background

The Puerto Rico trench is a subduction zone, where the North American (NOAM) plate is subducting under the Caribbean plate (Figure 1). Subduction is highly oblique (10°-20°) to the trench axis with a large component of left-lateral strike slip motion [*ten Brink et al.,* 2004b]. In addition to its extremely deep seafloor, the Puerto Rico trench is also characterized by the most negative free- air gravity anomaly on Earth, -380 mGal, which indicates a dynamic departure from isostatic equilibrium. It is unclear, however, whether the trench continues to subside or whether subsidence has recently ended, and the trench has not yet been filled with sediment. Subsidence of the trench probably caused the north shore of Puerto Rico and its offshore region, which are covered by carbonate rocks to tilt northward. The carbonate platform was horizontally deposited over Cretaceous to Paleocene arc rocks starting in the Late Oligocene. Then, at 3.3 Ma or later, the carbonate platform was tilted by 4° toward the trench such that its northern edge is at a depth of 4000 m and its reconstructed elevation on land in Puerto Rico is at +1300 m ([*ten Brink,* in press]; Figures 1, 2a). The tilting episode may have been very short , less than 40 kyr, [*ten Brink,* in press]. The subsidence and tilting have probably created gravitational instabilities and steep topographic slopes (Figure 1b), which are vulnerable to slope failures, particularly during earthquakes. Other tectonic motions, such as the opening of Mona Passage, the subduction of Main Ridge, and the subduction of the North American plate, continue to shape the area and can generate earthquakes that promote slope failure.

The 770-km long Puerto Rico trench from the Dominican Republic in the west to Anguilla in the east was mapped with multibeam sonar during three cruises in 2002 and 2003 [*ten Brink et al.*, 2004a]. Parts of the Puerto Rico trench were previously surveyed with side scan sonar [*Grindlay et al.*, 1997; *Scanlon and Masson,* 1996; *Schwab et al.*, 1991] and multibeam bathymetry [*Grindlay et al.*, 1997; *Dillon et al.*, 1998] often at lower resolution and with line orientation and spacing that did not provide complete bathymetric coverage. The data, presented here, were collected using the SeaBeam 2112 multibeam sonar system aboard the NOAA ship Ron Brown with sufficient swath overlap and proper line orientation for hydrographic survey. The data were gridded at 150 m grid size following resolution tests and at 50 m grid size at depths smaller than 5000 m. Vertical resolution is estimated to be 0.5-1% of the water depth(10-80 m; L. Mayer, oral comm., 2003). Backscatter mosaic images derived from the multibeam bathymetry data aided in the interpretation. The total mapped area with multibeam sonar is 100,000 km$^2$ (Figure 1). Seismic profiles from

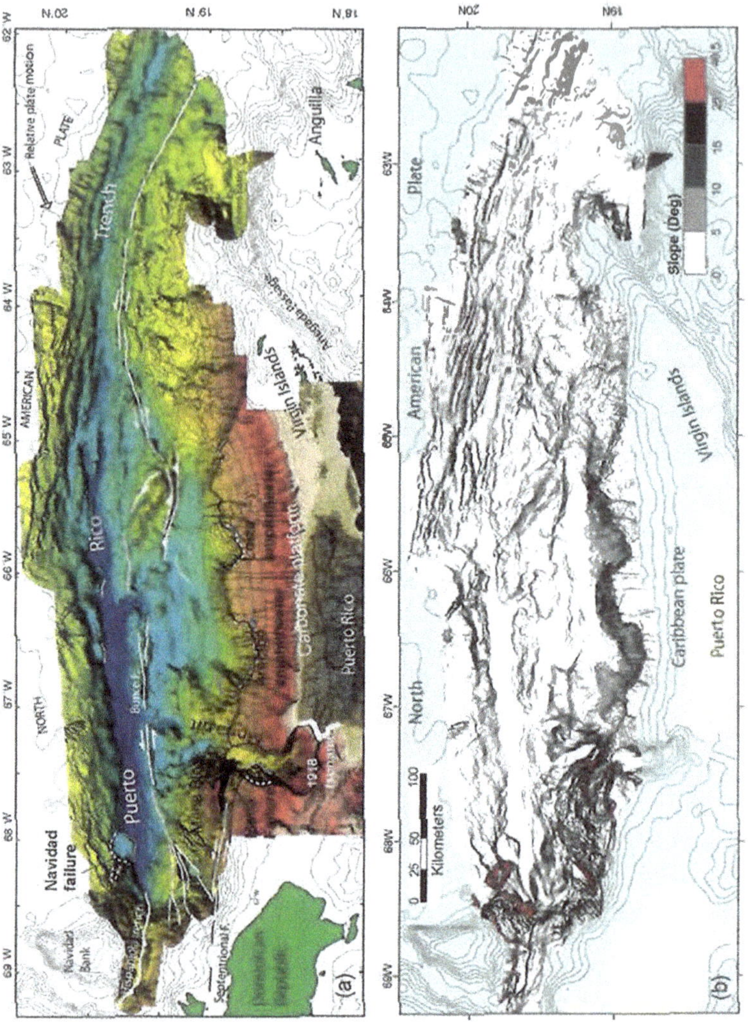


Figure 1. (a) Bathymetric map of the Puerto Rico trench and its vicinity from our new multibeam sonar survey [*ten Brink et al.*, 2004a], single-beam bathymetry compilation around Puerto Rico (A. Mercado, written communication, 2003), LIDAR data near shore (J. Brock, written communication, 2003), and topography of Puerto Rico. Contour interval is 500 m. Barbed white lines – thrust faults; white line – strike-slip fault; heavy black line – northern edge of tilted carbonate platform and southern edge on land; dashed lines – head scarp of slope failures; dotted line – possible debris toe; blue lines – fissures in the sea floor. (b) Map of the slope angle (in degrees) of the bathymetry of the Puerto Rico trench. The slope was calculated as the maximum horizontal slope on a grid interval of 300 m. A finer grid interval will produce higher slope angles. Thin contours – bathymetry from ETOPO2 at 500 m intervals.

the University of Texas Seismic data center (http://www.ig.utexas.edu/sdc) were migrated in time using the Stolt post-stack migration algorithm and provided vertical cross-section of some of the slope failures. We focus our discussion on three areas: The northern edge of the tilted carbonate platform, the Mona Rift, and the downgoing North American plate north of the trench (Figure 1).

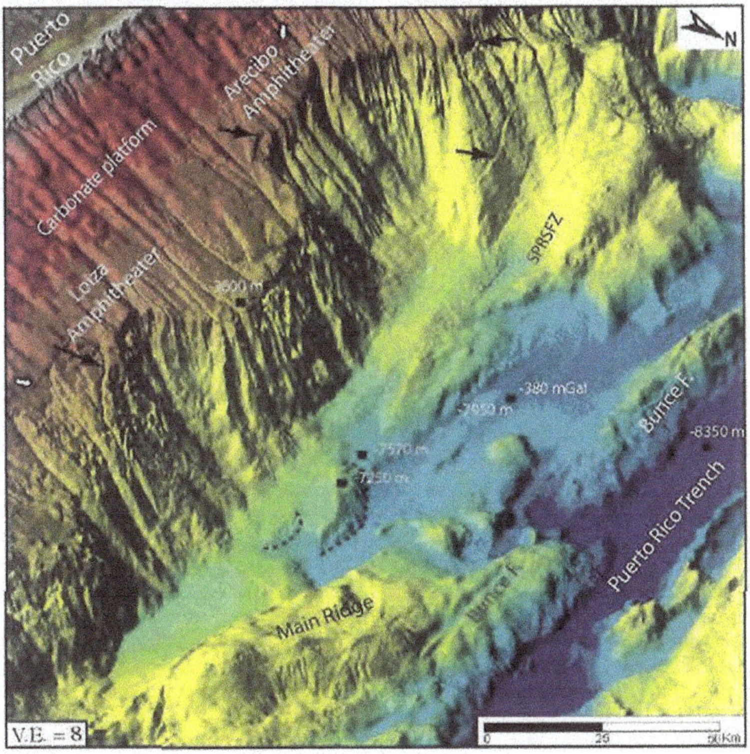


(a)

(b)

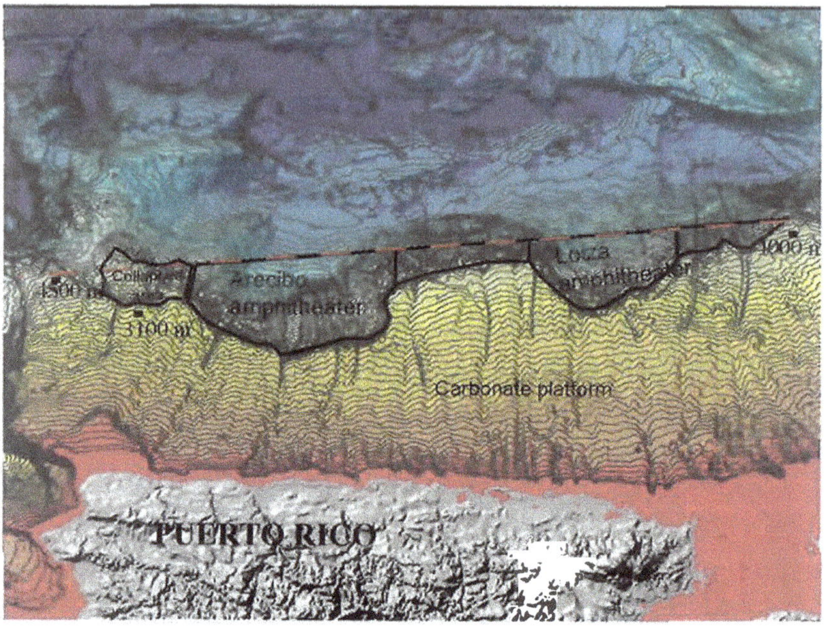


Figure 2. (a) Perspective view of the edge of the carbonate platform. Viewing direction is from the NE to the SW. White arrows point to the boundary between multibeam bathymetry and single-beam bathymetry. Black arrows point to fissures in the seafloor. Dotted lines are debris toes. The locations of a previously suggested strike-slip fault (SPRSFZ) and the largest free-air gravity anomaly on Earth (-380 mGal) are shown. Stippled textures (e.g., near SPRSFZ) are multibeam acquisition artifacts. (b) Bathymetry of the northern edge of the carbonate platform contoured at 100 m interval. The estimated edge of the platform before failure started is shown as a straight line located where a northward extension of the carbonate platform tilted at 5° would lie at a depth of 4500 m. The total estimated eroded area enclosed by black lines is 2120 km$^2$, and the total areas of the Arecibo and Loiza amphitheaters is 1147 and 692 km$^2$, respectively.

### 3. The northern edge of the carbonate platform

The northern edge of the carbonate platform north of Puerto Rico is characterized by two large amphitheater-shaped headwall scarps, the Arecibo amphitheater 37 km north of Arecibo, Puerto Rico [*Mercado et al.*, 2002; *Scanlon and Masson*, 1996; *Schwab et al.*, 1991] which is 50-km across, and the Loiza amphitheater, 48 km north of Loiza, Puerto Rico [*Scanlon and Masson*, 1996], which is 35-km across (Figures 2 and 3). Smaller failures are observed adjacent to these amphitheaters, and the westernmost edge of the platform appears to have collapsed but was translated very little. Several smaller scarps and slope failures were also identified all along the edge of the carbonate platform (Figures 2 and 3).

#### *3.1 Loiza amphitheater*

The backscatter image (Figure 3b) shows high reflectivity in the western half of the Loiza amphitheater, and lower reflectivity on the eastern side. The high reflectivity region extends downward into the abyssal plain and ends in a low hill (Figure 3a). We interpret this hill to be the toe of a debris field, which extended about 40 km from the head of the scarp. Seismic profile 1a across the failure scarp and the debris field has a classic concave shape (Figure 4a). The debris field itself is elevated by 40-50 m above the surrounding area, barely above the vertical resolution of our data. A layer of debris 200-250 msec thick can perhaps be identified on the seismic profile (Figure 4a). In contrast, the eastern sector of the amphitheater has low reflectivity. The part of seismic profile 7 (Figure 4b), which is below 6.5 s two-way travel time (TWTT), crosses the eastern sector obliquely and shows a hummocky surface and irregular reflectors. This sector of the amphitheater is interpreted to represent a series of smaller failures along the slope.

We propose that slope regions of high back scatter (such as failure A in Figure 3b) may be associated with a large debris flow over the slope, which has not yet been modified by secondary failure and may be relatively recent. Although high back scatter also results from sediment discharge along channels in the inclined carbonate platform, these narrow channels are clearly distinct from the slope reflectivity and they mostly terminate at the carbonate platform edge.

A 15-km long, several hundred meters wide and at least 100 m deep fissure runs at a distance of 2-5 km above the eastern headwall scarp of the Loiza

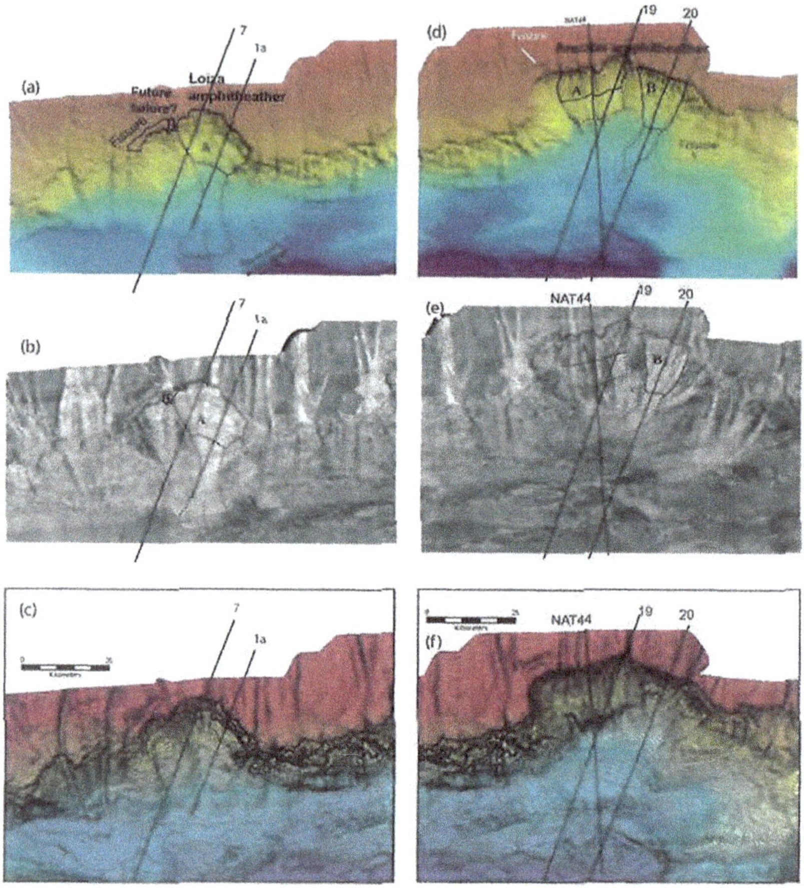


Figure 3 (a) View of the Loiza Amphitheater and the edge of the carbonate platform. Solid lines are locations of bathymetric profiles shown in Figure 4. Area A, which is enclosed by a solid line, is an interpreted slope failure, which produced the debris flow enclosed by a dashed line with a debris toe. Area B is a hypothesized area that would slide should the fissure SE of the headwall of the escarpment continue to open. Dashed line marked by 3 is a suggested old toe of a debris avalanche. (c) Backscatter image of the map shown in (a). Profiles and solid and dashed lines are as in (a). High reflectivity represents debris flow from Failure A and turbidity flows along the drainage gullies coming of the carbonate platform. (c) Contour map of Loiza Amphitheater and its vicinity. Contour interval is 50 m. Note the large fissure SE of the headwall, numerous other small failures outside the amphitheater. (d) View of the Arecibo Amphitheater and the edge of the carbonate platform. Solid lines are locations of bathymetric profiles shown in Figure 4. Area A enclosed by a solid line is interpreted as a rotational slide whose debris, enclosed by a dashed line that is perched on the lower slope. Area B enclosed by a solid line is interpreted as another slope failure, which produced the debris flow enclosed by a dashed line. (e) Backscatter image of the map shown in (d). Profiles and solid and dashed lines are as in (d). High reflectivity represents debris flow from slide B and turbidity flows along the drainage gullies coming of the carbonate platform. (f) Contour map of Arecibo Amphitheater and its vicinity. Contour interval is 50 m. Note the lack of debris toes at the base of the slope. Wiggly contours on a small scale are due to acquisition artifacts.

Amphitheater (Figures 3a and 3c). The fissure cuts obliquely to the slope of the carbonate platform, suggesting that it was not developed as subterranean drainage in the limestone layers. More likely, it represents an extension normal to the steep ($\leq 45°$) headwall of the amphitheater because of the horizontal tensile stresses that develop in an unsupported wall. If this interpretation is correct, then another future failure (marked B in Figure 3a), may take place, and depending on its acceleration, it may generate tsunami waves that will propagate toward the Virgin Islands.

A 300 m high and 25 km long ridge in the abyssal plane may be an old toe of a debris flow (Figure 3A) although a seismic profile, oriented obliquely to this ridge does not show a clear internal structure. If this ridge were a debris toe, it would have originated from an older failure on the eastern side of the Loiza Amphitheater, which was moving in a NW direction prior to the occurrence of Failure A.

### *3.2 Arecibo amphitheater*

The Arecibo amphitheater also appears to have been formed by several slope failure events. The amphitheater-shaped scar was previously postulated to have formed by a giant submarine slope failure with a displaced volume of over 910-1500 $km^3$ [*Grindlay*, 1998; *Schwab et al.*, 1991]. Maximum run-up of 30-55 m was calculated from this failure, with a run-up >10 m along the entire northern coast of Puerto Rico [*Mercado et al.*, 2002]. The backscatter image shows high reflectivity within the funnel-shaped area of high reflectivity on the western side (marked B in Figure 3e) of the amphitheater, which perhaps represents debris or turbidity flow. The cross-section of the area of high reflectivity area (Seismic profile 20 (Figure 4c) is concave (with the exception of the upper slope above 6.5 sec.), similar to the shape of the reflective slope failure A in the Loiza Amphitheater (Figure 4a).

The eastern amphitheater slope is much less reflective with the exception of narrow bands that can be traced to the drainage system higher up on the carbonate platform (Figure 3e). The slope in this region is probably the result of a rotational slide, where the upper slope slid about 8 km and the debris is perched on the lower slope (Figures 3f and 4d and 4e). The seismic stratigraphy on the slope shows remarkably coherent reflections suggesting that carbonate blocks slid coherently. The lower slope is steep and appears to be eroded by retrograde gullies (Figure 3f).

The abyssal plain in front of the Arecibo Amphitheater is 150-200 m shallower than the base of the slope east of the Arecibo amphitheater (Figure 2b

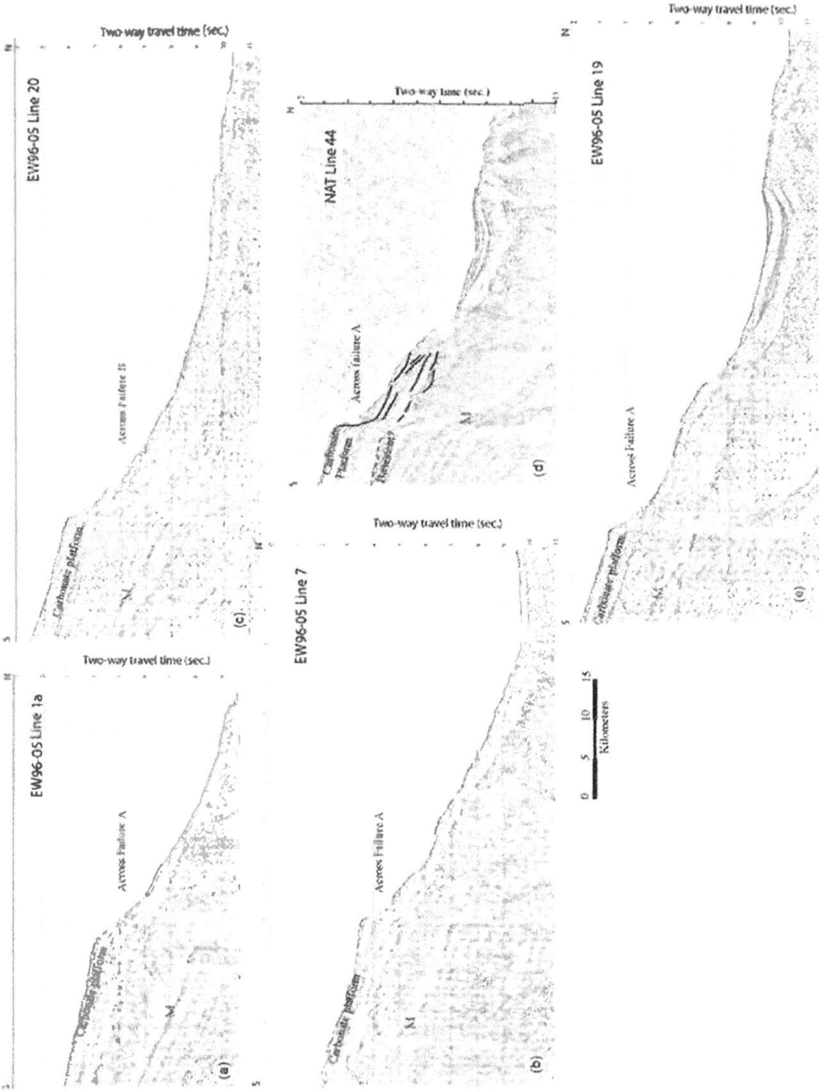


Figure 4. Reprocessed seismic reflection profiles (a and b) across the Loiza and (c, d, and e) the Arecibo amphitheater. See Figure 3 for location. Note that profiles 7 and 19 are oriented oblique to the slope. All the profiles, with the exception of (d) are single-channel profiles. All the profiles were migrated using the Stolt migration, filtered, and gained. Yellow – interpreted bottom of the carbonate platform. Green – interpreted boundaries between sequences of turbidity flows. M – water bottom multiple.

and 3f), perhaps because more sediment was deposited in front of the amphitheater than to the east. The abyssal plain is underlain by up to 2 s TWTT or over one km thick packet of sediments that dip and thicken to the north (Figure 3c, 3d, and 3e). Grindlay [1998] interpreted this sedimentary packet to be turbidities from the major slumping event that created the Arecibo amphitheater, but the internal stratigraphy indicates several distinct sedimentary packet, which were probably deposited during different failure events, or during massive sediment run-offs from the Puerto Rico after heavy storms.

A 7x3 km area along the eastern headwall of the Arecibo amphitheater appears to be surrounded by fissures (Figure 3d) and to be up to 200 m lower than the surrounding carbonate platform (Figure 3f). This may be an indication of the beginning of a new slope failure in part of the Arecibo amphitheater.

### *3.3 Mona rift*

A 20-km-long section of the western wall of the upper Mona Rift is scoured by a slope failure (Figures 5a and 5b). The headwall varies in height from 1250-3350 m below sea level. The failure scarp appears in map view to be <5 km wide and the debris field ahead of it is 2.5-10 km wide. In side view (Figure 5a), the failure appears almost as a rock fall. A cross-section of the failure (Figure 5c and d) shows an 11-km-long by 0.5 thick blocky section of debris field. The debris field is interpreted as a combination of rock fall and debris avalanche. Using seismic velocity of 2 km/s for the debris field and 2.75 km/s for the carbonate platform at the headwall [*van Gestel et al.*, 1998], the area of the debris field is the equivalent of a 4.1 km wide section of failed platform. The failure appears to be recent because the debris field comprises of discrete blocks and because does not cover the most recent sediments filling the rift.

The upper Mona rift (Figure 1a, 5a and b) cuts across the carbonate platform north of Puerto Rico. The rift walls are very steep (>25°) and are likely normal fault planes along which east-west extension between the Puerto Rico and the Hispaniola blocks [*Mann et al.*, 2002] takes place. The tsunami of 1918 which killed 40 people along the termination of the rift at the northwestern shore of Puerto Rico, probably initiated at the upper Mona rift [*Mercado and McCann*, 1998]. The tsunami followed almost immediately a magnitude 7.3 earthquake which caused extensive damage and additional 76 fatalities in NW Puerto Rico (ibid.). Eyewitness account (quoted in [*Mercado and McCann*, 1998]) described the sea receding before it returned to flood the shore with waves up to 6 m high. This indicates that the downward motion along the normal fault, on which the earthquake had likely occurred, was directed away from shore. Mercado and McCann (1998) therefore located the fault along the eastern wall of the rift.

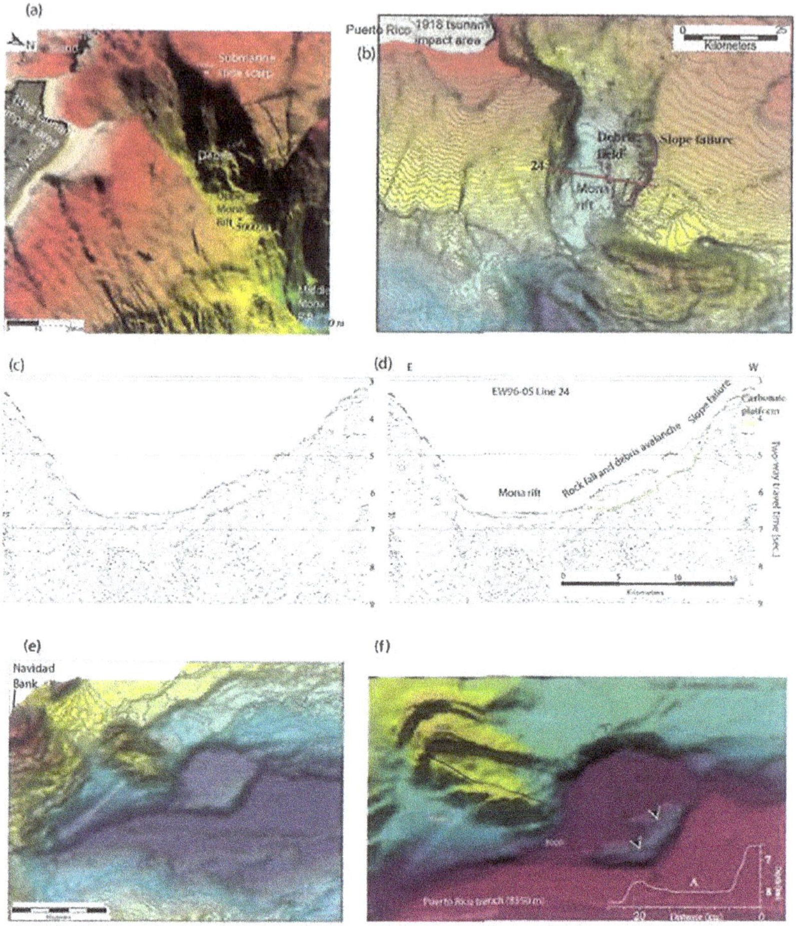


Figure 5. (a) Side view of the slide in the upper Mona rift. Back arrow marks the boundary between our higher resolution multibeam bathymetry data and the existing bathymetry compilation from single beam bathymetry (A. Mercado, written communication, 2002). (b) Contour map of the Upper Mona Rift. Contour interval is 100 m. Black arrow is as in (a). Solid line is the estimated area of the slide and dashed line is the estimated area of the debris field. Red line is the location of the profile in (c and d). (c) Seismic profile 24 across Mona failure. See Figure caption 4 for details of the processing sequence. (d) Interpretation of the seismic profile. (e) Bathymetric map of the Navidad Bank area. See Figure 1 for location. Contour interval is 100 m. There may be numerous slides on the slope of Navidad Bank that faces the Puerto Rico trench. (f) Bathymetric map of the largest failure east of Navidad Bank. Solid line labeled 1 perhaps encloses the slide area although there appear to be two debris toes, indicating perhaps repeated failure along the same escarpment. Dashed Line A marks the location of the profile in the inset. Solid line labeled 2 is perhaps another landslide that scooped part of the seamount.

Alternatively, Figure 5b shows that the slide scarp is curved toward the SW and therefore, the motion of the slide was perhaps directed to the northeast. A more accurate bathymetry of the shallower section of Mona rift is needed together with numerical simulations to determine what impact tsunamis generated from Mona rift slides has on the neighboring islands.

### *3.4 Northern side of the trench*

Slope failures are not limited to the carbonate platform north of Puerto Rico. Several scarps and deposits were observed in the multibeam bathymetry for the first time on the northern trench wall, where the North America plate drops into the trench by 2-2.5 km over a distance of 20-25 km. The largest scarp, located at a depth of 6500 m is 20-km wide and 1500-m high. A 600 m high mound 15 km in front of the scarp presumably is the deposit derived from the scarp (Figure 5f). The mound has two parts, which indicates that the failure occurred as at least two separate events. The circular region between the scarp and the toe is 300 m higher than the floor of the trench, and is likely filled with the products of the slope failures. There are no seismic profiles across this region to determine whether it is filled with debris avalanche or turbidity flows. The backscatter image shows low reflectivity from the scarp to the toe, with the exception of the area located within 5 km of the base of the scarp, perhaps because the failures are older and a veneer of pelagic sediments has covered their products.

The areas to the north and the west of that scarp, including what appears to be an excavated side of a seamount may also be associated with submarine slope failures. Slopes reaching 45° and vertical offsets of up to 8 km were also observed near Navidad Bank and Mona Block (Figure 1a), and along fault scarps on the descending NOAM plate NE of the Virgin Islands. These conditions may generate submarine slope failures. Slope failures on the northern wall of the trench should be evaluated for tsunami potential because the direction of slide movement and therefore the directivity of slide-generated waves are toward Puerto Rico.

## 4. Characteristics of slope failures around the Puerto Rico trench

Bathymetric slopes reaching 45° can be found along the northern edge of the Carbonate platform north of Puerto Rico, along the walls of the Upper and Middle Mona rift basins, around Mona block, along the eastern end of Septentrional Fault, around Navidad Bank, and along the descending NOAM plate north of the trench (Figure 1). Many landslide scarps along the Atlantic and Pacific margins of North America and in the Gulf of Mexico are, on the other hand, found on much smaller slope angles (as small as 0.5°, [*Prior and Coleman*, 1984]). The slope failures, which they represent, are thought to be induced by rapid sedimentation and tectonic activity, causing the slope to become steeper, and by elevated pore pressure, gas pressure, wave generated or tidal generated bottom pressure (e.g., [*Driscoll et al.*, 2000; *Locat et al.*, 2004; *Prior and Coleman*, 1984]).

The edge of the continental shelf along much of North America is covered with clastic sediments, which in high sedimentation areas may be unconsolidated. (The exceptions are the Blake Escarpment and the exposed lower continental slope from Florida to New Jersey, which are made of limestone and chalk [*Dillon et al.*, 1993; *Twitchell et al.*, 1993], and may be analogous to the Puerto Rico margin.) When the slope fails, these sediments may disintegrate into debris flows and a turbidity current, such as may have happened in the 1929 Grand Bank event [*Piper et al.*, 1999]. By contrast, slope failures around Puerto Rico involve either carbonate layers or oceanic crust. The north coast of Puerto Rico and the area surrounding the Upper Mona Rift are covered by an average of 1600 m thick layer of carbonate rocks (mostly limestone) [*van Gestel et al.*, 1998], which will tend to fail as a coherent rather than internally deformable rock mass. Navidad Bank is covered by several km of carbonate rocks [*Freeman-Lynde and Ryan*, 1987]. In all these areas we expect slope failures to occur as rock falls, block and debris slides, and to a lesser extent, debris avalanches, and debris flows (cf., *Varnes*, 1978). These failure modes are indeed observed in the multibeam imagery and seismic profiles, for example, Arecibo Failure A (Figures 3d, 4e and 4f) and Mona rift (Figures 5a and 5d).

The carbonate platform comprises a 1-2 km thick layer of highly massive or layered limestone, with some shales and shaly sandstone toward the bottom of the section (the San Sebastian Formation)[*Monroe*, 1980; *van Gestel et al.*, 1999]. It is underlain by volcanoclastic, volcanic, and igneous rocks of the arc and forearc, which may be less coherent during failure. The carbonate layer may act as a cap that protects the underlying less coherent layers from failing (Figure 4). When part of this cap detaches from the rest of the platform and either rotates (Figure 4d) or disintegrates (Figure 4a), it exposes the underlying layers to further failures, probably in the form of smaller more incremental events (e.g., Figure 4b). A carbonate block may detach from the platform by a combination of vertical fissures, in which carbonate dissolution may take place, and reduced basal friction on basal shale and clay layers. The fissures must be oriented subparallel to the edge of the platform to allow the blocks to detach, and the carbonate dissolution is probably aided by fresh water runoff from the island.

The process may therefore be limited by the rate at which these fissures can form and cut through the entire thickness of the carbonate layer.

Slope failures off Navidad Bank, Mona Block, and the walls of Mona rift are likely made of thick coherent limestone blocks. Slope failures on the NOAM plate are likely to comprise about 500 m of pelagic sediment and chert layers underlain by upper oceanic crust basalts, and these failures may develop differently than those along the carbonate platform.

The failure of coherent blocks on a steep slope appears to cause a high tsunami run-up. The 1992 Flores Island tsunami was accompanied by slope failures in the surrounding carbonate reef and generated a maximum run-up of 26 m [*Imamura et al.,* 1995], however, the submarine portion of the slide has not been mapped. The collapse of the Ritter Island volcano in 1888, which was associated with sliding of coherent blocks, had produced a maximum run-up of at least 15 m [*Ward and Day*, 2003]. The 1975 Kalapana earthquake in Hawaii was argued by some authors to be the result of a coherent slope failure of the flank of Kilauea volcano (e.g., [*Ma et al.,* 1999]) and it produced a maximum run-up of 15 m. While it appears that failure of coherent blocks on a steep slope may generate a high tsunami run-up, there is no evidence to suggest that the opposite is also true, namely, that the failure of less coherent blocks on lower slopes will generate small tsunamis. The 1929 Grand Banks tsunami with a maximum run-up of 13 m was triggered by a M=7.2 earthquake, and was probably amplified by large-scale slope failure and turbidity flow ([*Piper et al.,* 1999; *Hughes Clark,* 1990), but the exact contribution of the slope failure to the Grand Banks run-up is unknown.

For a slide of constant spatial dimensions, tsunami wave height is affected by the time evolution and terminal speed of the slide. These dynamic properties, in turn, depend on a complex interplay of forces (e.g., basal friction and drag) and physical properties (e.g., friction coefficient along joints and fractures) for coherent slides of the type studied here. (For comparable analysis of deformable slides, see *Locat et al.* [2004]) When performing the coupled landslide motion – hydrodynamic simulations, the impact of these factors on the generated tsunami is tested by systematically modifying the time duration of the slide.

## 5. Hydrodynamic Simulation

The effects of slope failure in the Puerto Rico trench on the resulting tsunamis was explored using a hydrodynamic simulation similar to that used by *Mercado et al.* (2002) for the entire Arecibo Amphitheater. A simplified representation of Failure A in the Arecibo Amphitheater (Figure 3d) is parameterized according to its effective width and total length of the slide slope in the area of evacuation (Figure 6). The dimensions of this failure are assumed to be 22 km wide (along strike) with a 3.5 km length (down-slope) of the depletion zone and a 4.5 km length of the debris field (total length 8 km). The average thickness of the slide is assumed to be approximately 1 km. The slide is then modeled as a region of depletion with a sharp head scarp and a down slope region of debris fan accumulation (cf., *Trifunac et al.*, 2003). Movement of the slide is specified according to a duration time ($t_d$) with smooth ramps used to simulate the accelerating (starting) and decelerating (stopping) phases of slide motion.

Slide movement is directly coupled with the hydrodynamic equations of motion through temporal and spatial derivatives of seafloor motion. As in *Mercado et al.*, (2002) hydrodynamic modeling is based on weakly nonlinear "extended" equations described by [*Lynett and Liu*, 2002]. Because of the large vertical motions associated with landslides, nonlinearity of the momentum equations may be more of a concern for landslide-generated tsunamis than for seismogenic tsunamis. When the maximum seafloor displacement at any particular time step is much smaller than the water depth above the slide, the weakly nonlinear equations of Lynett and Liu, [2002] can be used. Nonlinearity can also be important for accurately determining tsunami run-up, especially for large incident waves. As the tsunami propagates away from the source, frequency dispersion also becomes important. Landslide-generated waves are typically not the long waves characteristic of seismogenic sources, and so energy will be dispersed in the direction of wave propagation as different wave components (frequencies) travel at different velocities. Lynett and Liu (2002) use the arbitrary-level velocity computation [*Nwogu*, 1993] to "extend" the validity of frequency dispersion for the depth-integrated equations into the intermediate water regime allowing for accurate simulation of waves with lengths greater than two water depths. The WNL-EXT equations are implemented in the program COULWAVE using a finite-difference approximation with a high-order predictor-corrector scheme. The spatial grid size used for the computations is 266 m with a time step of 0.51 s. In addition, bottom friction is accounted for with a constant friction factor $f$=0.01 (cf., *Mercado et al.*, 2002), using the quadratic bottom friction formulation. A moving boundary condition [*Lynett et al.*, 2002] is implemented along the coast to represent run-up and overland flow. For the open-ocean boundary conditions, a sponge-layer absorption scheme is used.

## 6. Hydrodynamic Results

Because Arecibo Slide A interpreted from the multibeam bathymetry is smaller in volume by nearly an order of magnitude than the large Arecibo amphitheater slide modeled by [*Mercado et al.*, 2002], the associated maximum run-up of Slide A is also necessarily smaller (20 m instead of 55 m, Figure 7),

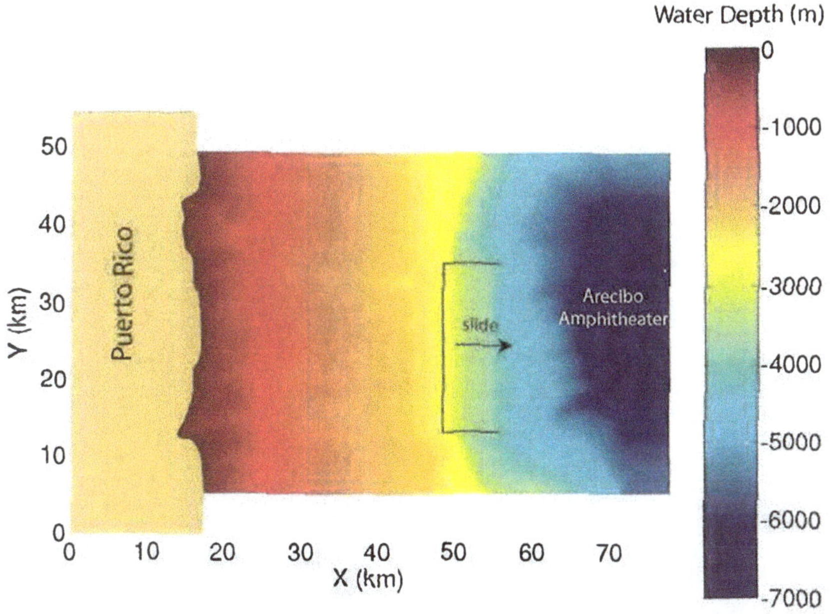


Figure 6. Bathymetry of part of the Arecibo amphitheater region along with a simple geometric representation of Arecibo slope failure A, interpreted from the multibeam bathymetry data (Figure 3a and c) and seismic profiles (Figure 4 d and e). Yellow region represents onshore Puerto Rico.

Figure 7a shows the wave field at $t \approx 155$ s and the maximum wave amplitude for Slide A, for $t_d = 133$ s. This duration time is smaller than the shortest duration time used by Mercado et al. (2002) for the much larger slide. It is reasonable to assume that the duration time scales with the dimensions of the slope failure. Similar plots for $t_d = 200$ s and $t_d = 400$ s are shown in Figures 7c-f, respectively, at the same scale as in Figures 7a and b. Increasing the slide duration dramatically decreases the maximum and near shore tsunami amplitude.

Because the slide is modeled as a region of progressive depletion and a down slope region of debris accumulation, it is difficult to assign an effective velocity for the slide as with simple block slides commonly used in tsunami studies (cf., [*Ward*, 2001]). Trifunac et al. (2003) describes composite slides of this type where different spreading velocities are assigned to the depletion and accumulation zones. In simpler, but perhaps less accurate terms, one can assign different length scales, such as run-out distance or horizontal displacement of the slide head, to calculate an effective velocity from $t_d$. Using8 km characteristic length scale for Slide A and $t_d = 200$ s (Figure 7c), the effective slide velocity is approximately 40 m/s. This compares to a tsunami phase velocity ($c = \sqrt{gh}$ for long waves) of 170 m/s at a water depth of $h$=3000 m

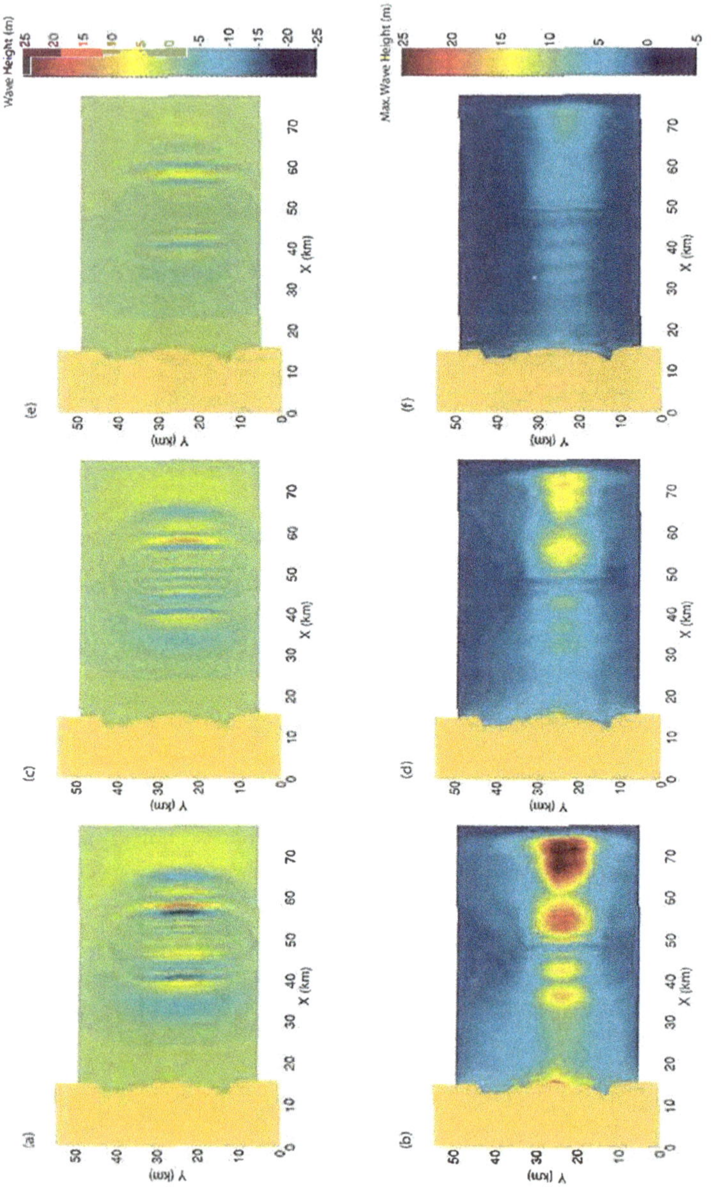


Figure 7. Hydrodynamic simulation for Slide A (Figure 6), using a slope failure duration $t_d$ =133 s. (a) Wave field at $t$=155 s; (b) Maximum wave height for the duration of model run (800 s). (c) Same as (a) for a failure duration of $t_d$ =200 s. (d) same as (b) for a failure duration of $t_d$ =200 s. (e) same as (a) for a failure duration of $t_d$ =400 s. (f) same as (b) for a failure duration of $t_d$ =400 s.

(approximate average water depth of the slide). For tsunamis propagating in the direction of slide motion (i.e., the outgoing tsunami), a slide velocity equal to the tsunami phase velocity is considered the ideal tuning velocity (*Ward*, 2001). As the duration decreases, the effective slide velocity is closer to the phase velocity and more efficient in terms of tsunami generation. For the tsunami propagating opposite the direction of slide motion (i.e., the incoming tsunami), generation efficacy follows a more complex relationship to both the initial acceleration of the slide and its terminal velocity. For the generation model used in this study, increasing the duration time does lead to a decrease in tsunami run-up on the near shore of Puerto Rico. We also examine the effect that a decrease in the friction factor $f$ from $f$=0.01 to $f$=0.001 has on the maximum tsunami wave heights. A decrease in $f$ results in a slight decrease in wave heights, similar to the effects noted by *Mercado et al.* [2002].

## 7. Recurrence time

A probabilistic assessment of tsunami hazard requires an estimate of recurrence time for slope failures. The absence of a historical record and clearly identified tsunami deposits on the northern coast of Puerto Rico, and the lack of dates on the observed slope failure, makes this task difficult. Nevertheless, the abundance of observed slope failures and hypotheses about the underlying causes of slope failures provide a starting point. A minimum recurrence time for slope failures along the edge of the carbonate platform was estimated assuming that the failure process was caused by the tilting of the platform at or after 3.3 m.y. ago, when the Puerto Rico trench collapsed and the island of Puerto Rico emerged (*ten Brink*, in press). We make the simplifying assumptions that slope failures were distributed evenly in time since the tilt event, and that the edge of the platform was initially straight (Figure 2b).

Unlike earthquakes, the size-frequency distribution of submarine slope failures is unknown. We therefore assume that the failures have a characteristic area and thickness similar to those observed. Three slope failure areas were calculated: Arecibo failure A with an area of 8x20 km, Loiza (future) failure B with an area of 3x15 km, and the Mona rift failure with an area of 4x20 km. Two other failures, Loiza failure A and Arecibo failure B are 22 and 15 km long, respectively, but their width is unknown. Based on the 3 calculated failure areas, the characteristic failure area along the edges of the carbonate platform is estimated to be 45-160 $km^2$. We calculate the number of large failures that occurred in the Arecibo and Loiza amphitheaters during the past 3.3 Ma to be the estimated lost area of carbonate platform divided by the characteristic failure area (Figure 2b). For the Arecibo Amphitheater, the calculated number of large failures is 7-25, and for the Loiza Amphitheater it is 4-13. The average recurrence time is therefore 470-132 kyr for the Arecibo amphitheater and 825-253 ky for the Loiza amphitheater. The combined landslide tsunami recurrence interval at the north coast of Puerto Rico from the entire edge of the carbonate platform is 253-70 kyr. This is probably a minimum estimate considering that there are other smaller failures which are probably capable of producing tsunamis.

Placement and configuration of turbidity layers can also help constrain the number of past slope failures within each amphitheater. Several events can be identified on the seismic profiles (Figure 4), but the resolution of these profiles is too poor to identify all the events. Additionally, some turbidity layers may be the result of massive sediment run-offs from the island after heavy storms, as represented in the backscatter images by the highly reflective channels coming down from the carbonate platform. A combination of near-bottom seismic reflection imaging and coring will help identify individual failure events.

## 8. Conclusions

Analysis of detailed multibeam bathymetry and coincident backscatter images from the edge of the carbonate platform north of Puerto Rico and the Virgin Islands combined with seismic cross-sections reveals numerous retrograde slope failures of various scales. Fissures, discovered in the ocean floor near the edge of the platform, indicate that the failure process is expected to continue in the future. The style of failure appears to be correlated with the presence of the carbonate layers at the headwall of the slide. Where the carbonate layer is thick, the failure often takes place as a rock fall or a rotated slump block perched on the lower slope, although debris avalanche or a debris flow may also take place. Secondary failure of the debris and of the exposed rocks that underlie the failed carbonate platform may lead to additional slope failures, probably in the form of debris flows and turbidity flows. Because the primary sliding material is carbonate rocks, which are expected to fail in coherent rock masses, and because the edge of the carbonate platform is much steeper than many continental slopes around the U.S., slides off Puerto Rico may result in a higher potential run-up of tsunamis than comparable slides elsewhere. However, a few conterminous U.S. margins such as the Blake Escarpment and the lower slope of the New Jersey margin have composition and slope angle similar to those of Puerto Rico [Dillon et al., 1993], hence their failure mode may be analogous to our observations.

The 50-km wide escarpment, known as the Arecibo amphitheater, previously thought to have been formed by a single giant voluminous failure, is in effect the result of several slope failures. A hydrodynamic simulation using the weakly-nonlinear "extended" wave equations in the COULWAVE model, predicts a maximum run-up < 20 m on the northern coast of Puerto Rico from one of the slides at this place. This run-up amplitude is only 36% of the previously estimated run-up, which was calculated for the entire amphitheater sliding at once. The run-up is predicted to be even smaller for a longer sliding duration. A recurrence interval for the largest observed slope failures, such as the one modeled here is estimated using many simplifying assumptions to be between 100-800 kyr. The lack of any established size-frequency relationship of submarine slope failures prevents us from extending this estimate to smaller failures.

Large submarine slides on the northern side of the Puerto Rico trench are found at greater depths (~6000 m), but can nevertheless be hazardous, because they aim toward Puerto Rico, have large horizontal and vertical dimensions and can excite lower frequency waves. These slope failures as well as a slide scarp discovered in the Upper Mona rift, possibly associated with the 1918 tsunami that hit northwestern Puerto Rico, should be further investigated.

This study demonstrates that high-resolution bathymetry, backscatter images, seismic profiles, and bottom samples, are necessary to evaluate the hazard from landslide tsunami. Emergency management cannot rely on tsunami warning for landslide tsunamis, because landslide tsunamis have short travel time from their source to the coast. Improved knowledge of the sea floor coupled with realistic simulations is needed to guide the preparation and mitigation of tsunami hazard.

## 9. Acknowledgements

The study was funded by NOAA Office of Ocean Exploration and the USGS Coastal and Marine Program and was conducted aboard the NOAA ship Ronald Brown. Shep Smith, Gene Parker, Toshi Uozumi, Bill Danforth, Chris Polloni, Brian Andrews, and Pilar Llanes were instrumental in collecting and processing the data. We thank Lisa Gahagan for help in retrieving the seismic profiles. We thank David Twitchell, Jim Robb, and Tip Meckel for their constructive reviews.

## 10. References

Burroughs, S.M., and S.F. Tebbens, Power law scaling and probabilistic forecasting of tsunami runup heights, Pure and Applied Geophysics, **162**, 331-342, (2005).

Dillon, W.P., J. Acosta, A. Uchupi, and U.S. ten Brink, Joint Spanish-American research uncovers fracture pattern in northeastern Caribbean, *EOS*, **79**, 336-337, (1998).

Dillon, W.P., J.S. Risch, K.M. Scanlon, P.C. Valentine, and Q.J. Huggett, Ancient crustal fractures control the location and size of collapsed blocks at the Blake Escarpment, east of Florida, in *U. S. Geological Survey Bulletin* **2002**, edited by W.C. Schwab, H.J. Lee, and D.C. Twichell, pp. 54-59, (1993).

Driscoll, N.W., J.K. Weissel, and J.A. Goff, Potential for large-scale submarine slope failure and tsunami generation along the U.S. Mid-Atlantic coast, *Geology*, **28**, 407-410, (2000).

Freeman-Lynde, R.P., and W.B.F. Ryan, Subsidence history of the Bahama Escarpment and the nature of the crust underlying the Bahamas, *Earth and Planetary Science Letters* **84**, 457-470, (1987).

Grindlay, N.R., Volume and density approximations of material involved in debris avalanche on the south slope of the Puerto Rico Trench, *A report to the Puerto Rico civil Defense and the University of Puerto Rico Sea Grant College Program*, 9 pp, (1998).

Grindlay, N.R., P. Mann, and J.F. Dolan, Researchers investigate submarine faults north of Puerto Rico, *EOS*, **78**, 404, (1997).

Heezen, B.C., and M. Ewing, Turbidity currents and submarine slumps, and the 1929 Grand Bank earthquake, *American journal of Science*, **250**, 849-873, (1952).

Hughes Clarke, J.E., Late stage slope failure in the wake of the 1929 Grand Banks earthquake, *Geo-Marine Letters*, **10**, 69-79, (1990).

Imamura, F., E. Gica, T. Takahashi, and N. Shuto, Numerical simulation of the 1992 Flores tsunami: Interpretation of tsunami phenomena in northeastern Flores Island and damage at Babi Island, *Pure and Applied Geophysics*, **144**, 555-567, (1995).

Locat, J., H.J. Lee, P. Locat, and J. Imran, Numerical analysis of the mobility of the Palos Verdes debris avalanche, California, and its implication for the generation of tsunamis, *Marine Geology*, **203**, 269-280, (2004).

Lynett, P., and P.L.-F. Liu, A Numerical Study of Submarine Landslide Generated Waves and Runup, *Proc. Royal Society of London A.*, **458**, 2885-2910, (2002).

Lynett, P., T.-R. Wu, and P.L.-F. Liu, Modeling wave runup with depth-integrated equations, *Coastal Engineering*, **46**, 89-107, (2002).

Ma, K.-F., H. Kanamori, and K. Satake, Mechanism of the 1975 Kalapana, Hawaii, earthquake inferred from tsunami data, *Journal of Geophysical Research*, **104**, 13,153-13,168, (1999).

Mann, P., E. Calais, J.-C. Ruegg, C. DeMets, P.E. Jansma, and G.S. Mattioli,

Oblique collision in the northeastern Caribbean from GPS measurements and geological observations, *Tectonics*, **21**, 1057, doi:10.1029/2001TC001304, (2002).

Mercado, A., N.R. Grindlay, P. Lynett, and P.L.-F. Liu, Investigation of the potential tsunami hazard on the north coast of Puerto Rico due to submarine landslides along the Puerto Rico trench, *Report submitted to Puerto Rico State Emergency Management Agency and Sea Grant College Program*, 432 pp., (2002).

Mercado, A., and W. McCann, Numerical simulation of the 1918 Puerto Rico tsunami, *Natural Hazards*, **18**, 57-76, (1998).

Monroe, W.H., Geology of the middle Tertiary formations of Puerto Rico, *U. S. Geological Survey Professional Paper*, **P593**, 93 pp., (1980).

Nwogu, O., Alternative form of Boussinesq equations for nearshore wave propagation, *Journal of Waterway, Port, Coastal, and Ocean Engineering*, **119**, 618-638, (1993).

Piper, D.J.W., P. Cochonat, and M. L. Morrison, The sequence of events around the epicenter of the 1929 Grand Banks earthquake: initiation of debris flows and turbidity current inferred from sidescan sonar, *Sedimentology*, **46**, 79-97, (1999).

Prior, D.B., and J.M. Coleman, Submarine slope instability, in *Slope instability*, edited by D. Brunsden, and D.B. Prior, pp. 419-455, Wiley and Sons, Chichester, (1984).

Satake, K. and Y. Tanioka, The July 1998 Papua New Guinea earthquake: Mechanism and quantification of unusual tsunami generation, Pure and Applied Geophysics, **160**, 2087-2118, (2003).

Scanlon, K.M., and D.G. Masson, Sedimentary processes in a tectonically active region: Puerto Rico North Insular Slope., in *Geology of the U.S. Seafloor: the View from GLORIA*, edited by J.V. Gardner, M. Field, and D.C. Twichell, pp. 123-134, Cambridge University Press, Cambridge, (1996).

Schwab, W.C., W.W. Danforth, K.M. Scanlon, and D.G. Masson, A giant submarine slope failure on the northern insular slope of Puerto Rico, *Marine Geology*, **96**, 237-246, (1991).

ten Brink, U.S., Vertical motions of the Puerto Rico Trench and their cause, *J. Geophys. Res.*, in press.

ten Brink, U.S., W.W. danforth, C. Polloni, B. Andrews, P. Llanes, S.V. Smith, E. Parker, and T. Uozumi, New sea floor map of the Puerto Rico trench helps assess earthquake and tsunami hazards, *EOS*, **85**, 349, 354, (2004a).

ten Brink, U.S., and J. Lin, Stress interaction between subduction earthquakes and forearc strike-slip faults: modeling and application to the northern Caribbean plate boundary, *Journal of Geophysical Research*, **109**, B12310,

10.1029/2004JB003031,(2004b).

Trifunac, M.D., A. Hayir, and M.I. Todorovska, A note on tsunami caused by submarine slides and slumps spreading in one dimension with nonuniform displacement amplitudes, *Soil Dynamics and Earthquake Engineering,* **23**, 223-234, (2003).

Twichell, D.C., P.C. Valentine, and L.M. Parson, Slope failure of carbonate sediment on the West Florida Slope, in *Submarine landslides; selected studies in the U.S. Exclusive Economic Zone*, edited by W.C.L. Schwab, H J; Twichell, D C, pp. 69-78, U. S. Geological Survey Bulletin, **2002**, (1993).

van Gestel, J.-P., P. Mann, J.F. Dolan, and N.R. Grindlay, Structure and tectonics of the upper Cenozoic Puerto Rico-Virgin Islands carbonate platform as determined from seismic reflection studies, *Journal of Geophysical Research*, **103**, 30,505-30,530, (1998).

van Gestel, J.-P., P. Mann, N.R. Grindlay, and J.F. Dolan, Three-phase tectonic evolution of the northern margin of Puerto Rico as inferred from an integration of seismic reflection, well, and outcrop data, *Marine Geology*, **161**, 259-288, (1999).

Varnes, D.J., Slope movement types and processes, in *Landslides: Analysis and Control*, edited by R. L Schuster and R. J. Krizek, pp. 11-33, National Academy of Sciences, Washington, D.C., (1978).

Ward, S., Landslide tsunami, *Journal of Geophysical Research*, **106**, 11201-11215, (2001).

Ward, S.N., and S. Day, Ritter Island Volcano; lateral collapse and the tsunami of 1888, *Geophysical Journal International*, **154**, 891-902, (2003).

# RISK ASSESSMENT OF TSUNAMI GENERATION FROM ACTIVE VOLCANIC SOURCES IN THE EASTERN CARIBBEAN REGION

George Pararas-Carayannis
*Honolulu, Hawaii, USA*

Earthquakes, volcanic eruptions, volcanic island flank failures and underwater slides have generated numerous destructive tsunamis in the Caribbean region. Convergent, compressional and collisional tectonic activity caused primarily from the eastward movement of the Caribbean Plate in relation to the North American, Atlantic and South American Plates, is responsible for zones of subduction in the region, the formation of island arcs and the evolution of particular volcanic centers on the overlying plate. The inter-plate tectonic interaction and deformation along these marginal boundaries result in moderate seismic and volcanic events that can generate tsunamis by a number of different mechanisms. The active geo-dynamic processes have created the Lesser Antilles, an arc of small islands with volcanoes characterized by both effusive and explosive activity. Eruption mechanisms of these Caribbean volcanoes are complex and often anomalous. Collapses of lava domes often precede major eruptions, which may vary in intensity from Strombolian to Plinian. Locally catastrophic, short-period tsunami-like waves can be generated directly by lateral, direct or channelized volcanic blast episodes, or in combination with collateral air pressure perturbations, nuéss ardentes, pyroclastic flows, lahars, or cascading debris avalanches. Submarine volcanic caldera collapses can also generate locally destructive tsunami waves. Volcanoes in the Eastern Caribbean Region have unstable flanks. Destructive local tsunamis may be generated from aerial and submarine volcanic edifice mass edifice flank failures, which may be triggered by volcanic episodes, lava dome collapses, or simply by gravitational instabilities. The present report evaluates volcanic mechanisms, resulting flank failure processes and their potential for tsunami generation. More specifically, the report evaluates recent volcanic eruption mechanisms of the Soufriere Hills volcano on Montserrat, of Mt. Pelée on Martinique, of Soufriere on St. Vincent and of the Kick'em Jenny underwater volcano near Grenada and provides an overall risk assessment of tsunami generation from volcanic sources in the Caribbean region.

## 1. Introduction

Tectonic deformation and active geo-dynamic processes in the Caribbean region have produced distinct seismic and volcanic activity sources capable of generating destructive tsunamis. The historic record documents numerous events in this region for the last 400 years (Lander et al. 2002, 2003; ETDB/ATL: Expert Tsunami Database for the Atlantic, 2002). The list includes tsunamis from distant sources, such as that generated by the 1755 Lisbon earthquake. Eighty-eight tsunamis from local and distant sources have been reported for the 1489 to 1998 time period. Several of these were generated by volcanic eruptions and by collateral volcanic flank failures, debris avalanches, and landslides. A rough evaluation of the cumulative frequency of tsunamis was done specifically for the Barbados and Antigua (Zahibo and Pelinovsky, 2001).

Additionally, recently found submarine debris avalanches on the sea floor around many islands in the Lesser Antilles suggest that large scale landslides and volcanic island flank collapses must have generated tsunamis in the distant past (Deplus et a 2001). Also, extremely large paleotsunami activity has been postulated for the Southern region of the Leeward Lesser Antilles, consisting of the islands of Aruba, Curacao and Bonaire (Scheffers, 2002) within the tectonically active Caribbean - South American plate boundary zone and the West Indies Island Arc. However, it remains to be determined whether the extremely large boulders and rocks found on these islands, some at high elevations, are indeed deposits from paleotsunamis - as speculated – and what the source regions may have been.

There have been several numerical modeling studies of tsunamis in the Caribbean and tsunami travel time charts for the region have been prepared (Weissert, 1990). Additionally, historical tsunamis of seismic origin have been extensively documented and numerically simulated. Some of the best-studied historic events are the 1867 Virgin Island earthquake and tsunami (Devill, 1867; Reid and Taber, 1920. Zahibo et al 2003), and the 1918 tsunami in Puerto Rico (Mercado and McCann, 1998). The heights of tsunami waves from distant sources – such as those generated by the 1755 Lisbon earthquake or from a postulated, massive landslide on La Palma, Canary Islands, have been estimated with numerical models (Mader, 2001a, 2001b). The threat of mega tsunami generation from the same postulated massive slope failure of the Cumbre Vieja stratovolcano on La Palma and the far field effects in the Eastern Atlantic from such an unlikely event, have been assessed (Pararas-Carayannis, 2002).

The flank instabilities of the island volcanoes in the Lesser Antilles are well known and adequately documented (Le Friant, 2001). In the last decade, tsunamis generated by landslides and flank failures that followed eruptions of the Soufriere Hills volcano on Montserrat Island and an increased level of activity of the submarine volcano known as Kick'em Jenny, north of Grenada, have raised concerns about the generation of destructive tsunamis from these and other volcanic sources in the region. Because of such concerns, water waves generated by potential debris avalanches and landslides on Montserrat were numerically simulated and evaluated (Heinrich et al., 1998, 1999a,b, 2001; Mangeney et al., 2000; Zahibo and Pelinovsky, 2001). Similarly, a gravitational dome collapse of the Soufriere Hills volcano on Montserrat Island and the resulting pyroclastic flows that could generate tsunamis were modeled (Hooper and Mattioli, 2001). Finally, numerical simulations of Kick'em Jenny's explosions were completed (Gisler et al 2003, 2004).

**Scope of Present Study:** The present evaluation extends these investigations by reviewing primarily the eruption mechanisms of some of the more active volcanoes of the Eastern Caribbean Region and the factors that contribute to their instabilities, massive flank failures, and tsunami generation. Active tectonics, subduction, block movements inferred from tectonic studies, seismological and geodetic data, and the evolution, geochemistry and eruption mechanisms of particular volcanic centers are reviewed for the purpose of evaluating volcanic explosivity factors – in general - and potential collateral mass edifice failures – in particular - that may result in tsunami generation. Specifically reviewed are historical eruptions and flank collapse events of Soufriere Hills on Montserrat, of Mt. Pelée on Martinique, of Soufriere on St.

Vincent, and of Kick'em Jenny near Grenada. Based on an analysis of the eruptive and mass failure mechanisms of these volcanoes, the risk of tsunami generation, from such sources, is evaluated for the entire Eastern Caribbean Region. Finally, through analogies and comparisons of differences and similarities of volcanic compositions and flank failure mechanisms, conclusions are drawn about the tsunamigenic efficiency of such processes, and the near and far field effects of tsunami waves generated by historical events, or that can be expected in the future from postulated massive edifice flank collapses of other volcanoes in the Caribbean region and around the world.

## 2. Tectonic Setting – Development Of Volcanic Centers In The Eastern Caribbean Region

Understanding tsunami generation mechanisms from volcanic sources in the Caribbean requires a brief review of the tectonic setting of the region and of the development of volcanic centers. The present review focuses on the presently volcanically active Eastern region, where most of the recent tsunamis of volcanic origin have occurred.

The Caribbean plate forms part of a region of great geologic and geographic diversity. It extends from southern N. America to northern S. America. It includes Central America, thousands of islands, and the oceanic areas in between (see Figure 1).

Geologically, the Caribbean has undergone many changes during its complex evolution (AAPG, 2003). There have been plate migrations, hotspot and mantle plume activity, island arc development and disappearance, subduction reversals, opening of young oceanic basins, major plate migration, and major block rotations. The Caribbean Plate is largely oceanic, although it carries large continental fragments in its western region.

Currently, the Caribbean plate is moving eastward in relation to the North and South American and the Atlantic Plates at a rate of approximately 20 millimeters per year (see Figure 2). The plate movement is responsible for the zones of subduction along the active boundaries and the formation of the West Indies Volcanic Island Arc on the overlying plate in the eastern region. (see Figure 3) These interactions lead to a moderate level of inter-plate seismicity and interplate and intraplate volcanic activity

Figure 1. The Caribbean tectonic plate and its volcanic island arc.

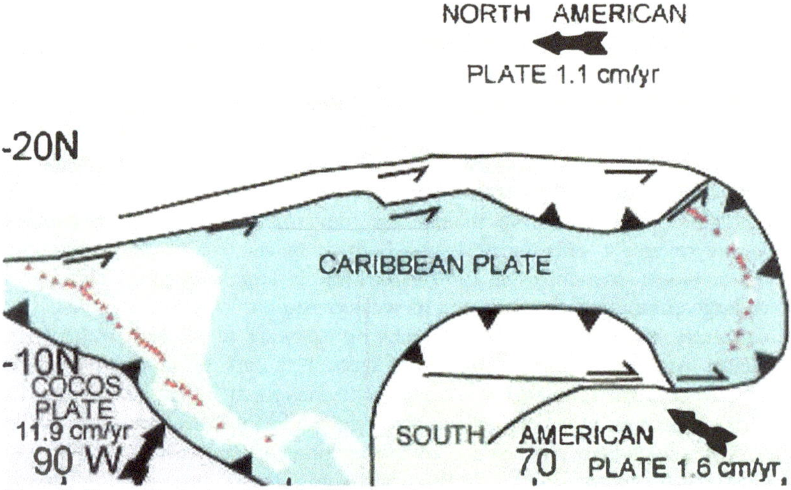


Figure 2. Present tectonics of the Caribbean region.

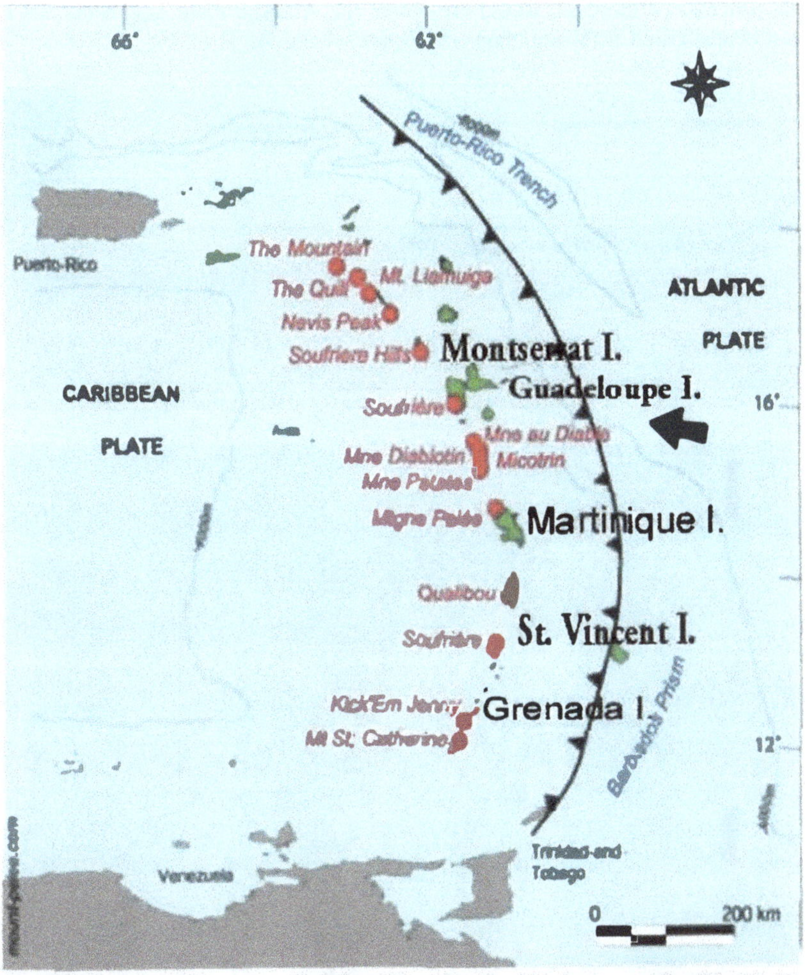


**Figure 3. Volcanoes of the eastern Caribbean Island Arc (modified web graphic of West Indies University)**

Presently, the Caribbean region is characterized by convergent, compressional and collisional tectonic activity, which results in frequent occurrences of earthquakes and volcanic eruptions. Often, localized landslides and volcanic island mass edifice failures are collaterally triggered. Most of these events occur near or along the geotectonically active plate boundaries and can generate local tsunamis with complex mechanisms, which represent the characteristics of each particular source. Seismic events in the Eastern Caribbean are principally associated with a subduction zone along a north-south line just east of the main island arc where the Atlantic Plate dips from east to west beneath the Caribbean Plate (see Figures 4 and 5).

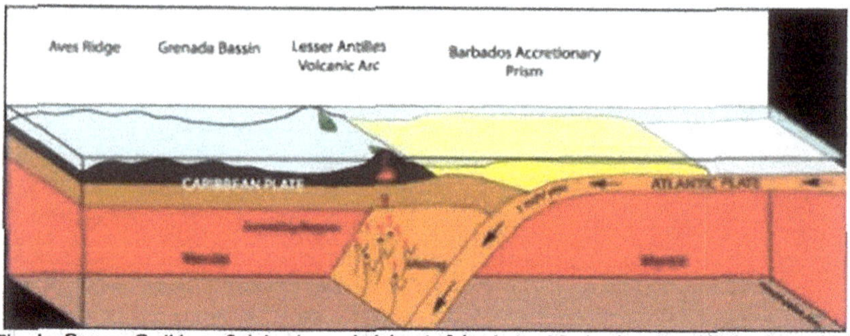


Fig. 4. Eastern Caribbean Subduction and Volcanic Island- Arc Mechanism (modified web graphic)

Fig. 5. Development of tensional volcanic fore-arc and back-arc by frontal subduction of the Atlantic plate

**2.1 Evolution of Volcanic Centers in the Eastern Caribbean Region**

Volcanic centers in the region are the by-product of active tectonic activity. Subduction is responsible for the evolution of the Lesser Antilles, an arc of small islands with active volcanoes characterized by both effusive and explosive activity (Martin-Kaye1969). Specifically, the down-dip compression on the Atlantic plate caused by tectonic movement has created the tensional volcanic back-arc, which is characterized by spreading and shallower seismic activity. As the fore-arc is driven by the mantle drag toward a trench – the zone of subduction - the resulting compression is balanced with the slab pull. This flow in the mantle causes the back-arc spreading (Seno and Yamanaka, 1998). Arc stresses and back-arc spreading result in increased volcanic activity in the region and in a greater potential for tsunami generation from subsequent volcanic mass edifice failures, and other mechanisms, which will be discussed and evaluated further in subsequent sections (Figure 6).

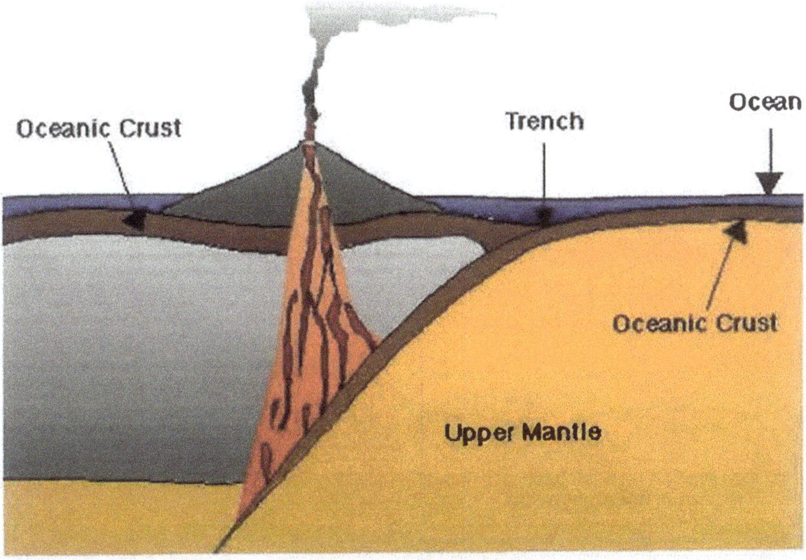


Fig. 6. Evolution of Island-Arc volcano (modified web graphic)

**3. Recent Tsunamis Of Volcanic Origin In The Lesser Antilles Region**

In recent times, Soufriere Hills on Montserrat, Kick'em Jenny near Grenada, Soufriere of St. Vincent, and Mt. Pelée on Martinique, are volcanoes in the Lesser Antilles region that have generated local tsunamis by renewed volcanic activity and associated flank failures and landslides (Lander et al 2002). Given the degree of violent volcanic activity and the flank instabilities of stratovolcanoes in the region, it is believed that the occurrences of tsunami waves have been under-reported in historical records, probably because the effects of such sea level disturbances were either localized or were overshadowed by greater catastrophes caused by violent volcanic eruptions The following is a brief overview of some of the reported historical tsunami events.

### *3.1 Montserrat Island Tsunamis*

The recent historic record documents several tsunamis at Montserrat Island. Earthquakes in the area generated some of these, while others were generated by pyroclastic flows of the Soufriere Hills stratovolcano, by debris avalanches, and by major flank failures and landslides (Figure 7). Also, the coastal geomorphology of the eastern part of Montserrat near the Chance Peak of the Soufriere Hills volcano indicates that massive landslides must have generated local tsunamis in the distant past. According to the more recent historic record, an earthquake in the region on September 13, 1824, resulted in a remarkable rise and fall of sea level at Plymouth. Another major earthquake near Antigua, reportedly triggered landslides into the sea in Antigua, Montserrat and Nevis Islands. However, most of the noteworthy tsunamis were generated recently as a result of renewed activity of the Soufriere Hills volcano.

Fig. 7. Debris avalanches and pyroclastic (lava) flows associated with the 1999 eruption of the Soufriere Hills volcano on the island of Montserrat reached the sea and generated a tsunami. Photo of landslide scar (Montserrat Volcanic Observatory).

On December 26, 1997, following a major eruption of Soufriere Hills a landslide - assisted by pyroclastic flows - reached the sea along the southwestern coast of the island and generated significant tsunami waves. (Heinrich et al., 1998, 1999a,b, 2001; Figure 8). Maximum runup of the waves, about ten kilometers away from the source region, was about 3m with inland inundation of about 80 meters. Similar debris avalanches and pyroclastic flows associated with a 1999 eruption of Soufriere Hills reached the sea and generated another local tsunami. The height of the waves in the immediate area ranged froml–2m but attenuated rapidly. By the time they reached the islands of Guadeloupe and Antigua the maximum runup heights were only about 50 cm. The most recent tsunami occurred on July 12, 2003, following a major collapse of a lava dome (Pelinovsky et al 2004). A pyroclastic flow reached the sea and generated a tsunami, which was reported to be about 4 meters on Montserrat and about 0.5-1 m at Guadeloupe.

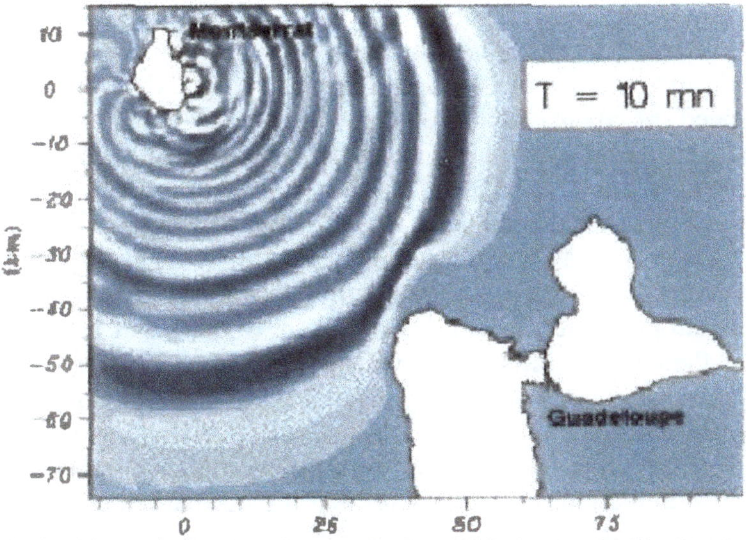


Fig. 8. Travel time chart of the tsunami generated by the 1999 debris avalanche at Montserrat Island.

### *3.2 Martinique Island Tsunamis*

Mt Pelée on Martinique is another active stratovolcano with unstable flanks composed primarily of pyroclastic rocks. As such, it must have generated numerous tsunamis in the distant geologic past. The first reported violent eruption of Mt Pelée occurred in 1792. The record does not indicate whether the eruptions caused flank failures on the island or generated tsunamis. Such events could have occurred but not been reported. Even the recent historic record is unclear. For example, there are reports of observed sea level agitations on Martinique in 1767, but it is not known whether these were tsunami waves generated by a distant earthquake or an island flank failure. On 30 November 1823 an earthquake in the area generated a tsunami, which caused damage to St. Pierre Harbor. In 1824, another earthquake near St. Pierre was probably responsible for a very "high tide" that reportedly grounded several ships in the harbor.

In the spring of 1902, Mt Pelée began erupting again. According to historic records, as the summit eruptions intensified, the water of the Etang Sec crater lake heated to near boiling point. On May 5, the crater rim broke, and extremely hot water cascaded down River Blanche. The hot water, mixed with loose pyroclastic debris and mud, formed a massive 35-meter high lahar that reached a speed of about 100 kilometers per hour. The hot volcanic mudflow buried everything in its path. Near the mouth of River Blanche, north of St. Pierre, it hit a rum distillery and killed 23 workers. The lahar continued into the sea, where it generated 4-5 meter tsunami waves, which flooded the low-lying areas along the waterfront of St. Pierre. Subsequently, on May 8, 1902 (Figure 9), a catastrophic nuée ardente cascaded for about 6 km down-slope from the central crater of the volcano, at a velocity of more than 140 km per hour, completely destroying St. Pierre, and killing 29,000 of its inhabitants. According to the historic record there were only two survivors – one in a prison dungeon. There is not much information on the tsunami that the nuée ardente must have generated, as the immensity of St. Pierre's destruction overshadowed everything else.

### *3.3 St. Vincent Island Tsunamis*

There is not much information about tsunamis generated from eruptions or flank failures of the Soufriere volcano on St. Vincent Island, although several must have occurred. The historic record shows that the volcano erupted violently in 1718, 1812, 1902, 1971-1972 and in 1979. The 1902 eruption was the most catastrophic and killed 1,600 people. The record shows that, on May 7, 1902, a day before the most violent eruption of Mt Pelée on Martinique,

Fig. 9. Mt. Pelée's eruption of May 8, 1902 killed 29,000 people and destroyed the city of St. Pierre. Local destructive tsunamis were triggered by a lahar, a nuée ardente and by flank failures. (Photograph by Heilprin. taken on May 26, 1902)

tsunami-like disturbances of up to 1 meter were reported for the harbors of Grenada, Barbados and Saint Lucia. Although the origin of these waves is not known with certainty, the most likely source could have been pyroclastic flows reaching the sea from the violent eruption of Soufriere volcano on St. Vincent. Alternatively, the sea level disturbances could have been generated by a most likely unreported flank failure of Mt Pelée, which was also erupting at that time. The historic record documents that on May 7, 1902 the submarine communication cables from Martinique to the outside world were cut.

### *3.4 Grenada Island Tsunamis*

Kick-'em Jenny is an active submarine volcano about 8 km off the North side of the island of Grenada (Figure 10), which erupted frequently during the 20th Century (Smithsonian Institution, 1999). There have been several local tsunamis generated by these eruptions. The volcano's first recorded eruption reportedly occurred in 1939, but undoubtedly there were many unreported occurrences before that date. Since 1939 there have been at least ten more eruptions. The better known are those that occurred in 1943, 1953, 1965, 1966, 1972 and 1974. The 1974 eruption was major. The last known major eruption occurred in 1990.

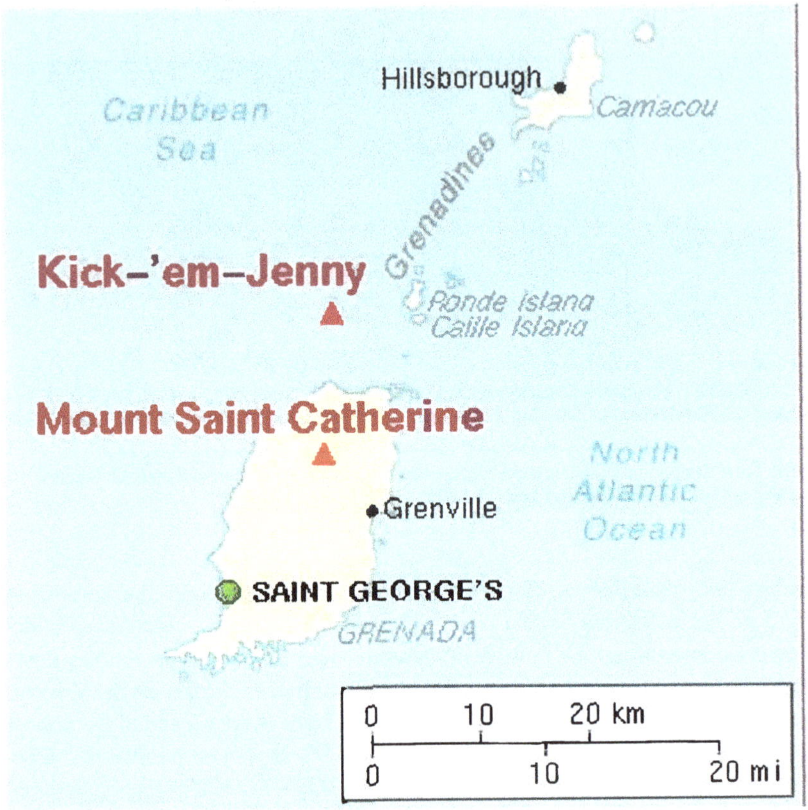


Fig. 10. Volcanoes of Grenada (USGS graphic)

The 1939 and 1974 eruptions ejected columns above the sea surface. At the peak of the July 24, 1939 eruption - which lasted more than 24 hours - a cloud rose 275 meters above the sea surface (Tilling, 1985; Seismic Research Unit Website, Univ. of West Indies, 2001). The event was witnessed by a large number of people in northern Grenada. Kick-'em Jenny's 1939 eruption also generated a series of tsunami-like waves, which had amplitudes of about 2 meters in northern Grenada and the southern Grenadines. The waves probably reached the west coast of the Barbados, but were not noticed as their heights had attenuated significantly.

## 4. Mechanisms of Tsunami Generation From Volcanic Sources

Based on what appear to be debris avalanches or toes of large scale landslides on the ocean floor, it has been postulated that mega-tsunamis were generated in the distant geologic past by massive volcanic flank failures in the Canary, Cape Verde and Hawaiian islands, as well as elsewhere in the Atlantic, Pacific and Indian oceans. Pararas-Carayannis (1992, 2002, 2003) evaluated mega-tsunami generation from prehistoric and postulated massive landslides and flank failures of oceanic basalt shield stratovolcanoes such as Kilauea, Mauna Loa, Cumbre Vieja, Cumbre Nueva, Taburiente and Piton De La Fournaise, and from the explosions/collapses of continental stratovolcanoes linked to catastrophic phreatomagmatic episodes of Plinian and Ultra-Plinian intensities, such as those that occurred at Krakatau and Santorin. Some volcanic source mechanisms of tsunami generation were realistically modeled in estimating the near and far field wave characteristics (Mader, 2001; Le Friant, 2001; Gisler 2004). However, there has not been adequate evaluation of tsunami generation from eruptions and flank failures of unstable basalt/andesitic Caribbean volcanoes, which have different geochemistry and eruption styles and processes.

### *4.1 Tsunami Generation Mechanisms of Shield Volcanoes*

Oceanic, basalt shield volcanoes have different styles of eruption, thus their mechanisms of flank failures and of tsunami generation differ from those of volcanoes along continental boundaries. Most of the basaltic stratovolcanoes have Hawaiian styles of eruptions, which usually involve less explosivity, and passive lava flows because of lower silica, gas content and ejecta viscosity. Occasional sudden gas releases may produce explosive lava fountains and unstable pyroclastic deposits. Small scale hydromagmatic explosions can also occur near the coast. Examples would be those that formed the Diamond Head and Coco Head craters on the island of Oahu, in Hawaii. However, most of the eruptions of shield volcanoes are usually confined near summit calderas or along flank craters and vents. Resulting slides from unconsolidated pyroclastics usually involve relatively small volumes of material, which rarely reach the sea to generate waves of any consequence. However, destructive tsunamis can be generated from massive volcanic edifice failures of larger blocks which may be triggered by large scale magmatic chamber collapses, erosion, gaseous pressure, phreatomagmatic and forced dike injection, or by isostatic and gravity induced kinematic changes (Pararas-Carayannis 2002). Any of these processes can trigger large volcanic mass failures of shield volcanoes, alone or in combination with other mechanisms.

### *4.2 Tsunami Generation Mechanisms of Caribbean Volcanoes*

There is a plethora of geologic evidence indicating that volcanoes in the Caribbean region have generated tsunamis by a variety of mechanisms, recently and within the last 100,000 years. Destructive tsunami waves were generated by violent sub-aerial and submarine eruptions and accompanying earthquakes, by caldera and submarine flank collapses, by subsidences, by atmospheric pressure waves, by lahars, nuées ardentes - a type of pyroclastic avalanche mixture of gas, dust, ash and other hot glowing incandescent solid particles and lava fragments -, pyroclastic flows and debris avalanches. Also, tsunamis must have been generated from gravitational mass edifice failures due to the characteristic flank instabilities of the volcanoes in this region - even in the absence of obvious triggering events. For example, earth tides could have triggered such failures.

Evaluation of flank instabilities of Caribbean stratovolcanoes and their potential for tsunami generation requires a closer examination of the styles, intensity and geometry of eruption mechanisms, of precursor events, of the time history of volcanic episodes, of the geochemistry and composition of the lava and ejecta, as well as an assessment of tectonic processes in the region which result in volcanic arc stresses, back-arc spreading and an increased level of volcanic activity. Small scale flank collapses which result in tsunami generation are a standard phase in the evolution cycles of Caribbean volcanoes (Young 2004.). The following is a review of different factors that contribute to eruptions episodes of Caribbean volcanoes, which may range in style and intensity from Strombolian to Vulcanian/Plinian, but are not as catastrophic as the Plinian and Ultra-Plinian episodes of the Krakatoan/Santorin variety. Specifically reviewed in the next section are precursor events and eruption processes of some of the previously mentioned active volcanoes of the Lesser Antilles – processes that result in moderate flank failures and the generation of tsunamis or tsunami-like waves. Although the present analysis focuses on mechanisms of flank failures of active Caribbean stratovolcanoes, it should be pointed out that the processes and mechanisms that are described are similar to those occurring at many other volcanoes around the world. Furthermore, it should be pointed out that, in contrast to tsunami generation from seismic sources which cannot be predicted, the generation of tsunamis from volcanic sources can be forecast with proper monitoring of precursor events, of volcanic activity and of flank instabilities. The following sections provide an analysis of factors that contribute directly to eruptions, flank instabilities and failures of Caribbean volcanoes and, indirectly, to the generation of destructive sea waves.

## 5. Factors Contributing To Volcanic Explosivity, Structural Flank Instabilities, Mass Edifice Failures, Debris Avalanches And Tsunami Generation In The Caribbean Region

Tsunamis can be generated by volcanic caldera and lava dome collapses, by vertical, lateral or channelized explosive activity and by the associated atmospheric pressure perturbations, pyroclastic flows, lahars, debris avalanches or massive volcanic edifice failures. Contributing tectonic factors include island arc volcanism that overlies a subduction zone and which can result in the most catastrophic types of eruptions. Many additional specific factors determine the eruption style, higher explosivity, the generation of pyroclastic flows, the structural flank instabilities, the slope failures and the debris avalanches that characterize the mainly andesitic Caribbean volcanoes. The following is a description of these factors for the most active volcanoes in the Caribbean region. However, the same factors apply to all the basaltic/andesitic volcanoes around the world that border tectonic boundaries.

### *5.1 Geochemical Factors*

Whether a volcano will have effusive eruptive activity or explosive type of bursts will depend primarily on geochemical factors. The build up of pressure of volatile gases within the magmatic chambers determines a volcano's eruption style, explosivity, flank instability and potential for subsequent tsunami generation.

Variations in the chemical composition of volcanic eruptive effluents result from differences in the mineralogy and the bulk composition of mantle material, differences in depth at which melting occurs, differences in percentages of melting at the source, and alterations in composition as the magma rises to the surface (Wright and Helz, 1987). The chemical composition of molten rocks in the magmatic chambers of a volcano and the different abundances of elements, particularly silica, determine the viscosity of effluents that can rise to the surface. Magmas, which are low in silica and rich in iron and magnesium, like the basalts of Hawaiian types of volcanoes, are very fluid. Basalt contains anywhere from 45% to 54% silica, and generally is rich in iron and magnesium. Dissolved gases escape prior to an eruption, thus resulting in a subsequent effusive eruptive activity associated with relatively gentle lava flows. Because of the low viscosity, lava flows travel great distances from the eruptive vents, and produce broad, shield-shaped volcanoes. Kilauea, in Hawaii, is an example of such a volcano.

By contrast, the chambers of stratovolcanoes along continental margins contain magma which is composed primarily of andesite or dacite, that are relatively high in silica and low or moderate in iron and magnesium. Andesite may contain anywhere from 54 to 62 percent silica. Thus magma material tends to be very viscous. Gases are trapped and cannot escape until the magma enters the volcanic conduits leading to the surface. The reduction in pressure allows the gas bubbles in the magma material to nucleate and expand. When the outward pressure exerted by the gas bubbles exceeds the strength of the lava material above it, the volcano erupts violently. Because of the high pressure of the expanding volatiles, the lava is fragmented to produce volcanic ash and pyroclastics that are ejected out of the volcanic vent at high velocity. Most of the ash travels far but pyroclastics accumulate near the vent, thus producing a steep-sided stratovolcano with very unstable flanks. An example of such a volcano is Mount St. Helens in the State of Washington.

The volcanoes of the Lesser Antilles in the Caribbean region are mainly andesitic stratovolcanoes, characterized by both effusive and explosive activity (Brown et al. 1977). However, even effusive activity may culminate into explosive activity at later stages of an eruption, as the more viscous magma material reaches the surface. Explosive eruptions of Caribbean volcanoes may last for hours and will result in greater volcanic cone instabilities. Intensities and types of explosive episodes may range from Strombolian, Vulcanian to even Vulcanian/Plinian. During such eruptions, destructive sea waves may be generated directly by pyroclastic flows, lahars and nuées ardentes reaching the sea, and by atmospheric pressure waves of blast events. Debris avalanches and massive volcanic flank failures can also generate destructive waves during an eruption or subsequently.

### *5.2 Growth and Collapses of Lava Domes*

As a result of the geochemistry and the higher viscosity of the mainly andesitic magma, Caribbean volcanic activity often results in the formation and growth of lava domes near a volcano's summit or along its flanks (Voight, 2000). Rapid lava dome growth, after a quiet volcanic activity period, is indicative that pressure is building up within a volcano. Subsequent collapses of lava domes often trigger a flurry of volcanic eruptions which will vary in intensity and may be associated with catastrophic pyroclastic flows, lahars, debris avalanches and large scale flank failures – which can generate directly destructive local tsunamis. In fact the collapse of a growing lava dome has the potential of generating directly destructive pyroclastic flows, lahars or debris avalanches, which, upon reaching the sea, can generate destructive local tsunamis. Indirectly, such volcanic processes may weaken the flanks of the volcano, thus flank failures may be retrogressive and time-dependent. Thus, lava dome growth, chemical and geometric factors, and the mechanisms of weakening that result in eventual collapses, are important factors in assessing a volcano's overall instability and in forecasting a major eruption or even the generation of a tsunami.

### *5.3 Volcanic Explosivity Factors*

As already indicated, the sudden releases of gases are responsible for a volcano's explosivity and flank instability and thus trigger flank failures which can generate destructive tsunami waves. Additionally, the direct explosive outflux of volcanic volatiles can create sudden atmospheric pressure disturbances which, coupled with the sea surface, can also generate destructive waves.

Significant atmospheric pressure perturbations that are propagated by buoyancy forces during violent volcanic eruptions can have periods greater than 270 seconds (Beer, 1974). The periods of the atmospheric waves is dependent on the volumes of volatiles and the duration of volcanic bursts. For example, spasmodic volcanic bursts of Caribbean volcanoes during a violent eruption may produce extremely large movements of air parcels similar to those produced by the 1980 Mount St. Helens eruption (Mikumo and Bolt, 1985; Tilling 1984; Tilling, et al 1990) or by the 1992 Pinatubo eruption (Tahira et al., 1996). Eruptive phases consisting of numerous eruptive pulses may last from a few hours to days. Explosive pulses may last from a few seconds to minutes and may vary in intensity depending on the impulsivity of the degassing source (Newhall & Self, 1982). Such volcanically generated atmospheric pressure waves can generate several destructive tsunami-like waves of longer periods with heights which will not attenuate as rapidly as those from other wave generation mechanisms.

### *5.4 Blast Geometry Factors*

Additionally, the geometry of eruption blasts is a factor that will also determine a volcano's flank instability and its potential for massive failures, avalanches or subareal or submarine landslides and subsequent tsunami generation. Generally, volcanic blasts mechanisms can be vertical, lateral-direct or channelized. If the blasts directly affect a body of water, then more destructive local tsunamis can be expected.

### *5.5 Vertical Blast Mechanisms*

More typical for Caribbean volcanoes are the vertical type of blast mechanisms, which will deposit unconsolidated pyroclastic debris and build up cones with unstable flanks. Gravity forces may eventually lead to cone collapse or other failures, which in turn may trigger lahars, larger flank landslides or debris avalanches, usually on the steep-sided stratovolcanoes. As previously discussed, strong blasts can also generate atmospheric pressure perturbations which, depending on duration and intensity, may couple with the sea surface to form additional destructive waves of varying periods.

### *5.6 Lateral and Channelized Blast Mechanisms*

The extensive distribution of deposits from the 1980 eruption of Mt. St. Helens (Figure 11) is indicative that lateral and channelized blasting effects can be far reaching. A large lateral blast from an erupting Caribbean island volcano could be extremely destructive but could also generate much more destructive local tsunami waves. The May 8, 1902, explosion of Mt. Pelée that destroyed the city of St. Pierre on the island of Martinique was such a lateral blast. Lahars and nuées ardentes generated destructive local tsunamis. The distribution of pyroclastic deposits on the western flank of the underwater volcano Kick'em Jenny indicates that such lateral blasting occurred there in the past. There is also a major escarpment that suggests other types of massive slope failures.

## 6. Mechanisms Of Volcanically- Induced Tsunami Generation In The Lesser Antilles Islands Of The Caribbean

In view of the above-described complex factors, proper numerical modeling of tsunami generation from volcanic sources requires consideration of explosion dynamics, of the time history of blast events, of the geometry of sudden explosions, and of water displacements caused by both movements of mass and by atmospheric pressure waves.

`Recent historical tsunamis generated at Montserrat, Martinique, St Vincent, and Grenada Islands in the Lesser Antilles region were briefly reviewed. Obviously, the mechanisms of tsunami generation from volcanic sources can vary significantly even within the same geotectonic region. In the following sections, the eruptive processes at work for active volcanoes such as Soufriere Hills, Mt. Pelée, Soufriere and Kick-'em Jenny will be reviewed and analyzed,

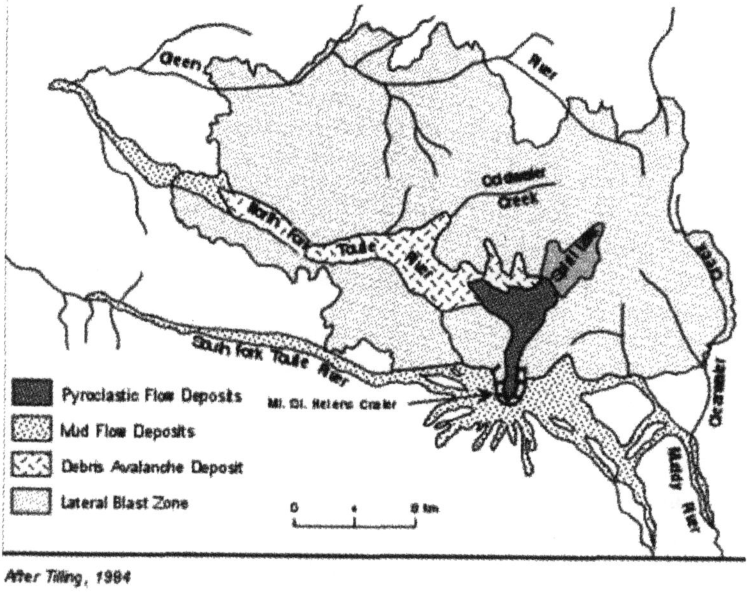


Fig. 11. Distribution of deposits from the May 18, 1980 lateral and channelized blast of Mt. St. Helens (after Tilling, 1984)

as well the time history of major specific eruptions that resulted directly or indirectly in tsunami generation.

### *6.1 Soufriere Hills Volcano on Montserrat Island - Eruptive Processes and Mechanisms of Tsunami Generation*

The Soufriere Hills, located on the southern part of Montserrat Island (Figure 12), is a very active, primarily andesitic stratovolcano (Rowley1978), which is the predominant type of explosive volcano in the world. Its present elevation is 915 m. The first known historical eruption was in 1995. Since then there have been several more eruptions in the late 1990s (Hooper and Mattioli, 2001), including one that destroyed the former capital Plymouth. The volcano is presently very active. All of its eruptions have been associated with earthquake swarms, lava dome collapses, steam explosions, ash falls, pyroclastic flows and debris avalanches.

The volcano's flank instability and its potential for landslides and tsunami generation result from the composition of its magma which is very sticky and has a high content of dissolved water. In fact, eruptions appear to be linked with rainstorms and high earth tides. When the volcano erupts it tends to form a steeply sloped peak made of alternating layers of lava, block, and ash. Thus, the slopes of the volcano become unstable and susceptible to massive landslides and debris avalanches, some of which can reach the sea and generate local tsunamis. In fact several have occurred in the last few years.

### *6.1.1 Eruptive Processes*

To understand better the tsunamigenic potential of the Soufriere Hills volcano on Montserrat, we must review further its eruptive processes. Two types of eruptive mechanisms characterize this volcano, both of which have the potential to generate local tsunamis. In both cases, magma inside the volcano is driven up by buoyancy and gas pressure, which may vary depending on its viscosity.

In one type of eruption, the volcano will explode and shoot molten rock violently into the air in the form of dense clouds of lava fragments. The larger and heavier fragments tend to fall back around the volcanic vent, which may become increasingly unstable as the eruption progresses. Often, the accumulation may run down slope as ash flows, while some of the finer particles may be carried by the wind. Often, both ash and pyroclastic flows can trigger debris avalanches and larger landslides and slope failures on the volcano (DeGraff 1988) that can generate small local tsunamis.

In the second type of eruption, the molten rock - which is lighter than the surrounding solid rock - breaks through the weaker stratigraphic layers and rises closer to the surface forming a lava dome. When the dome becomes too steep, or if pressure within builds up, it collapses and disintegrates, spewing lava and hot ash down the side of the volcano. Also, it may form a mushroom cloud of ash that can spread over the island. The collapse of such lava domes releases pressure and frequently causes subsequent massive eruptions that can also affect the volcano's flank stability, thus generating pyroclastic flows, debris avalanches, landslides and massive flank failures. In fact a combination of the above-described eruptive processes occurred during eruptions in the summer of 1995. Specifically, on July 18, 1995, Soufriere Hills had its first recorded eruption in historic times. It begun with a small phreatic eruption. Periods of intense seismic activity were associated with ejection of steam and ash, shortly after a new vent opened southwest of an old volcanic dome known as Castle Peak. The eruption culminated into a major event on August 21, 1995, when the volcano begun spouting molten ash, rock, and gas over the island, killing 19 and incinerating the capital city of Plymouth. A strong burst of steam carried a cloud of ash to an altitude of 7,000 feet. The eruption triggered several landslides that reached the sea, but there was no report of any unusual waves being generated. (Mangeney et al. 1998; Calder et al. 1998).

Fig. 12. The Island of Montserrat, the Soufriere Hills volcano, and the abandoned town of Plymouth.

### *6.1.2 The Eruption and Tsunamis of 26 December 1997 and 1999 and 2003:*

As already mentioned, the Soufriere Hills volcano either erupts by exploding and expelling lava or by dome collapse. Both types of eruption can be destructive as they can produce dangerous ash hurricanes and pyroclastic flows, trigger landslides and debris avalanches and thus generate tsunamis. Although the 1995 eruption and other volcanic processes that occurred subsequently did not generate a tsunami, apparently they weakened Soufriere Hills' flanks. This weakness contributed to the subsequent volcanic flank failures associated with the eruptions of 1997, 1999 and 2003 – which generated tsunamis.

Specifically, on June 25, 1997, after two years of precursory swelling and micro earthquake activity, Soufriere Hills volcano erupted again. A damaging pyroclastic flow of ash, gas, and rock killed at least ten people and destroyed nine villages. A lava dome was subsequently observed which built up steadily in the volcano's crater for over two months. On 26 December 1997, following the collapse of this lava dome, a major eruption occurred. The eruption generated ash hurricanes, which destroyed Plymouth. Both the ash hurricanes and a landslide - possibly assisted by pyroclastic flows triggered by the dome-collapse - reached the sea, along the southwestern coast of the island and generated significant tsunami waves. (Heinrich et al., 1998, 1999a,b, 2001). The maximum runup of the waves was about 3m. about ten kilometers away from the source region, with inland penetration of about 80 meters. The volume of the landslide debris, which generated this tsunami, was estimated to be about 60 million cubic meters (Lander et al., 2003).

Similar debris avalanches and pyroclastic flows associated with the 1999 eruption of Soufriere Hills reached the sea and generated another local tsunami. The height of the waves in the immediate area ranged from 1–2m but attenuated rapidly. By the time the waves reached the islands of Guadeloupe and Antigua their heights attenuated considerably. Maximum runup heights were only about 50 cm.

The most recent tsunami was produced by the eruption of July 12, 2003 (local date; Figure 13) following a major collapse of a lava dome (Pelinovsky et al 2004; Young 2004). Pyroclastic flows and a debris avalanche reached the sea at the end of Tar River Valley on the east coast and generated this tsunami, which was reported to be about 4 meters at Spanish Point on Montserrat Island and about 0.5-1 m at Deshaies and near Plage de la Perle on Guadeloupe where it caused some damage to fishing boats.

That debris avalanches and extensive landslides of andesitic volcanoes will only generate local destructive tsunamis, is supported by the April 20, 1988 massive flank failure of the northeast flank of the volcano La Fossa on the Island of Vulcano in the southern Tyrrhenian Sea, in Italy (Figure 14). According to modeling studies - which were based on photogrammetric techniques conducted in 1981 and 1991 - the large 1988 flank failure of La Fossa involved a mass with

Fig 13. Pyroclastic flow from the 2003 eruption of Soufriere Hills volcano on Montserrat reaching the sea.("Copyright Montserrat Volcano Observatory/Government of Montserrat and British Geological Survey; photo used by permission of the Director, MVO")

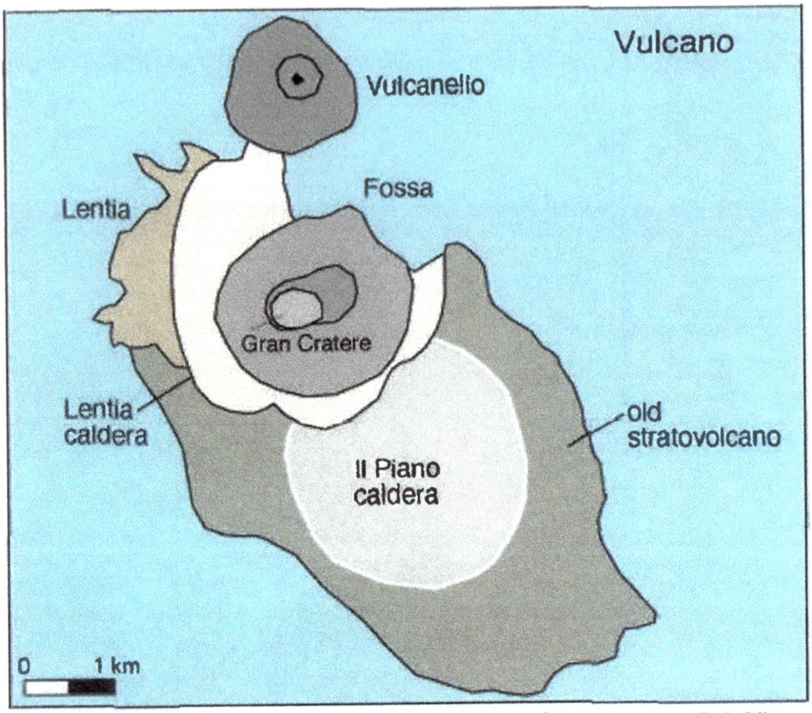


Fig. 14. Map of the Island of Vulcano in Italy where a 200,000 cubic meter massive flank failure on the northeast side generated local tsunami (after Imbo, 1965 and Keller,1980)

a volume estimated to be about 200,000 cubic meters. The mass that was detached fell into the sea for about 10 seconds. A small tsunami was generated in the bay between P●int Nere and Point Luccia on the island. Maximum observed runup height of the waves was about 5.5 meters at Porto di Levanteand presumably even at Monterosa on Lipari Island. (Barberi, et al 1990; Lander et al. 2003).

### *6.2 Mt. Pelée Volcano on Martinique Island – Eruptive Processes and Mechanisms of Tsunami Generation*

Mt. Pelée on Martinique is a very active island-arc stratovolcano with unstable flanks made mostly of pyroclastic rocks (Smith and Roobol 1990). Its summit elevation is 1397 m. It undergoes similar eruptive processes as other Caribbean volcanoes and can also generate destructive local tsunamis by pyroclastic flows, flank failures or debris avalanches. However, what makes Mt. Pelée unique has been its unusual lava dome formations, the intensity and styles of its eruptions and the unusual and violent pyroclastic flows it can generate (Fisher and Heiken1982). The volcano has a long history of eruptions in the last

5,000 years (Westercamp and Traineau 1983). In more recent historic times the volcano erupted in 1635, 1792, in 1851-1852, in 1902- 1905 (Heilprin 1908) and in 1929-1932 (Perret 1937).

The historic record documents two extremely violent eruptions in 1792 and in 1902 - associated with numerous other phenomena that followed dome collapses – and by which local tsunamis were generated. The eruptive processes of Mt. Pelée and the tsunami generation mechanisms that are described in subsequent sections are based on what occurred on Martinique in May of 1992 and whatever little is known about the violent volcanic eruption of 1792.

### *6.2.1 Eruption Processes of Mt. Pelée*

Eruptions of Mt. Pelée range in volcanic explosivity intensity from severe Vulcanian (VEI = 3) - which can occur yearly - to cataclysmic Vulcanian-Plinian events (VEI = 4) separated in time by many decades. The Peléean eruptions - as they are now termed because of their unique characteristics - are extremely violent eruption events that often include collapses of ash columns, and unique pyroclastic flows known as "nuées ardentes", and debris avalanches containing large amounts of ignimbrites (ash flow tuffs). These unusual pyroclastic flows are usually triggered after a lava dome collapse.

### *6.2.2 The Eruption and Tsunamis of May 1902*

A previously stated, an extremely violent volcanic eruption occurred on Mt. Pelée in 1792. It is very probable that a tsunami was generated at that time as a result of a flank failure or pyroclastic flow, but there are no reports documenting it. However, the 1902 eruption (Figure 15) and its associated unusual phenomena are well documented in the literature (Lacroix, 1904; Heilprin 1908; Fisher et al 1980). The May 1902 tsunamis were generated by a lahar and a subsequent nuée ardente of a violent eruptive phase.

In early 1902, a large dome of very viscous lava had grown on Mt. Pelée's flank near its summit, largely by expansion from within. As the lava dome grew, its outer surface cooled and hardened. There is not much information on the size of this particular lava dome, but it could have been as big as that of the Katmai volcano in Alaska, which collapsed and triggered an eruption in 1912. That dome had been circular and measured about 250 meters across and 60 meters in height. However, what was reported about Mt. Pelée's lava dome is that it had cut a large V-shaped notch through the cliffs that surrounded the volcano's summit crater. According to reports, the "notch was like a colossal gun sight pointing directly at the town of St. Pierre".

According to historic records, on May 5, 1902, a 35-meter lahar cascaded down the flank of the volcano and reached the sea. The lahar generated a local tsunami wave of about 4-5 meters in height, which killed one hundred people in St. Pierre. Subsequently, at approximately7:50 a.m. on May 8, 1902, the pressure from within the volcano reached a critical level. Suddenly, the summit lava dome collapsed and shattered with a deafening roar, spilling loose fragments down- slope. The sudden release of pressure triggered

Fig. 15. The 1902 Eruption of Mt. Pelée on the island of Martinique. The destruction of the town of St. Pierre was caused by a nuée ardente. (Photograph by Heilprin, 1902).

by the dome collapse resulted in an extremely violent eruptive phase of Mt. Pelée.

A large nuée ardente cascaded from the central crater for about 6 km down the south flank, at a velocity of more than 140 km per hour. In less than one minute it struck the coastal town of St. Pierre, destroying it completely and killing 29,000 of its inhabitants (Figure 16). Only two people are known to have survived. According to reports (Heilprin 1908; Fisher et al 1980) the directional blast was so strong that it carried a three-ton statue sixteen meters from its mount. One-meter thick masonry walls were blown into rubble. "Supporting girders were mangled into twisted strands of metal". The heat of the nuée ardente nuee was immense and ignited huge fires. Thousands of barrels of rum that was stored in the city's warehouses exploded and burned in the streets.

There is not much direct information on the tsunami that the nuée ardente must have generated, as the immensity of St. Pierre's destruction overshadowed everything else. However, it was reported that the nuée ardente continued seaward toward the harbor where it destroyed at least twenty ships that were anchored offshore. The American sailing ship "Roraima", which had arrived only a few hours earlier, burned and all its crew and passengers perished. According to reports, the steamship "Grappler" was capsized by the force of the nuée ardente. However, it is more likely that it was capsized by the wave generated by the nuée ardente within the harbor.

Mechanisms of tsunami generation involving cascading volcanic gases and rapidly moving pyroclastics flows are not confined to Caribbean volcanoes or to Mt. Pelée, in particular. There is evidence that similar hot glowing avalanches of hot gas, dust, ash and pyroclastics have generated several tsunamis in the distant past in New Zealand and elsewhere around the world.

### *6.3 La Soufrière Volcano on St. Vincent Island – Eruptive Processes and Mechanisms of Tsunami Generatio*

La Soufrière is an active and dangerous stratovolcano on the island of St. Vincent in the Windward Islands of the Caribbean (Figure 17), with a well-documented history of violent eruptions (Robson, 1965a, Shepherd and Aspinall. 1982). The present elevation of its summit is at 1220 m. There is a lake within the summit crater. La Soufrière should not be confused with a volcano by the same name on the island of Guadaloupe.

There is evidence of activity on Soufrière for the last 650,000 years (Hay 1959; Rowley1978). In recent times, major eruptions occurred in 1718, 1784,1812, 1814, 1880, 1902-03 (Anderson 1784; Anderson, T. 1903; Flett 1902,1908; Anderson and Flett 1903; Sapper 1903; Anderson 1908, Carey and

Fig. 16. Devastation of the town of St. Pierre on Martinique Island by a Nuee Ardente of the 1902 eruption of Mt. Pelée. (Photograph by Heilprin, 1902).

Sigurdsson 1978). In the twentieth century there were major eruptions, in 1971-72 (Aspinall et al 1972; Baker, 1972; Tomblin et al. 1972; Aspinall 1973; Aspinall et al 1973) and in 1979 (Shepherd et al 1979; Shepherd et al 1982; Barr and Heffter 1982; Brazier et al 1982; Fiske and. Sigurdsson. 1982; Graham and Thirlwall. 1981) (see Figure 18). The 1812 eruption resulted in many deaths. However, the 1902 eruption was the most catastrophic of all resulting in the loss of 1,600 lives.

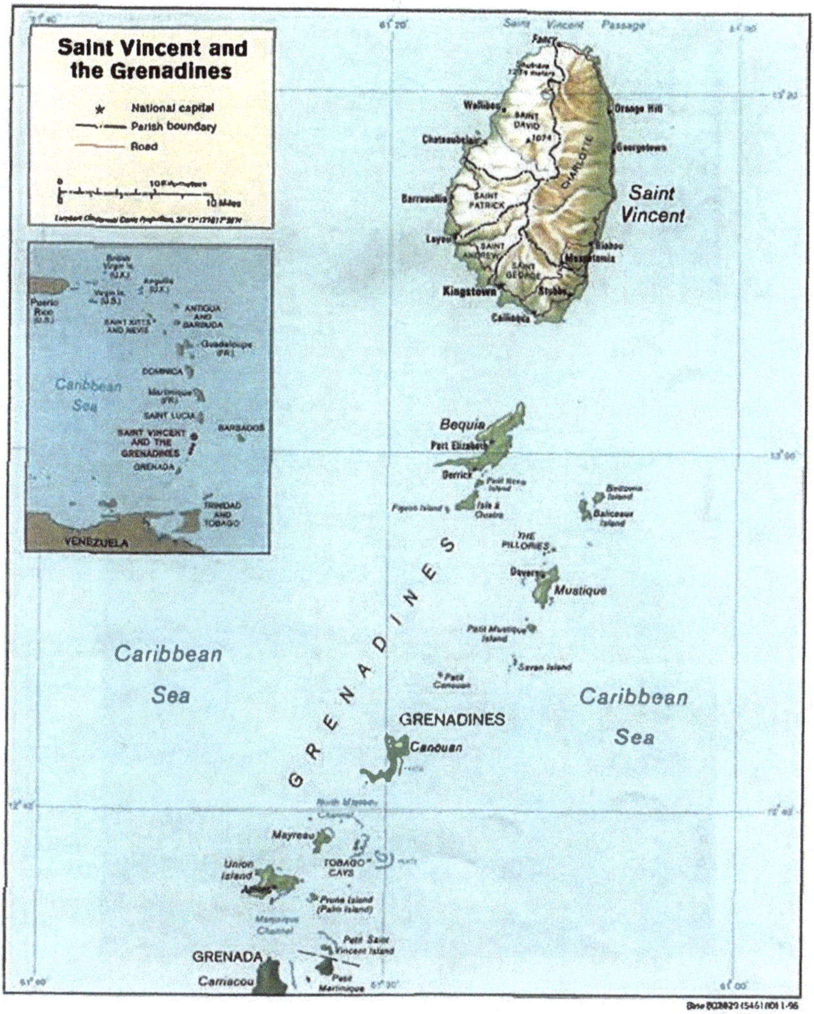


Fig. 17. St. Vincent Island and the Grenadines (web graphic)

### *6.3.1 Eruption Processes of Soufrière*

Geologic evidence indicates that for the past 4,000 years the Soufrière volcano's eruptions have alternated between explosive episodes associated with the forceful ejection of fragmented material and pyroclastic flows to quiet effusion of slow moving lava that forms summit domes (Earle 1924; Hay1959; Heath et

Fig. 18. The 1979 eruption of La Soufrière on St Vincent Island (Photograph by Richard Fiske)

al 1998). The 1979 eruption is typical of such variation. It begun quite suddenly with less than 24 hours of precursor activity. The mechanism of its subsequent explosive eruption has been well documented (Shepherd and Sigurdson, 1982). The first eruptive episode was Vulcanian in character. It sent a plume of steam and tephra to a height of 20 km. and lasted a little less thantwo weeks (Sparks and Wilson. 1982). The second episode consisted of a quiet extrusion and growth of a basaltic andesite lava dome (Huppert et al. 1982).

### *6.3.2 The Eruption and Tsunamis of May 7, 1902*

There is not much information about tsunamis generated from eruptions or flank failures of St. Vincent's Soufrière stratovolcano, although several must have occurred in the geologic past - and even more recently. As mentioned previously, on May 7, 1902, a day before the most violent eruption of Mt Pelée on Martinique, tsunamis like disturbances were reported for the harbors of Grenada, Barbados and Saint Lucia. The most likely source could have been air pressure waves from the violent eruption of Soufrière on that day, or pyroclastic flows and debris avalanches reaching the sea.

Coincidentally, the historic record also shows that on the same day - May 7, 1902 - the submarine communication cables from the island of Martinique to the outside world were cut. The exact area where the cables failed is not known. Thus, it is difficult to determine what caused the cable failures and whether the sea level disturbances observed at the harbors of Grenada, Barbados and Saint Lucia had the same source. The waves could have been generated by an unknown flank failure of Mt Pelée, and the cable failures by an underwater debris avalanche. On May 5, Martinique had already experienced a destructive local tsunami generated by a lahar.

## *6.4 Kick-'em Jenny Submarine Volcano near the Island of Grenada – Eruptive Processes and Mechanisms of Tsunami Generation*

Kick-'em Jenny is a growing submarine volcano about 8 km off the north side of the island of Grenada. It is the southernmost active volcano in the Lesser Antilles volcanic arc and has erupted frequently during the 20th Century (Smithsonian Institution, 1999). Presently, the volcano has a circular base of about 5000 m, its main cone has reached a height of about 1300 m above the sea floor, and its summit is about 190 m below the sea surface (Figure 19). The volcano is expected to reach the surface and form an island in the future – if there is no flank subsidence or cone collapse.

Kick-'em Jenny's first recorded eruption occurred in 1939, but many unreported eruptions must have occurred prior to that date. Since 1939 there have been at least twelve or more events. Most of the historical eruptions were documented by acoustic measurements, since submarine volcanoes generate strong acoustic signals that are recorded by seismographs. Known eruptions occurred in 1939,1943, 1953, 1965, 1966, 1972, 1977, 1988, and in 1990. The better-known events are those that occurred in 1943, 1953, 1965, 1966, 1972 and 1974. The last major eruption occurred in 1990. Earthquake swarms in late 2001 indicated renewed activity. The latest eruption occurred on March 15, 2003.

In 2003, during a survey of Kick- 'em Jenny, an inactive underwater volcano was discovered about 3 km away. It is, now known by the name of Kick-'em Jack.

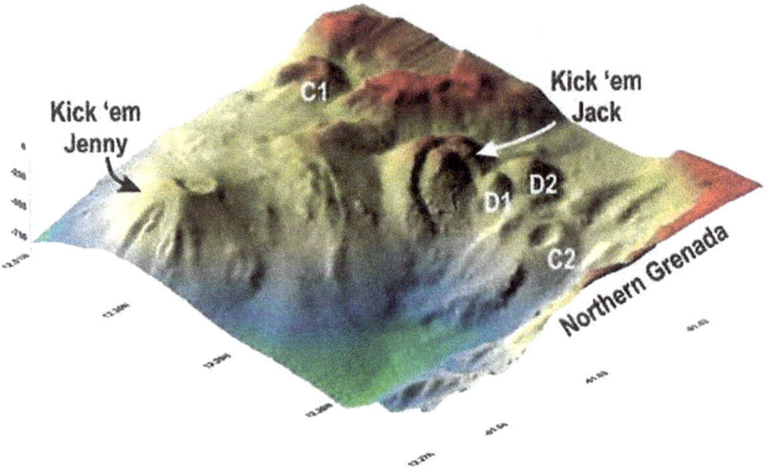


Fig. 19. Two-minute topography of the sea floor north of the island of Grenada, showing the geomorphology of the calderas, cones and domes, generally known as the Kick'em Jenny volcano (web graphic)

### *6.4.1. The Eruption and Tsunamis of 1939 and 1974*

According to historical accounts and eyewitness reports from northern Grenada, the July 24, 1939 eruption of Kick-'em Jenny was major and lasted for at least 24 hours. The eruption ejected a cloud plume above the sea surface. Furthermore at the peak of the eruption, the cloud plume rose 275 meters above the sea surface (Tilling, 1985; Univ. of West Indies, 2001). The eruption generated numerous tsunami-like waves of short period. These waves had maximum amplitudes of about 2 meters in northern Grenada and the southern Grenadines, but were almost imperceptible when they reached the west coast of the Barbados.

### 6.4.2 Eruption Processes of Kick'em-Jenny

The underwater topography of the sea floor north of Grenada indicates that Kick-'em Jenny comprises of three small craters and two lava domes – all of which probably share the same magmatic chambers.

As most of the Caribbean volcanoes, Kick-'em Jenny has had both violent and effusive eruptive episodes. Eruptions of the volcano have been associated with magmas, which have ranged in compositions from basalt to basaltic andesitic. Thus, gently extruded submarine pillow lavas and domes as well as tephra and other pyroclastics from minor phreatomagmatic explosions, are present in submarine deposits around the volcano.

### *6.4.3 Flank instability*

The distribution and orientation of pyroclastic deposits on the sea floor, primarily to the west side of Kick-'em Jenny, indicate that many volcanic eruptions must have occurred that have been lateral or channelized blasts, possibly following the collapse of lava domes (Figures 20 and 21). Furthermore NOAA surveys in 2003 demonstrated the presence of deposits from a debris avalanche. The geomorphology of the sea floor indicates that this debris avalanche extends west for 15 km and perhaps as much as 30 km from the volcano, into the Grenada Basin (Sigurdsson et al 2004). Also, earlier multibeam surveys of the sea floor discovered the existence of an arcuate fault escarpment - of yet unknown age - to the east of the active cone. Because of its shape and length, this escarpment cannot be related to caldera subsidence and collapse. Its configuration and the overall geomorphology of the sea floor suggest that a larger scale subsidence or volcanic mass edifice collapse occurred in the distant past. It also suggests that Kick'em Jenny volcano might have been at or above sea level in the past.

Overall, the volcano's present rapid upward growth towards the surface of the sea is indicative of active vertical summit eruptions and the build up of a cone by deposition of pyroclastics. However, the flanks of this cone must be very unstable and subject to collapses and the generation of future debris avalanches, which could slow the volcano's present rate of growth. Additionally, hydromagmatic explosions associated with future eruptions could also result in greater flank instability and might also slow down the rate ofgrowth. Future major eruptions can be expected to be more violent and to eject sizeable columns above the sea surface to heights much greater than those of the 1939 and 1974 events. Major future eruptions can be expected to have considerably higher plume clouds, because of the greater strength of hydromagmatic episodes as the summit approaches the sea surface and the inclusion of a higher content of molecular water – in the form of superheated steam - along with ejected tephra and other fine pyroclastic materials.

### *6.4.4. Assessment of the Tsunamigenic Potential of Future Eruptions of Kick'em Jenny*

The frequency of Kick-'em Jenny's eruptions and the volcano's rapid growth toward the sea surface have raised concerns that future eruptions will generate tsunami waves with far reaching destructive effects on Caribbean islands and along the coast of Venezuela. Earthquake swarms in late 2001 added to concerns that Kick-'em Jenny will again have a major eruption.

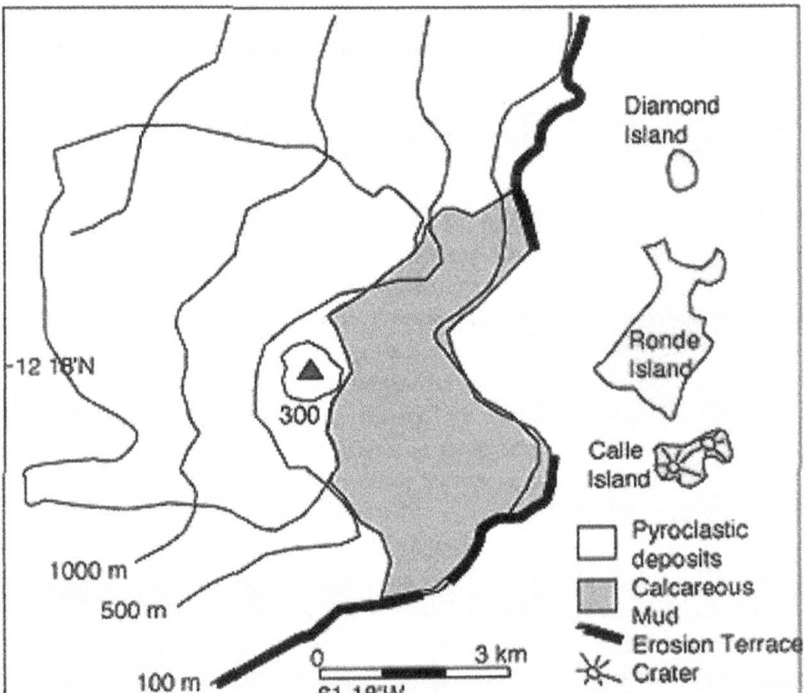


Fig. 20. Bathymetry and distribution of volcanic deposits from eruptions of Kick'em Jenny volcano (Web graphic at http://volcano.und.edu/vwdocs/volc_images/north_america/kick.html)

Although there is a good probability that several eruptions will occur in the near future – and in fact the latest occurred on March 15, 2003 - the potential tsunami risk from a future eruption has been highly exaggerated by the

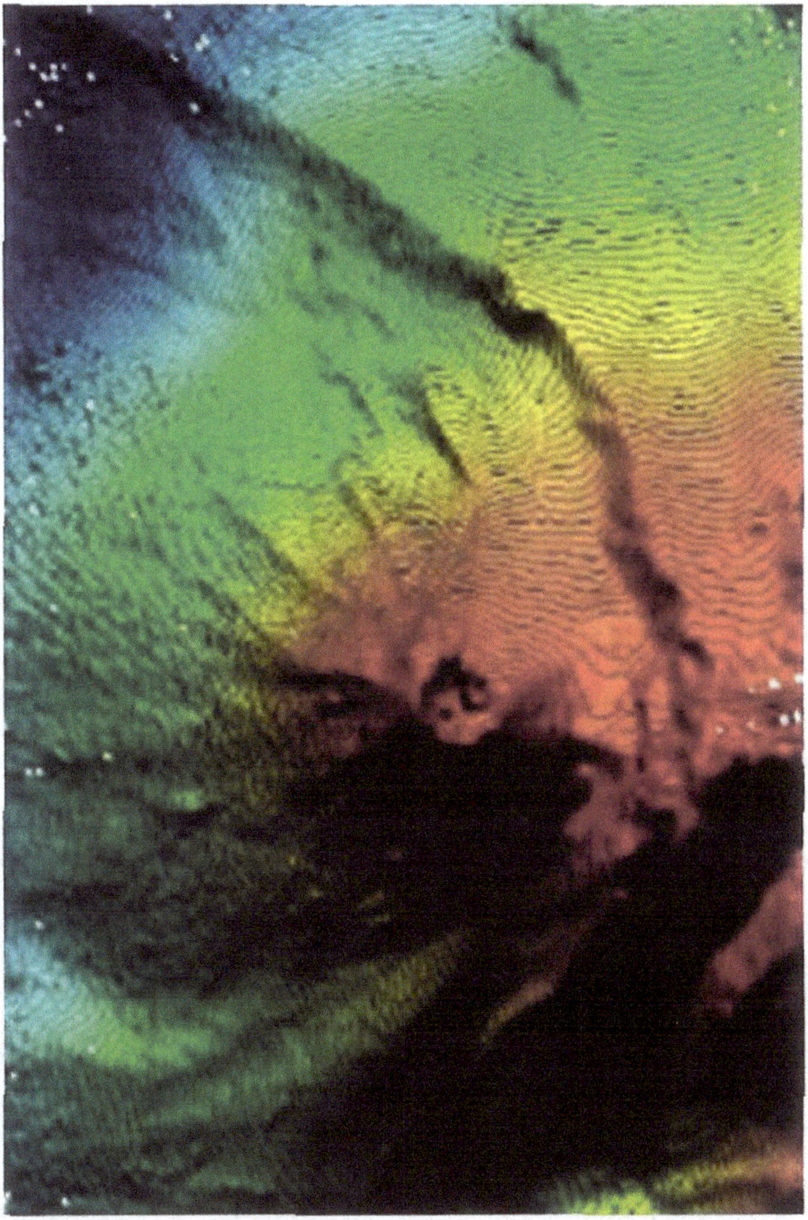

Figure 21. Morphology of Kick'em Jenny volcano and of an extensive slope failure of unknown age. (NOAA multibeam submarine survey)

introduction of speculative and highly unrealistic "worst case" scenarios. Kick-'em Jenny is not Krakatau and does not pose the purported potential tsunami danger that has been misreported by the media. Kick-'em Jenny is a much smaller volcano than Krakatau and has much smaller crater dimensions and magmatic chambers. The tectonic interactions that have produced this volcanic center in the Caribbean are substantially different than those of Krakatau, which erupted in 1883 and generated a destructive tsunami, which killed nearly 37,000 people in Indonesia (Pararas-Carayannis, 2003). Kick-'em Jenny's magmatic geochemistry is substantially different. Its magma composition ranges from mainly basalt to basaltic andesite. At the present stage of its development, Kick-'em Jenny volcano's small dimensions and geochemistry prevent eruptions of Vulcanian or Plinian intensity or extremely massive volcanic edifice collapses. The following is a realistic analysis of Kick-'em Jenny's tsunamigenic potential and future risks.

### *6.4.5 Tsunami Generation from Submarine Explosive Eruptions*

Even a major explosion at a peak phase of Kick-'em Jenny's eruption would be expected to generate tsunami-like waves, not as a single event but spread over a period of 24 hours or more. The periods of these waves will be relatively short and will range from 1-4 minutes at the most. Because of the short periods and wavelengths, the wave heights will decay rapidly with distance. As in 1939, the waves from future eruptions will be of significance along the north coast of Grenada and along the western coasts of Isla de Ronde and Isla Calle (Grenadine Islands), and possibly Tobago, St. Vincent and Barbados, but not anywhere else in the Caribbean. This conclusion is further supported by the numerical modeling studies that were conducted at the Los Alamos National Laboratory (Gisler et al 2004). Specifically, numerical simulations of Kick-'em Jenny's explosions with the same 3-D compressible hydro code used for asteroid impacts - and injecting as much as 20 kilotons of thermal energy at the apex of Kick-'em Jenny's volcanic cone, confirmed that only short period tsunami-like waves can be generated and that the waves will attenuate rapidly away from the source. Accordingly, it is concluded that explosive eruptions do not couple well to water waves. The waves that are generated from such eruptions are turbulent and highly dissipative, and don't propagate well.

At the present time, the depth of Kick-'em Jenny's summit and the hydrostatic pressure above it dampen the energy of eruptive explosions - although both the 1939 and 1974 events were violent enough to break through the sea surface. As the volcano keeps on growing towards the surface, the hydrostatic column pressure above the eruptive vents will decrease significantly. Future eruptions can be expected to be more explosive. However, even such future eruptions will only generate waves of short period and their heights will decay rapidly.

### *6.4.6 Tsunami Generation from Submarine Crater Collapses*

Even if one or all of Kick-'em Jenny's three small craters collapse, no major waves will be generated. For example, when the summit of the submarine volcano Loihi collapsed in Hawaii during the summer of 1996, the wave that was generated was of short period and decayed very rapidly (Mader 2004). The cavity generated by the Loihi collapse was 1000 meters wide and 300 meters deep, which is much greater than any potential cavity that could be expected from collapses of any or all of Kick-'em Jenny's craters. The fact that the top of the Loihi cavity was at 1050 meters depth while the top of Kick'em- Jenny's main crater is at 190 meters will not be much of a factor in tsunami wave generation, since the waves will be of short periods and will behave as short waves.

### *6.4.7 Tsunami Generation from Submarine Volcanic Dome Collapses*

Similarly, submarine dome collapses on Kick-'em- Jenny will probably trigger major eruptions – perhaps lateral blasts - with the associated pyroclastic flows and debris avalanches. However, it is expected that the volume of the ejecta and gases will be relatively small and that any resulting tsunami-like waves that will be generated will not be greater than those generated by the 1939 or 1974 eruptions.

### *6.4.8 Tsunami Generation From Future Subaerial Volcanic Collapses, Flank Failures And Massive Volcanic Edifice Failures*

When Kick-'em Jenny breaks through the sea surface and begins to build in height, it is expected that its eruptions will be more violent and that its flanks will be even more unstable than they are now. As with other active Caribbean volcanoes, waves may be generated by violent eruptive episodes, from caldera and dome collapses, from pyroclastic flows, landslides, flank failures, debris avalanches or even massive volcanic edifice failures. However the tsunami waves will be of relatively short periods (1-4 minutes at the most). Although the waves may be significant locally, they will decay rapidly.

### *6.4.9. Worst Possible Scenario*

Based on the pattern of Kick-'em Jenny's eruptive activity, a "worst case scenario" at the present time would be a repeat of the 1939 eruption, but at the shallower depth of the present summit. The waves from the 1939 event were about 2 meters in northern Grenada and the southern Grenadines but substantially lower on the west coast of the Barbados. A large violent eruption similar to the 1939 event, at the present depth of summit, can be expected to generate waves with a probable maximum runup of about 3 meters in Northern Crenada and the Grenadines, and as much as 1-2 meters along the west coast of the Barbados, Trinidad, and St. Vincent. The wave heights along the nearest coastline of northern Bonaire and Venezuela may be up to 1 meter at the most. When the volcano breaks through the surface of the sea, the probable maximum runup could be as much as 4 meters on Northern Grenada and as much as 2 meters along the west coast of the Barbados, Trinidad, and St. Vincent.

## 7. Assessment Of Future Risks Of Tsunamis From Volcanic Sources – Tsunami Forecasting And Preparedness For The Caribbean Region

The historic record indicates that Caribbean volcanoes pose a serious threat for several islands in the region (Robertson, 1992, 1995). The1902 eruptions of Mt Pelée on Martinique and of La Soufrière on St Vincent, and the more recent eruptions of Soufriere Hills on Montserrat, and of Kick-'em Jenny in Grenada increased awareness that tsunamis generated from volcanic sources represent an additional hazard that needs to be addressed individually for each of the region's active volcanoes. Fortunately - and in contrast to the unpredictability of tsunamis of seismic origin - tsunamis generated form volcanic sources can be forecast for the Caribbean region and appropriate measures can be taken. Preparedness for the tsunami hazard should include the monitoring of precursory eruptive processes as ongoing presently (Sigurdsson 1981; Shepherd and Aspinall 1982; Shepherd 1989) but, additionally, studies of geomorphologies and flank instabilities of each individual volcano and the mapping of risk areas that can contribute to massive volcanic edifice failures – with or without a volcanic triggering event – and thus in the generation of destructive tsunami waves.

Already, as a result of greater awareness and concerns about the threat of volcanic hazards in the Caribbean region, several scientific organizations have already established monitoring stations on several islands. For example, following the devastating 1902 eruption of La Soufrière volcano on St. Vincent Island, a surveillance program was initiated. In 1952, a Seismic Research Unit was established on the island and a sustained program of volcano monitoring was undertaken (Fiske and Shepherd 1990). Similarly, following the 1995 eruption of the Soufriere Hills volcano on the island of Montserrat, a very effective monitoring program was established by the Montserrat Volcano Observatory. Additionally, the Universities of Puerto Rico and of the West Indies have undertaken extensive monitoring functions and programs. Present volcano monitoring operations include routine measurements of geological, geophysical and geochemical parameters and assessments of precursory-to-an-eruption phenomena. With some small additional effort, these existing volcano monitoring programs in the Caribbean region can easily assess future risks for the collateral tsunami hazard, develop micro-zonation maps of potential tsunami hazard sites along the coast and help establish programs of tsunami preparedness for the public. The following sections summarize briefly the importance of monitoring some of the precursory-to-an-eruption phenomena as they relate to potential tsunami generation.

### *7.1 Micro-earthquake Activity*

Routine monitoring of a volcano's micro-earthquake activity helps forecast eruptions (Hirn et al1987). For example, before a major eruption occurs on Soufriere Hills on the island of Montserrat, the increasing pressure within the volcano generates a flurry of micro-earthquakes, which are indicative of magma movement. When this activity peaks and the focus of micro earthquakes becomes shallower, it becomes evident that the pressure within the volcano has reached a critical phase and that a fairly imminent eruption can be expected. Such monitoring is presently in effect for several islands with active volcanoes.

### *7.2 Monitoring Lava Dome Formation and Rate of Growth*

Measuring the swelling of the volcano with tilt meters and other geodetic and photogrammetric means may also indicate if there is intrusion by a lava dome, the dome's rate of growth, and its potential for collapse. Mature lava domes, which grow slowly, are usually non-explosive. Similarly non-explosive are post eruption lava domes, such as the felsic lava dome known as the Tower of Pelée - extruded in the waning stages of the 1902 eruption of Mt. Pelée on Martinique.

However, younger, fast growing, pre-eruption extruded domes that contain lava which has not been completely degassed, may explode or collapse. The eruption and explosion of Mt. St. Helens in the State of Washington were preceded by the rapid development of a lava dome over a three-year period from 1980 to 1983. Also, a lava dome formed rapidly on the Soufriere Hills volcano's crater on Montserrat Island over a two-month period prior to the major eruption of 26 December 1997. The dome formation served as a natural warning for the residents of Plymouth to evacuate, thus there was no losses of life. Stations on several Caribbean islands with active volcanoes, routinely monitor lava dome formation and rates of growth.

### *7.3 Evaluation of Potential Lava Dome Collapses*

The periodic explosion or gravitational collapses of the viscous masses of lava domes can sometimes generate deadly pyroclastic flows that can reach the sea and generate tsunami waves. Lava dome collapses were associated with the 1902 eruption and the nuée ardente of Mt. Pelée on Martinique, the 1902 eruption of Soufriere on St. Vincent Island, and the 1997, 1999 and 2003 eruptions, pyroclastic flows and debris avalanches of Soufriere Hills on Monteserrat Island -- the latter generating significant tsunami waves along the southwestern coast of the island.

In view of the above, it is important to monitor changes of lava domes and their potential for collapses. Furthermore, since lava dome collapses, particularly near a volcanic summit may be followed by violent eruptions, pyroclastic flows and debris avalanches, the expected path of destruction and potential flank failure sites can be determined by careful evaluation of the local topography and geomorphology. Based on such assessments, coastal areas subject to the tsunami hazard could be identified, microzonation maps can be drawn and appropriate warning signs be posted for the protection of the public

## 8. Summary and Conclusions

Historical tsunami events from volcanic sources in the Caribbean Region have been under-reported as the immensity of destruction from volcanic events has overshadowed them. Small scale flank failures are quite common for most of the active volcanoes in the Caribbean. Such volcanic sources have the potential of generating destructive local waves in confined bodies of water and in the near field environment of an open coast. Local tsunamis can also be generated by gravity-induced flank failures, even in the absence of eruptive triggering events. Heavy rains and earth tides appear to play a significant role in small scale flank failures of unstable volcanic slopes.

Tsunami or tsunami-like waves can be generated by a variety of volcanic mechanisms, pyroclastic flows, debris avalanches, and volcanic edifice mass failures and by aerial or submarine landslides. Impulsively generated waves from such complex source mechanisms behave non-linearly and change significantly away from the source, with varying near and far field effects and terminal run up heights. However the wave periods are short and range from 1-4 minutes at most. The heights of these waves attenuate rapidly with distance because of relatively smaller source dimensions and shorter wave periods and do not pose a significant danger at great distances from the source. Caribbean volcanoes and their associated flank failures can be forecast with careful monitoring and programs of preparedness need to be established.

At the present time, the Soufriere Hills volcano on the island of Montserrat poses the greater threat of local tsunami generation in the Eastern Caribbean Region. Its eruptive activity in the last decade, the rapid rates of lava dome formations and growth and the associated collapses and eruptive style, indicate ongoing active volcanic processes that will continue for many years. Tsunamis can be expected in the near future from both pyroclastic flows reaching the sea and by flank collapses.

The historic record supports that Mt. Pelée on the island of Martinique will continue to pose a threat for a repeat of a Vulcanian-Plinian episode in the future. When this will happen is not known. However, given the sophistication of present monitoring programs, any future activity of the volcano will be properly forecast and cautionary measures will be taken. Local tsunamis may be expected around the island by flank failures of unstable slopes, even in the absence of a triggering volcanic event.

The historic record supports that the stratovolcano "La Soufrière" on the island of St. Vincent poses a very significant threat for renewed activity in the future. Given the fact that there is a lake at the summit, there is also a potential danger that even an eruption of moderate activity may cause a breach on the crater's rim and trigger a dangerous lahar which may be destructive and may even generate a local tsunami if it reaches the sea. Also the instability of La Soufrière flanks pose a threat of failures and of local tsunami generation – even in the absence of a volcanic eruption. Heavy rains, gravitational forces and earth tides may be significant triggering factors.

Kick-'em Jenny volcano will continue to rise towards the surface and eventually will form an island volcano. It is possible that its present rate of growth may be slowed down by cone collapses and subsidence. The dimensions of the volcano and the style of expected eruptions and intensities limit the size of tsunamis that can be generated from future events. A large violent eruption of the Kick'em Jenny volcano at the present depth of the summit, can be expected to generate waves with a probable maximum runup of about 3 meters in Northern Crenada and the Grenadines, and as much as 1-2 meters along the west coast of the Barbados, Trinidad, and St. Vincent. The wave heights along the nearest coastline of northern Bonaire and Venezuela may be up to 1 meter at the most. When the volcano breaks through the surface of the sea, the probable maximum runup of a tsunami from a major eruption could be as much as 4 meters on Northern Grenada and as much as 2 meters along the west coast of the Barbados, Trinidad, and St. Vincent.

**9. References**


AAPG International Meeting,, 2003.*Caribbean Plate Origin*. Caribbean Tectonics, Barcelona, Spain, September 21-24.

Anderson, J., 1784. *An account of Morne Garou, a mountain in the island of St. Vincent with a description of the volcano on its summit*, Philosophical Transactions of the Royal Society 125:32.

Anderson, T., 1903. *Recent volcanic eruptions in the West Indies*. The Geographical Journal.

Anderson, T., 1908. *Report on the eruptions of the Soufrière in St. Vincent i n 1902, and on a visit to Montagne Pelèe in Martinique - The changes in the district and the subsequent history of the volcanoes*. Philosophical Transactions of the Royal Society Series A, 208 (Part II):275-352.

Anderson, T., and Flett S. J.,1903. *Report on the eruption of the Soufrière of St. Vincent in 1902 and on a visit to Montagne Pelèe in Martinique*. Part I. Royal Society Philosophical Transactions Series A-200:353-553.

Aspinall, W.P., 1973. *Eruption of the Soufrière volcano on St. Vincent island, 1971-1972*. Science 181:117-124.

Aspinall, W.P., H. Sigurdsson, and Shepherd. J.B.,1973. *Eruption of the Soufriere volcano on St. Vincent island,* 1971-19721. Science 181:117-124.

Aspinall, W.P., Sigurdsson, H., Shepherd, J.B., Almorales, H. and P.E. Baker. 1972. *Eruption of the Soufrière Volcano on St. Vincent island, 1971-72.* In Smithsonian Institute for Short-Lived Phenomena.

Baker, P.E., 1972. *The Soufrière volcano, St. Vincent and its 1971-72 eruption.* Journal of Earth Sciences, Leeds 8 (Pt. 2):205-217.

Barberi, F., Blong R. et al. 1990. *Reducing volcanic disasters in the 1990's,* Volcanol. Soc. Japan Bull., 35, 80

Barr, S., and J.L. Heffter. 1982. *Meteorological analysis of the eruption of Soufrière in April 1979*. Science 216:1109-1111.

Beer, T., 1974. *Atmospheric Waves*. Wiley, New York, 300 pp.

Brazier, S. A., Davis, N., Sigurdsson, H. , and R. S. J. Sparks. 1982. *Fallout and deposition of volcanic ash during the 1979 explosive eruption of the Soufrière of St. Vincent.* Journal of Volcanology and Geothermal

Research 14 (3-4):335-359.

Brown, G.M., Holland, J.G., Sigurdsson, H., Tomblin, J. F. and R.J. Arculus. 1977. *Geochemistry of the Lesser Antilles volcanic island arc.* Geochimica et Cosmochimica Acta 41:785-801.

Carey, S.N., and Sigurdsson H.. 1978. *Deep-sea evidence for distribution of tephra from the mixed magma eruption of the Soufrière on St. Vincent, 1902: Ash turbidites and air fall.* Geology 6:271-274.

DeGraff, J.V., 1988. *Landslide hazards on St. Vincent and the Grenadines, West Indies.* Washington, D.C.: Dept. Reg. Div., Organisation of American States.

Deplus, C., Le Friant, A., Boudon, G., Komorowski J. -C., Villemant, B., Harford, C., Segoufin, J., and Cheminee, J. -L., 2001. *Submarine evidence for large-scale debris avalanches in the Lesser Antilles Arc,* Earth and Planetary Science Letters, 192, 2, 145–157.

Devill, S-C. 1867. *Sur le tremblement de terre du 18 Novembre 1867 aux Antilles,* Comptes Rendus Acad. Sci. Paris, 65, 1110–1114.

Earle, K.W. 1924. *The Geology of St. Vincent and the neighbouring Grenadines.* Kingstown: Geological Survey of the Windward & Leeward Islands.

ETDB/ATL 2002. *Expert Tsunami Database for the Atlantics,* Version 3.6 of March 15, 2002. Tsunami Laboratory, Novosibirsk, Russia.

Fisher, R.V., and Heiken, G., 1982. *Mt. Pelee, Martinique; May 8 and 20, 1902, pyroclastic flows and surges*: Journal of Volcanology and Geothermal Research, v. 13, p. 339-371.

Fisher, R.V., Smith, A.L., and Roobol, M.J., 1980. *Destruction of St. Pierre, Martinique by ash-cloud surges, May 8 and 20, 1902*: Geology, v. 8, p. 472-476.

Fiske, R.S., and Sigurdsson, H., 1982. *Soufriere volcano, St. Vincent: Observations of its 1979 eruption from the ground, aircraft, and satellites*: Science, v. 216, p. 1105-1126.

Fiske, R. S., and J. B. Shepherd. 1990.*12 Years of Ground-Tilt Measurements On the Soufrière of St- Vincent, 1977-1989.* Bulletin of Volcanology 52 (3):227-241.

Flett, J.S. 1902. *Note on a preliminary examination of the ash that fell on Barbados after the eruption at St. Vincent, West Indies.* Quarterly Journal of the Geological Society of London 58:368-370.

Flett, J.S. 1908. *Petrological notes on the products of the eruptions of May 1902, at the Soufriere in St. Vincent.* Philosophical Transactions of the Royal Society of London 208 (Series A):225-253.

Gisler, G., Weaver, R ., Mader C. and M. Gittings 2004. *Two-Dimensional*

*Simulations of Explosive Eruptions of Kick-em Jenny and other Submarine Volcanoes*. Los Alamos National Laboratory. . NSF Caribbean Tsunami Workshop. Puerto Rico March 30-31, 2004

Graham, A. M., and M. F. Thirlwall. 1981. *Petrology of the 1979 Eruption of Soufrière Volcano, St. Vincent, Lesser Antilles*. Contributions to Mineralogy and Petrology 76 (3):336-342.

Hay, R.L. 1959. *Formation of the crystal-rich glowing avalanche deposits of St. Vincent, B.W.I.* Journal of Geology 67:540-562.

Heath, E., R. MacDonald, H. Belkin, C. Hawkesworth, and H. Sigurdsson. 1998..*Magmagenesis at Soufrière Volcano, St Vincent, Lesser Antilles arc*. Journal of Petrology 39 (10):1721-1764.

Heilprin, A., 1908. *The eruption of Pelee*: Philadelphia Geographic Society, 72 p.

Heinrich, F., Mangeney, A., Guibourg, S., and Roche, R. 1998. *Simulation of water waves generated by a potential debris avalanche in Montserrat, Lesser Antilles*, Geophys. Res. Lett., 25, 9, 3697–3700,

Heinrich, F., Guibourg, S., Mangeney, A., and Roche, R. 1999a. *Numerical modelling of a landslide-generated tsunami following a potential explosion of the Montserrat Volcano*, Phys. Chem. Earth (A), 24, 2, 163–168,

Heinrich, F., Roche, R., Mangeney, A., and Boudon, G. 1999b. *Modeliser un raz de maree cree par un volcan*, La Recherche, 318, 67–71,

Heinrich, F., Boudon, G., Komorowski, J. C., Sparks, R. S. J., Herd, R., and Voight, B. 2001. *Numerical simulation of the December 1997 debris avalanche in Montserrat*. Geophys. Res. Lett., 28, 13, 2529–2532.

Hirn, A., Girardin, N., Viode, J.-P., and Eschenbrenner, S., 1987. *Shallow seismicity at Montagne Pelee volcano, Martinique, Lesser Antilles*: Bulletin of Volcanology, v. 49, p. 723-728.

Hooper, D. M. and Mattioli, G. S. 2001. *Kinematic modelling of pyroclastic flows produced by gravitational dome collapse at Soufriere Hills*. Natural Hazards, 23, 65–86.

Huppert, H. E., J. B. Shepherd, H. Sigurdsson, and R. S. J. Sparks. 1982. *On Lava Dome Growth, With Application to the 1979 Lava Extrusion of the Soufrière of St-Vincent*. Journal of Volcanology and Geothermal Research 14 (3-4):199-222.

Imbo, G., 1965. *Catalogue of the active volcanoes of the world including solfatara fields*, Part XVIII, Italy: International Association of Volcanology.

Keller, J., 1980 .*The island of Vulcano*: Soc. Italiana Min. Petr., 36, p. 368-413.

Lacroix, A., 1904. *La Montagne Pelee et ses eruptions*: Paris, Masson et Cie,

622 p.

Lander, J. F., Whiteside, L. S., and Lockridge, P. A 2002. *A brief history of tsunami in the Caribbean Sea*, Science of Tsunami Hazards, 20, 2, 57–94.

Lander James F., Whiteside Lowell S., Lockridge P A 2003. *TWO DECADES OF GLOBAL TSUNAMIS 1982-2002*. Science of Tsunami Hazards, Volume 21, Number 1, page 3.

Le Friant, A. 2001. *Les destabilisations de flanc des volcans actifs de l'arc des Petites Antilles: origines et consequences,* These de Doctorat, Universite de Paris VII, 377p.

Mader C. L., 2001. *Modeling the La Palma Landslide Tsunami*. Science of Tsunami Hazards, Volume 19, pages 150-170 (2001)

Mader, C .L. 2004. *The Loihi Cone Collapse*. In Numerical Modeling of Water Waves - Second Edition, CRC Press, pages 130-132.

Mangeney A., Heinrich F., Roche, R., Boudon, G., and J. L. Chemin̂e. 2000. *Modeling of debris avalanche and generated water waves*. Application to real and potential events in Montserrat. Phys, Chem. Earth 25(9-11), 741-745.

Martin-Kaye, P.H.A. 1969. *Summary of the geology of the Lesser Antilles*. Overseas Geology & Mineral Resources 10:172-206.

Mercado, A. and McCann, W. 1998. *Numerical simulation of the 1918, Puerto Rico tsunami*, Natural Hazards, 18, 1, 57–76.

Mikumo, T. and Bolt, B.A. 1985. *Excitation mechanism of atmospheric pressure waves from the 1980 Mount St. Helens eruption.* Geophysical Journal of the Royal Astronomical Society, 81(2), 445-461.

Newhall, C.G. and Self, S. 1982. *The volcanic explosivity index (VEI): An estimate of explosive magnitude for historical volcanism*. Journal of Geophysical Research, 87(C2), 1231-1238.

Pararas-Carayannis, G. 1992. *The Tsunami Generated from the Eruption of the Volcano of Santorin in the Bronze Age*. Natural Hazards 5:115-123.

Pararas-Carayannis, G. 2002. *Evaluation of the threat of mega tsunami generation from postulated massive slope failures of island stratovolcanoes on La Palma, Canary Islands, and on the Island of Hawaii*, Science of Tsunami Hazards, Vol. 20, 5, 251–277.

Pararas-Carayannis, G. 2003. *Near and Far-Field Effects of Tsunamis Generated by the Paroxysmal Eruptions, Explosions, Caldera Collapses and Slope Failures of the Krakatau Volcano in Indonesia, on August 26-27, 1883,* Journal of Tsunami Hazards, Vol. 21, Number 4.

Pelinovsky E., Zahibo N., Dunkley P., Edmonds M., Herd R., Talipova T. ,

Kozelkov A., and I. Nikolkina, 2004. *Tsunami Generated by the Volcano Eruption on July 12-13, 2003 at Montserrat, Lesser Antilles.* Sciences of Tsunami Hazard Vol. 22, No. 2, pages 44-57.

Perret, F.A., 1937. *The eruption of Mt. Pelee, 1929-1932*: Carnegie Institute of Washington Publication, v. 458, 126 p.

Reid, H. F. and Taber, S. 1920. *The Virgin Islands Earthquakes of 1867– 1868,* Bull. Seismol. Soc. America, 10, 9–30.

Robertson, R.E.A. 1992. *Volcanic Hazard and Risk Assessment of the Soufrière Volcano, St. Vincent, West Indies.* MPhil, Earth Sciences, The University of Leeds, Leeds.

Robertson, R.E.A. 1995. *An Assessment of the Risk From Future Eruptions of the Soufrière Volcano of St. Vincent, West Indies.* Natural Hazards 11 (2):163-191.

Rowley, K.C. 1978. *Stratigraphy and geochemistry of the Soufriere Volcano, St. Vincent, West Indies.* PhD, Seismic Research Unit, University of the West Indies, St. Augustine.

Scheffers A. and Kelletat D. 2004. *New Evidence and Datings of Paleo-Tsunami Events in the Caribbean,* Essen University (Germany). NSF Caribbean Tsunami Workshop. Puerto Rico March 30-31.

Univ. of West Indies 2001. Seismic Research Unit Website.

Seno T. and Yamanaka Y.1998. *Arc stresses determined by slabs: Implications for back-arc spreading,* Earthquake Research Institute, University of Tokyo, Geopys. Res. Lett., 3227-3230.

Shepherd, J.B., Aspinall, W.P., Rowley, K.C., and others, 1979, *The eruption of Soufriere volcano, St. Vincent, April-June, 1979*: Nature, v. 282, p. 24-28.

Shepherd, J.B. 1989. *Eruptions, eruption precursors and related phenomena in the Lesser Antilles. In Volcanic hazards:* IAVCEI Proceedings in Volcanology 1, edited by J. H. Latter. Berlin, Heidelbery: Springer-Verlag.

Shepherd, J.B., and Sigurdsson, H., 1982. *Mechanism of the 1979 explosive eruption of Soufriere volcano, St. Vincent*: Journal of Volcanology and Geothermal Research, v. 13, p. 119-130.

Shepherd, J.B., and W.A. Aspinall. 1982. *Seismological studies of the Soufrière of St. Vincent, 1953-1979: Implications for volcanic surveillance in the Lesser Antilles.* Journal of Volcanology and Geothermal Research 12:37-55.

Sigurdsson, H. 1981. *Geological observations in the crater of Soufriere volcano,* St. Vincent: University of the West Indies.

Sigurdsson H., Carey S and Wilson D. 2004. *Debris Avalanche Formation at*

*Kick'em Jenny Submarine Volcano.* NSF Caribbean Tsunami Workshop. Puerto Rico March 30-31, 2004

Smith, A.L., and Roobol, M.J., 1990. Mt. Pelee, Martinique; *A study of an active island-arc volcano*: Boulder, Colorado, Geological Society of America Memoir 175, 105 p.

Smithsonian Institution 1999. - Global Volcanism Program Kick-'em-Jenny Website, August.

Sparks, R. S. J., and L. Wilson. 1982. *Explosive Volcanic-Eruptions .5. Observations of Plume Dynamics During the 1979 Soufrière Eruption, St Vincent.* Geophysical Journal of the Royal Astronomical Society 69 (2):551-570.

Tahira, M., Nomura, M., Sawada, Y., and Kamo, K. 1996. *Infrasonic and acoustic-gravity waves generated by the Mount Pinatubo eruption of June 15, 1991. Fire and Mud.* University of Washington Press, Seattle, 601- 614.

Tilling, Topinks, and Swanson, 1990, *Eruptions of Mount St. Helens: Past, Present, and Future:* USGS General Interest Publication.

Tomblin, J.F., H. Sigurdsson, and W.A. Aspinall. 1972. *Activity at the Soufrière Volcano, St. Vincent, West Indies, between Oct. 31-Nov. 15, 1971.* Nature 235 (5334):157-158.

Voight B. 2000. *Structural stability of andesite volcanoes and lava domes* Philosophical Transactions: Mathematical, Physical and Engineering Sciences Vol 358, No 1770, Pages: 1663 - 1703 / May 15.

Weissert, T. P.1990. *Tsunami travel time charts for the Caribbean*, Science of Tsunami Hazards, 8, 2, 67–78.

Westercamp, D., and Traineau, H., 1983. *The past 5,000 years of volcanic activity at M. Pelee, Martinique (F.W.I.); Implications for assessment of volcanic hazards:* Journal of Volcanology and Geothermal Research, v. 17, p.159-185.

Wright, T.L., and Helz, R.T., 1987. *Recent advances in Hawaiian petrology and geochemistry,* in Decker, R.W., Wright, T.L., and Stauffer, P.H., eds., Volcanism in Hawaii: U.S. Geological Survey Professional Paper 1350, v. 1, p. 625-640.

Young R. S., 2004. *Small scale edifice collapse and tsunami generation at eastern Caribbean volcanoes; a standard phase of the volcanic cycle.* NSF Caribbean Tsunami Workshop, Puerto Rico March 30-31.

Zahibo, N. and Pelinovsky, E. 2001. *Evaluation of tsunami risk in the Lesser Antilles*, Natural Hazard and Earth Sciences, 3, 221–231.

# TWO-DIMENSIONAL SIMULATIONS OF EXPLOSIVE ERUPTIONS OF KICK-EM JENNY AND OTHER SUBMARINE VOLCANOS

GALEN GISLER, ROBERT WEAVER, CHARLES MADER
*Los Alamos National Laboratory*

MICHAEL GITTINGS
*Science Applications International*

Kick-em Jenny, in the Eastern Caribbean, is a submerged volcanic cone that has erupted a dozen or more times since its discovery in 1939. The most likely hazard posed by this volcano is to shipping in the immediate vicinity (through volcanic missiles or loss-of-buoyancy), but it is of interest to estimate upper limits on tsunamis that might be produced by a catastrophic explosive eruption. To this end, we have performed two-dimensional simulations of such an event in a geometry resembling that of Kick-em Jenny with our SAGE adaptive mesh Eulerian multifluid compressible hydrocode. We use realistic equations of state for air, water, and basalt, and follow the event from the initial explosive eruption, through the generation of a transient water cavity and the propagation of waves away from the site. We find that even for extremely catastrophic *explosive* eruptions, tsunamis from Kick-em Jenny are unlikely to pose significant danger to nearby islands. For comparison, we have also performed simulations of explosive eruptions at the much larger shield volcano Vailulu'u in the Samoan chain, where the greater energy available can produce a more impressive wave. In general, however, we conclude that explosive eruptions do not couple well to water waves. The waves that are produced from such events are turbulent and highly dissipative, and don't propagate well. This is consistent with what we have found previously in simulations of asteroid-impact generated tsunamis. Non-explosive events, however, such as landslides or gas hydrate releases, do couple well to waves, and our simulations of tsunamis generated by sub-aerial and sub-aqueous landslides demonstrate this.

## 1. Introduction

Water and magma make a highly explosive combination, particularly at water depths less than about 130 meters. The explosive vaporization of water, heated by contact with magma at 1200 C or hotter, produces an instantaneous pressure of ~50 kBar, that can have extremely dangerous consequences. The August 1883 explosion of Krakatau is thought to have been caused by this hydromagmatic mechanism. This event produced tsunami that killed many thousands of people, and propagated around the world.

It is of interest to discuss whether there is a significant danger of tsunami from the submarine volcano Kick-em Jenny in the eastern Caribbean. We conclude here that there is not. The simulations that we have performed, in an axisymmetric geometry resembling Kick-em Jenny, suggest that only for very much more energetic events are significant waves generated, and that even these waves do not propagate as classical tsunami. These results are consistent with conclusions we have drawn from simulations of other explosively-generated waves.

Kick-em Jenny, located 8 km north of the island of Grenada in the volcanic arc of the Lesser Antilles, is one of the most active volcanoes in the region. It has erupted a dozen times since 1939, and is a known hazard to shipping, marked on navigation charts. The principal dangers caused by Kick-em Jenny are from volcanic missiles projected to altitudes of a few hundred meters, and from gases emitted into the seawater from the volcano, reducing the density of the water and causing ships to lose buoyancy. No significant tsunami have been observed to arise from these eruptions, which have generally been of magnitude between 0 and 1 on the Volcanic Explosivity Index (VEI) scale (Simkin et al 1981), though there were early reports of minor waves.

Smith and Shepherd (1993, 1995) investigated the tsunami hazard posed by the Kick-em Jenny volcano. While the top of the volcanic cone is at a depth such that the hydrostatic water pressure confines the explosive effects of the eruption, it was thought that the cone was building towards the surface, and might eventually pose a hazard due to the explosive vaporization of sea water. They used linear theory to calculate initial amplitudes, dispersion, and propagation, and shoaling, given a spectrum of potential events and their probability. Their worse case scenario included run-ups as high as 46 meters on the northern shore of Grenada for a VEI = 6, or Krakatau-like event, considered as likely on a 1000-year scale, or as high as 8 meters for a more realistic, 100-year, VEI = 3 event.

Recent bathymetry conducted since the publication of the Smith and Shepherd papers has led to the understanding that the depth of the Kick-em Jenny summit, now at 190 m, has not significantly diminished since its discovery, so the danger from tsunami caused by this volcano is now generally thought to be insignificant. Nevertheless, it is of interest in the general case to study the coupling of underwater volcanic explosive eruptions to the generation of tsunami.

Accordingly, we have conducted a series of two-dimensional axisymmetric simulations of explosive underwater volcanic eruptions to study the coupling of these events to the production of water waves.

## 2. The SAGE hydrocode

The SAGE hydrocode is a multi-material adaptive-grid Eulerian code with a high-resolution Godunov scheme originally developed by Michael Gittings for Science Applications International (SAIC) and Los Alamos National Laboratory

(LANL). It uses continuous adaptive mesh refinement (CAMR), by which we mean that the decision to refine the grid is made cell-by-cell and cycle-by-cycle continuously throughout the problem run. Refinement occurs when gradients in physical properties (density, pressure, temperature, material constitution) exceed user-defined limits, down to a minimum cell-size specified by the user for each material in the problem. With the computing power concentrated on the regions of the problem which require higher resolution, very large computational volumes, and substantial differences in scale, can be simulated at low cost.

SAGE can be run in several modes of geometry and dimensionality, explicitly 1-D Cartesian and spherical, 2-D Cartesian & cylindrical, and 3-D Cartesian. The RAGE code is similar to SAGE but incorporates a separate module for implicit, gray, non-equilibrium radiation diffusion. Both these codes are part of LANL's Crestone project, in turn part of the Department of Energy's program in Advanced Simulation and Computing, or ASC.

Because modern supercomputing is commonly done on machines or machine clusters containing many identical processors, the parallel implementation of the code is supremely important. For portability and scalability, SAGE uses the widely available Message Passing Interface (MPI). Load leveling is accomplished through the use of an adaptive cell pointer list, in which newly created daughter cells are placed immediately after the mother cells. Cells are redistributed among processors at every time step, while keeping mothers and daughters together. If there are a total of M cells and N processors, this technique gives very nearly M/N cells per processor. As neighbor-cell variables are needed, the MPI gather/scatter routines copy those neighbor variables into local scratch.

In a multi-material code like SAGE, every cell in the computational volume can contain all the materials defined in the problem, each with its own equation of state (and strength model, as appropriate). There are a number of equations of state available, analytical and tabular. In the calculations reported here, we use the LANL SESAME tables for air and basalt, and for water we used a somewhat more sophisticated table (including a good treatment of the vapor dome) from SAIC. For the strength of basalt, we used a simple elastic-plastic model with pressure hardening (with depth) for the basalt.

The boundary conditions we use in these calculations are designed to allow unhindered outflow of waves and material. This is accomplished by the use of "freeze regions" around the edges of the computational box, which are updated normally during the hydrodynamic step, then quietly restored to their initial values of pressure, density, internal energy, and material properties before the next step. This technique has proven to be extremely effective at minimizing the deleterious effect of artificial reflections. But by far the best technique for dealing with unwanted boundary effects is to put the boundaries very far away from the regions of interest or to place the boundary beyond a material interface that truly exists in the problem and might be expected to interact with waves in an appropriate way (i.e. through reflection, transmission, and absorption).

### 3. Simulations of Kick-em Jenny

For the sake of simplicity, we perform our simulations in two-dimensions only, ignoring the very real three-dimensional character of the Kick-em Jenny volcano. Its cinder cone is embedded in a horseshoe-shaped slump caused by slope failure from a larger mound to its east. The event that caused this slump must have produced significant tsunami, and we shall argue that the danger from future such slumps is greater than from eruptive events on the cinder cone. The geometry of the slump and other three-dimensional features would provide collimation and amplification of eruptively produced waves that propagate in certain directions while attenuating waves that propagate in other directions. We ignore these effects in order to focus solely on the generative mechanism.

Accordingly, we model the volcano as a simple geometrical frustum, with a base of 5 km diameter, a top of 100 m diameter, and a height of 1.4 km (see Fig. 1). The cone has a hot magma core of 20 m diameter. We take the

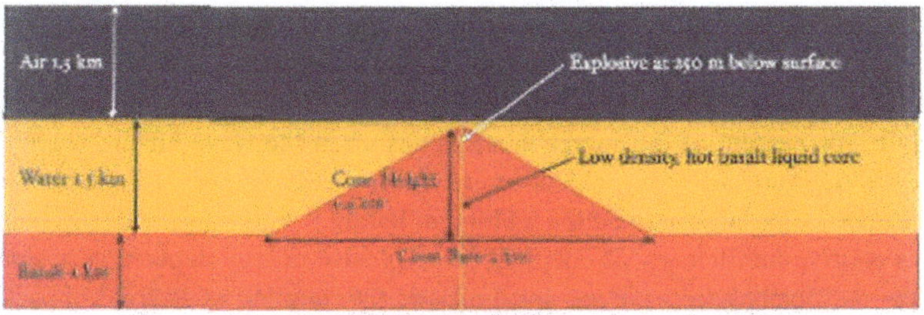


Figure 1. Geometry of SAGE set-up for Kick-em Jenny simulations.

water depth to be 1.5 km, so that the submerged top of the frustum is only 100 m below the water surface, thus significantly shallower than the true cone summit, and above the threshold depth for pressure confinement of a hydromagmatic eruption. We use three materials in the problem, air for the atmosphere, water for the ocean, and basalt for the seafloor, the volcanic cinder cone, and the hot magma core. We use tabular equations of state for these from the LANL Sesame Library except for water, which we take from a high-quality PacTech table. We use a simple elastic-plastic strength model for the solid basalt, and no strength for the magma, water, or air.

To model an explosive eruption, we take the extreme (and admittedly unlikely) case of an instantaneous explosion near the top of the cone. Because we anticipate that the strongest coupling to the water motion will be through the motion of rock, we do not place the explosion at the summit, but some depth (usually 150 m) below the summit.

We present here three sample runs that span the range of interest. The parameters of these runs and the resulting wave heights are presented in Table 1.

We chose these parameters to span a major portion of the range of interest, as an exploration of what waves might possibly be generated by significant explosive eruptions at Kick-em Jenny. Our extrapolated wave heights at 10 km, the distance are significantly less than reported by Smith and Shepherd (1993).

We illustrate these three runs by showing the final density configuration in Figure 2. The explosive energy is sourced in instantaneously at the beginning of the calculation. A hot crater quickly opens in the basalt, and the explosive vaporization of the water in contact with this crater produces a large transient water cavity. A "debris curtain" or rim wave makes a precursor tsunami that dies off very quickly. The main wave is produced by the collapse of the transient water cavity, and the strong water currents modify the shape of the basalt crater produced in the explosion. The wave that is produced by the cavity collapse is very turbulent and dissipative, and it propagates slowly. When the main wave leaves the computational domain (6 km from the center), we terminate the simulation, though the center is still hot and turbulent.

We find that the coupling of explosive energy to wave energy is much less efficient than the coupling of slower mechanisms of energy release. At most only a few percent of the source energy ends up in the wave kinetic energy, as shown in Figure 3 for the middle run of Table 1. Much more energy goes into the vaporization of water and the melting of basalt.
This relative inefficiency of coupling for explosive energy deposition is similar to the inefficiency we have previously found for coupling of asteroid impacts to tsunami wave energy. In the latter case we have found that impact- generated waves decline rather more steeply with distance than waves from seismic or landslide events, and their speeds and wavelengths are correspondingly lower (Gisler et al. 2002).

For comparison, we have performed a simulation of an underwater granular basalt landslide event with free energy ~20 MTons, comparable to our kej12 explosive eruption simulation, and find in this event a wave energy coupling efficiency of about 16%.

TABLE 1. SUMMARY OF IMPORTANT RUNS

| run name | VEI | kT explosive | wave height at 3 km | extrapolated wave height at 10 km |
|---|---|---|---|---|
| kej11 | 5.5 | 233,000 | 300 m | 37 m |
| kej12 | 4.5 | 21,200 | 130 m | 21 m |
| kej13 | 3.7 | 3,700 | 20 m | 2.7 m |

We also performed simulations of explosive eruptions on the massive shield volcano Vailulu'u, at the end of the Samoan chain. While such eruptions are even less likely in that case than in the case of Kick-em Jenny, we thought it useful to examine a case that was larger in dimension and energy availability. In that case also we find that the coupling to water waves is of very poor efficiency, but we notice that if we artificially weaken the strength parameters for basalt so that the rock deformation is significantly greater than in the nominal case, the wave heights and wave kinetic energy are considerably enhanced.

We conclude that the efficient production of a tsunami requires a disturbance that covers a substantial distance or lasts a considerable time. Earthquakes or landslides, or more generally a movement of the seafloor or a pressure pulse communicated by the seafloor, produce tsunamis efficiently. Explosions or impacts do not couple to water motion as efficiently as do slower motions of rock.

Specifically, the tsunami danger from explosive eruptions of Kick-em Jenny is much less important than the danger that might result from a slope failure at that volcano, similar to that which caused the horseshoe-shape cleft in which the volcano currently nestles.

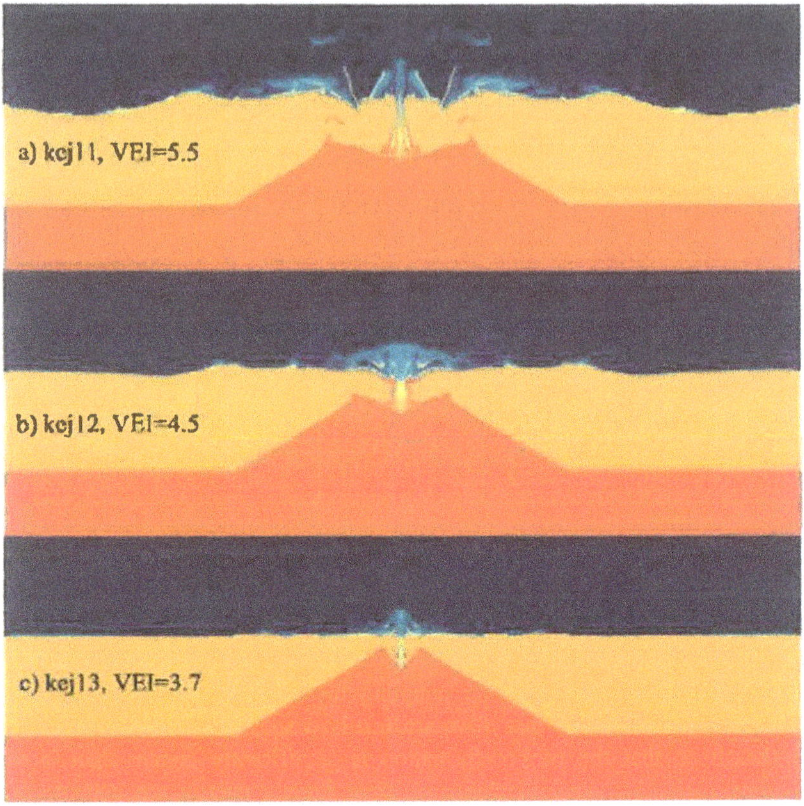


Figure 2. Final wave profiles for the three representative runs.

## 4. References

Galen Gisler, Robert Weaver, Charles Mader, Michael L. Gittings, 2003, "Two and Three-Dimensional Simulations of Asteroid Ocean Impacts", *Science of Tsunami Hazards*, **21**:, 119-134.

T. Simkin, L. Siebert, L. McClelland, D. Bridge, C. Newhall, and J. H. Latter, 1981, *Volcanoes of the World*, Smithsonian Institution, Hutchinson Ross, Stroudsburg.

Martin S. Smith and John B. Shepherd, 1993, "Preliminary Investigations of the Tsunami Hazard of Kick 'em Jenny Submarine Volcano," *Natural Hazards*, **7**:257-277.

Martin S. Smith and John B. Shepherd, 1995, "Potential Cauchy-Poisson Waves Generated by Submarine Eruptions of Kick 'em Jenny Volcano," *Natural Hazards*, **11**:75-94.

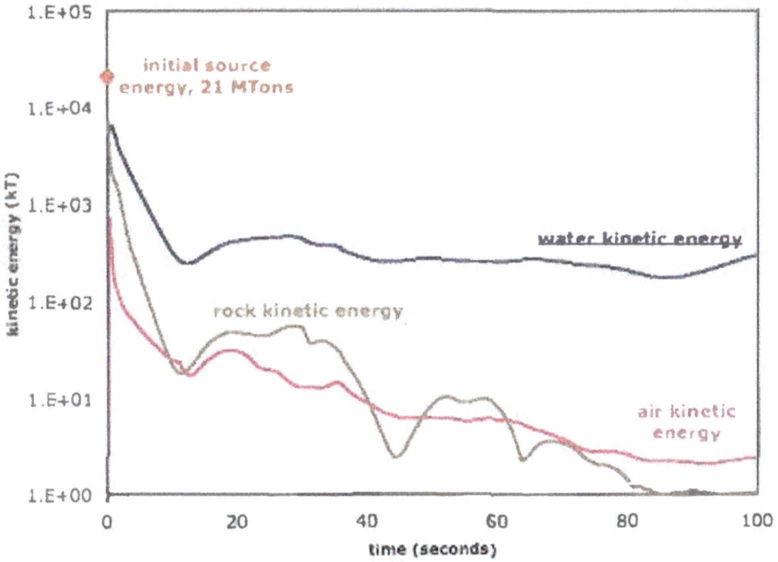


**Figure 3.** History of material kinetic energies for our run kej11, with an initial explosive energy of 21 MTons.

# THE TSUNAMI EVIDENCE IN THE CARIBBEAN: PALEOTSUNAMI STUDIES

# GEOMORPHOLOGIC AND STRATIGRAPHIC INVESTIGATIONS ON HISTORIC AND PRE-HISTORIC TSUNAMI IN NORTHWESTERN PUERTO RICO: IMPLICATIONS FOR LONG TERM COASTAL EVOLUTION

JUAN C. MOYA
*1308 Summoners Tale Ct. Austin, TX 78748.*

AURELIO MERCADO
*Department of Marine Sciences, University of Puerto Rico, P.O. Box 9013, Mayagüez P.R. 00681-9013.*

In 1918 an earthquake of magnitude M7.3 affected the northwestern portion of Puerto Rico. This earthquake was generated on the Mona Passage and produced an important tsunami. The event was associated with the rupture of the Mona Canyon Fault, which according to its Seismic Moment ($M_o$) had a rupture of about 3.5 meters of vertical displacement with an approximate length of 70 km. The waves generated by the tsunami reached up to 6 m along the Puerto Rican coast. Descriptions from survivors and historic records led us to reconstruct the occurrence of this event. Some stratigraphic and geomorphologic investigations were developed following the reconstruction of this tsunami. Eight cores were collected on the swamps of northwestern Puerto Rico looking for historic and pre-historic evidence of tsunami. Also, a trench was opened on the coastal plain. The geomorphologic analysis included the search of tsunami impacts along the coast, which consist of a series of different flat, swampy and low lands beaches, separated by prominent cliffs. Historical data and interviews of survivors suggest also that the tsunami impacted all the different geomorphic settings including swamps, coastal cliffs and coastal dunes. Field investigations searched for the location of the historic observations and descriptions of eyewitnesses in different areas including the swamps of Aguada and Aguadilla; the dune areas at the beach mainly in Aguadilla; and along the cliffs in Aguadilla.

The stratigraphy and sedimentologic record in the cores show some layers appearing as parallel cross bedding laminations at different depths. These layers are different from the massive deposition defined above and below, which consist mainly of massive layers of white and gray calcareous and quartz fine sand. These laminations appear to be related to single and sudden events that cut-crossed the massive layers being deposited. The records show evidence of events prior to the 1918 tsunami. Preliminary interpretation concluded that the laminations found are positively associated with tsunami deposition. Two radiocarbon dates were obtained from the layers with laminations. The dates showed an event occurred at 1270–1410 AD, and a second event dated 820–400 BC. The upper layers located between 0.30 to 0.50 meters deep in almost all the pipes have been assumed to be associated with the 1918 tsunami. Finally, the data shows that tsunami events combined with Neotectonic deformation have been very important in the morphologic evolution of northwestern Puerto Rico.

## 1. Introduction

In 1918 an earthquake of magnitude M7.3 affected western Puerto Rico. This seismic event produced an important tsunami that killed several people [*Reid and Taber,* 1919]. These authors described evidence for a run-up of up to 5 meters and possibly higher in some places. The earthquake was associated with the rupture of the Mona Canyon Fault, which according to its Seismic Moment ($M_0$) had a rupture of about 3.5 meters of vertical displacement with an approximate length of 70 km [*Mercado and McCann, 1998*]. This earthquake is just one of the examples of the potential that many faults in the Mona Passage have to generate tsunami along the coast of western Puerto Rico.

This work contributes to the technical knowledge of the earthquake and tsunami hazards in Puerto Rico and in the northeastern portion of the Caribbean Plate. Data related to the evidence of the 1918 tsunami is presented, including long- and short-term interpretations on the effects of tsunami in the northwestern coast of Puerto Rico. Analysis and interpretation of pre-historic seismic events is also presented, plus a discussion of future events associated with the oblique subduction and the extension occurring in Mona Passage, western Puerto Rico, and eastern Dominican Republic.

## 2. Background

This study covers the northwest of Puerto Rico between Aguadilla and Rincón, mainly because this was the area most affected by the 1918 tsunami (Figures 1 and 2). The study concentrates on suggested evidence of pre-historic and historic tsunami in this region, and analyzes the potential effect of future seismic events accompanied by tsunami as a contribution to a response plan for the island. The study presents suggested evidence of previous earthquakes generating tsunami based on the stratigraphic records registered by earthquake induced sedimentation. Finally, this study includes interpretations based on tsunami effects taken from the descriptions of *Reid and Taber*, [1919], supplemented by the modeling investigations developed by *McCann and Mercado [1997]*, interviews from survivors of the 1918 tsunami, and geologic and geomorphologic analysis of the evidence left by the 1918 tsunami. This is one of the first attempts to define the importance of tsunami in the evolution of the coastal morphology of northwestern Puerto Rico.

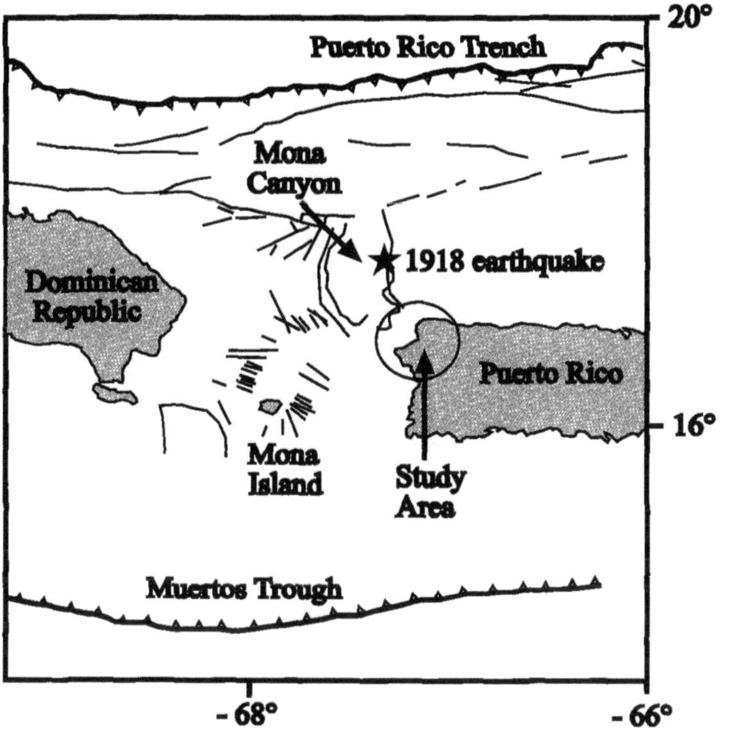


Figure 1. Map showing the location of the study area and the regional tectonic features close to the northwestern portion of Puerto Rico. The star represents the location of the 1918 earthquake located on Mona Canyon.

### 3. Tsunami records

Historically, tsunami have left geologic evidence in different environments depending on the local characteristics [*Minoura and Nakaya,* 1991; *Bryant and Young,* 1996]. From a stratigraphic point of view, places where sudden geomorphic and sedimentologic changes have occurred during and after a tsunami [*Minoura and Nakaya,* 1991; *Minoura et al., 1997*] have proved to be good locations to search for tsunami evidence. These environments are those where inputs of sand, marine organisms, or other geologic materials were carried from the sea and deposited in a low-energy coastal environment, such as swamps, tidal zones, sandy beaches, and highly organic soils.

From a geomorphologic point of view, the search for tsunami is more challenging due to the variety of settings where they occur. For example, *Bryant,*

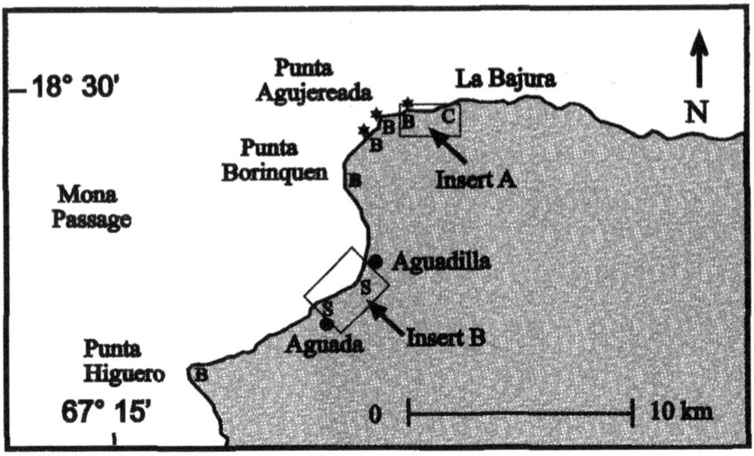


★ Marks of impacted rocks on the cliff

B Boulders carried out by the wave

C Coral carried out

S Sandy sediments in the swamps

Figure 2. Location of the field investigations in northwestern Puerto Rico, including the field evidence of potential tsunami related features.

*et al.* [1992, 1996] and *Bryant and Young* [1996] have described structures and morphologies observed in areas where tsunami have impacted specific geomorphic settings conformed by coastal cliffs. For cliff areas, rock impacts produced by tossed boulders have been common evidences of tsunami. Also, marine rocks, corals, marine shells and organisms which were ripped out by the sea waves, have been carried along coastal creeks, rivers, or water paths located close to the coast, reversing the sedimentation process from the sea toward land.

In sandy coasts, historical descriptions from tsunami mention that long-term formed coastal dunes have been broken by sea waves [*Minoura and Nakaya,* 1991]. These dunes have formed fan-delta morphologies associated with their disintegration. These sedimentologic [*Minoura and Nakaya,* 1991] and geomorphologic features [*Bryant et al.,* 1996] may support research using these

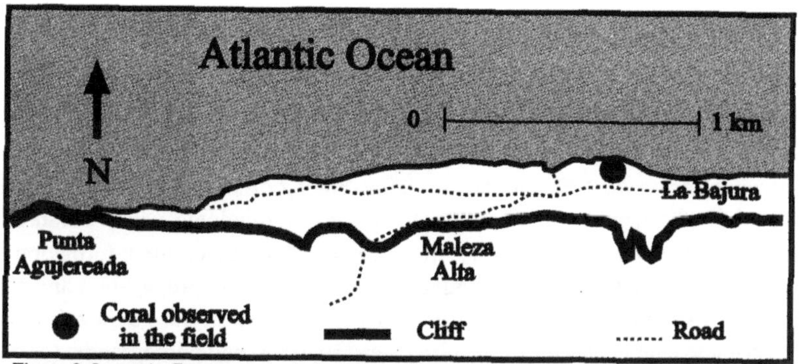


Figure 2. Insert A. Geomorphologic features and location of the coral reef observed in La Bajura.

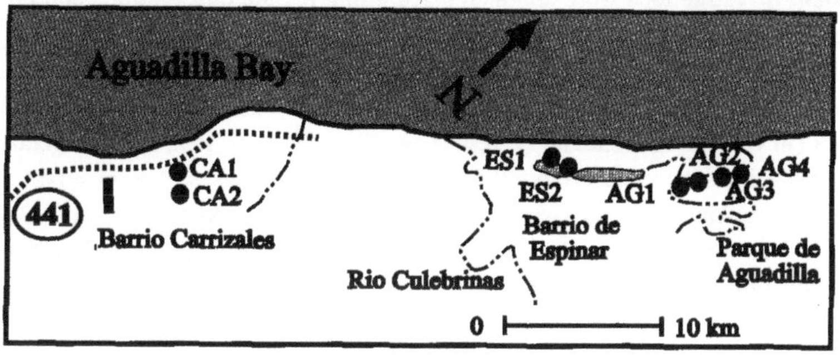


● Areas sampled for tsunami deposits

▮ Location of the trench opened

(441) State road 441

Figure 2. Insert B. Areas sampled for tsunami deposits and location of the trench in Carrizales, Playa de Aguada.

criteria applied to tsunami evidences in coastal zones such in the Caribbean Islands.

In northwestern Puerto Rico the presence of sand dunes along the beaches, swampy environments, and cliffs with loose boulders present good conditions for tsunami investigations, mainly next to the towns of Aguadilla and Aguada, where the 1918 tsunami was more catastrophic.

**4. Methodology**

This study was based on four main lines of evidence. First, compilation of the historic descriptions of the 1918 tsunami [*Reid and Taber,* 1919]. Second, interviews with 14 survivors and eye-witnesses of the 1918 tsunami. The interviews [*Acevedo,* 1996] helped to define among other things, the areas most affected by the tsunami and some potential geomorphologic or environmental changes observed by people during the event.Third, based on the information of steps one and two, the data was compared with the tsunami propagation and run-up model used by *McCann and Mercado [1997]*. Fourth, the geomorphologic and geologic characteristics of the coast were analyzed and photographic (aerial) and field investigations were conducted searching for the observations and other potential evidence of the 1918 tsunami.

From the sources mentioned, it is confirmed that the area most affected by the 1918 tsunami was the zone between Punta Higuero (in Rincón) and La Bajura (in western Isabela and eastern Aguadilla) (Figures 1, 2, and 3). Detailed field investigations focused on these areas. In most cases, the observations described by the eyewitnesses matched well the data from *Reid and Taber* [1919]. This information helped in searching for tsunami deposits and evidence in the swamps of Aguada and Aguadilla, where some cores were taken from the swampy areas (Figure 2). As complementary work, a trench was opened in the Playa de Aguada. The stratigraphy obtained was described and correlated using sedimentologic descriptions.

We also looked for evidence of the tsunami path in the sand dunes in La Bajura, northwest of Aguadilla, following the criteria suggested by *Minoura and Nakaya* [1991]. Finally, we searched along the cliffs north of Aguadilla (Figure 2), using data from the tsunami runup defined by *McCann and Mercado [1997]*, which was compared with the historical data from *Reid and Taber* [1919] and the eyewitnesses. People interviewed mentioned that they heard the noise of rocks being tossed on the cliffs and debris against the shores during the tsunami. *Reid and Taber* [1919] described boulders carried out and left on the beach by the big waves, and some were left on the railroad tracks on the cliff of Aguadilla. This combined data provided significant elements for analysis of the event.

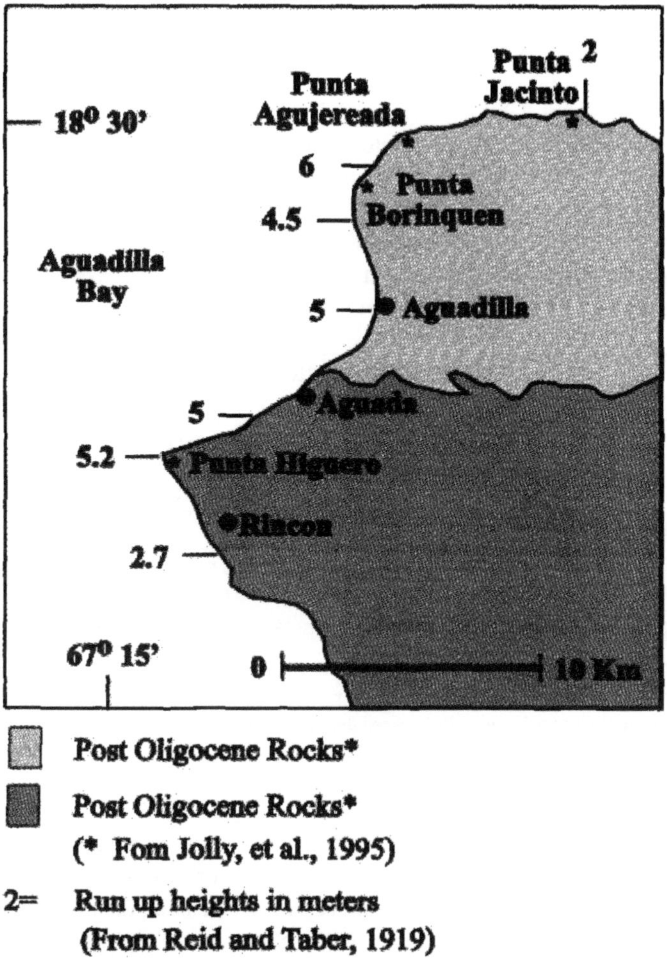


*Figure 3. General geology and runup heights observed in northwestern Puerto Rico after the* 1918 tsunami. Geology adapted from *Jolly et al.*, [1998].

## 5. General Tectonic Setting of Puerto Rico

Puerto Rico is located in the northeastern portion of the Caribbean Plate, which slowly advances to the east with respect to the North and South American Plates. Puerto Rico and other Caribbean islands occupy the easternmost part of a 2000-km long, 100-250 km wide Plate Boundary Zone, absorbing very oblique-sinistral convergence between the Caribbean and North American Plates

(Figures 1 and 4). The island is part of the Puerto Rico Microplate [*Byrne et al.*, 1985; *McCann,* 1995], the easternmost of the blocks in the Plate Boundary Zone. It is bounded by four different tectonic elements [*McCann et al.*, 1996; *Moya and McCann,* 1996]. To the north lies the subduction zone where the North American Plate descends into the mantle along the Puerto Rico Trench. To the south the Muertos Trough, a structure of probable oblique compressional origin, is the contact between the Caribbean Plate with the Puerto Rico Microplate. To the east and southeast is the Anegada Passage, a zone of active extensional deformation. To the west is an ill-defined zone of extension affecting the western part of the island of Puerto Rico, the eastern portion of the

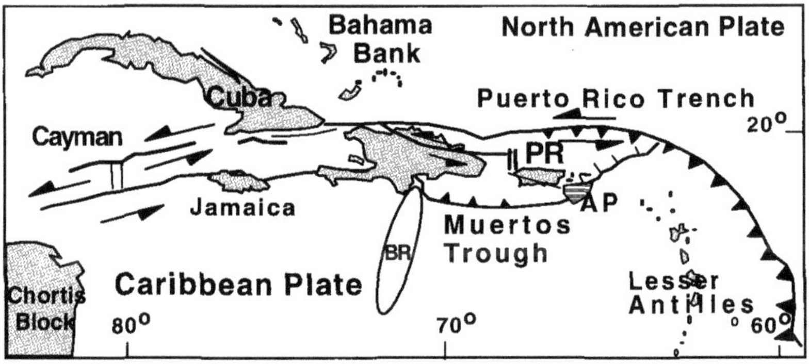


Figure 4. Plate Boundary Zone limits between the Caribbean and North American Plates next to the western portion of Puerto Rico.

Dominican Republic, and the Mona Passage [*Moya*, 1998]. Northwestern Puerto Rico would be located within this extensional zone as suggested by seismic events [*Moya*, 1998]. This extension includes offshore and onshore zones (Figure 4). In western Puerto Rico and eastern Hispaniola, the extension is being developed mainly onland. However, in the center of the region, the extension is occurring on the Mona Passage and very well expressed at the Mona Canyon.

This extensional zone has produced important earthquakes, such as the 1918 earthquake that occurred on the Mona Canyon Fault [*Mercado and McCann, 1998*]. These authors have shown the importance of many faults located in the Mona Passage and other areas as tsunamigenic sources, which present an important hazard for western Puerto Rico.

### *5.1 Seismicity*

The northeastern Caribbean is one of the most seismically active segments of the Caribbean Plate boundary as expressed by the historic data [*McCann,* 1985; *Panagiotopoulos,* 1995]. Large earthquakes have been fairly frequent in this island arc segment during this century. Recent seismicity shows the distribution and limits of the small microplates composing the Plate Boundary Zone. West of the Puerto Rico Microplate, in the eastern part of the Dominican Republic, there is a vast area that is nearly devoid of shallow seismicity. This quiet region is the El Seibo Microplate, as defined by *McCann et al.* [1996]; and *Moya and McCann* [1996]. To the north, higher levels of activity reflect motion along the subduction zone and faults at the northern flank of the platelet. To the west, the seismic activity reflects motion between the Central portion of Hispaniola and its eastern part (Figure 4).

Also, data on recent seismicity [*Moya and McCann, 1991*; *McCann et al.,* 1996] shows that southwestern Puerto Rico and Mona Passage have higher levels of microearthquakes than surrounding areas. *Asencio* [1980] and *Moya and McCann* [1991] show that for Puerto Rico many events can be identified in the historical record as having originated mainly in the Mona Passage. The 1918 earthquake was located in the Mona Canyon now known as the Mona Rift, west of Aguadilla, with an assigned Richter Magnitude of 7.3 in 1918. One hundred and fourteen people died and damage was estimated at $ 4 million (1918 dollars). These events show the importance of the seismic activity in the region associated with the separation between eastern Hispaniola and western Puerto Rico.

### *5.2 Geology and geomorphology of northwestern Puerto Rico*

The geology of northwestern Puerto Rico is documented as Cretaceous-Paleocene volcanoclastic sediments and volcanic rocks, deformed in the Eocene and Early Oligocene. The zone most affected by the 1918 tsunami is the area between Aguadilla and Rincón in northwestern Puerto Rico. Here, the geology consists of Eocene and Post Oligocene rocks and Late Quaternary sediments (Figure 3) [*Monroe;* 1967, 1977, 1980].

The geomorphic characteristics of the coast along northwestern Puerto Rico consists of a series of different flat swampy and low lands beaches, separated by prominent cliffs. The swampy low lands are associated with older courses of rivers, as is the case of the Culebrinas River. This river has had different paths in the Late Quaternary as expressed by abandoned meanders, which are observed in aerial photos and maps. The coastal plain at the mouth of this river consist of Holocene beach, alluvial, and swamp deposits [*Monroe*, 1969]. The morphology of this area has been associated to a Late Holocene fluvial-coastal interaction as expressed by the morphological and sedimentological features.

## 6. Interpretation of the 1918 earthquake: Observed Run up and effects

According to *Reid and Taber* [1919] and eyewitnesses, the 1918 tsunami had different runup heights along the shores of northwestern Puerto Rico (Figure 3). The people interviewed mentioned that the sea receded, came and receded, and came again a few times. It was not a single wave, but a set of waves, sometimes with more than five minutes of difference in the arrival time. In some areas survivors described that the sea water was ponded in the plains and swamps for days. In the Aguada beach at Carrizales (Figures 2 and 3), for example, human bodies were taken later from the swamp, after the water receded. Bad odors remained in the zone for more than a week, associated with dead marine organisms and dead bodies, particularly in the area of Espinar and La Bajura (see Figure 2). One person interviewed showed that in Carrizales the sea wave affected his family property and killed some relatives. There the tsunami penetrated in the swamps at least 50 m and in other areas by more than 150 meters.

According to some of the survivors, the sea level receded about 500 to 700 meters before the first wave arrived, and then the sea came back with a runup of up to 3 to 5 meters high. From different sources we established that the church of Aguadilla was not affected by the wave as has sometimes been stated. However one block to the south the streets were flooded with ponded water a few inches deep. According to the Civil Defense Office in Aguada, after the 1918 tsunami event some storm wind waves (called locally "bravatas" and generally due to extra-tropical winter storms) have had enough energy to affect the coast with high waves. But the runup from these storm waves has not been high enough to penetrate the swamps of Aguada or as far as the 1918 tsunami runup did in the city of Aguadilla, so the possibility of recent storm deposits located 50 to 150 meters from the beach on these swamps after the 1918 tsunami has been eliminated. However, in 1928 the eye of Hurricane San Felipe, a category 5 hurricane, crossed the area when it was leaving Puerto Rico. In the interviews nobody mentioned that the effects of the storm surge during this event penetrated the swamps in Aguadilla and Aguada. It is important to mention this mainly because tsunami effects can be similar to storm surge effects, something that will be considered later. Storm surge model simulations (Mercado, 1994) show that surge elevations along the open northwestern coast of Puerto Rico are relatively low, being on the order of 1.2 m for a category 5 hurricane like San Felipe. This does not include the wave setup component, which can add an additional meter or so. This is basically due to the extremely narrow shelf existing there.

### *6.1 Bathymetry*

One of the possible causes of the high impact of tsunami on the northwestern part of Puerto Rico is the presence of a small canyon formed by a secondary fault which connects with the Mona Canyon Fault [*Mercado and McCann, 1998*; *Moya*, 1998] (Figure 5). This secondary canyon points out directly towards the northwestern part of Puerto Rico, which causes the seismic induced waves (tsunami) coming from the Mona Canyon or from the Mona Passage to arrive faster and directly to the Aguadilla beach, as predicted by the simulations of *McCann and Mercado [1997]*. The bathymetry of this canyon goes from 4000 m in the Canyon to zero along the coast in less than 15 km (Figure 5). The bathymetry of the rest of the Mona Passage to the south is shallower, mainly south of Rincón. The impact of the 1918 tsunami was less

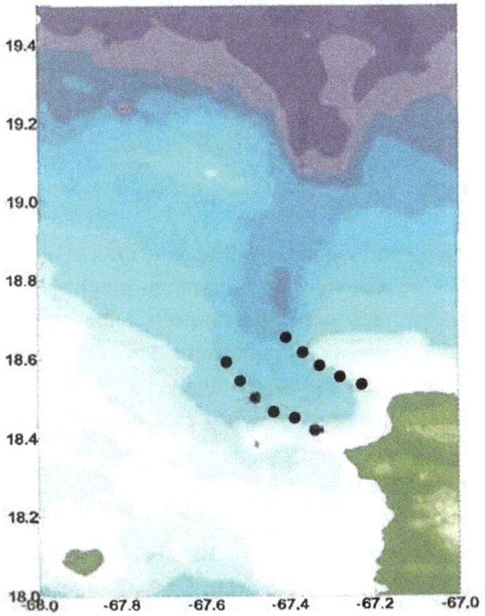


Figure 5. Bathymetry of the Mona Canyon and the Mona Passage. Dotted lines show the tectonic depression connecting Mona Canyon with the northwestern coast of Puerto Rico.

catastrophic south of Rincón. This may be caused, among other things, by the dissipation of the energy due to the shallow bathymetry and the fact that the fault had a strike angle close to 0° T, and its southern limit extended only to the latitude of Aguadilla (*Mercado and McCann*, 1998). This places all coastal areas south of Punta Higuero in Rincón approximately along the axis of the fault, along which the radiated tsunami energy is known to be small.

### *6.2 Predicted run up*

*McCann and Mercado [1997]*, using the Japanese TIME tsunami model, did a numerical simulation of the 1918 tsunami, including estimation of runup. The model suggests that runup waves with heights up to 3.4 meters occurred during the 1918 tsunami close to the city of Aguadilla. This is consistent with the data observed by *Reid and Taber* [1919] which mention runup varying between 2.4 and 3.4 m. Their model also predicts several important sets of waves rather than a single one, which is also consistent with data from the people interviewed. In addition, the simulated time of arrival of the first wave matched very well with the findings of *Reid and Taber* [1919]. These results guided the field investigations.

## 7. Field Investigations

### *7.1 Cores in the swamps*

In the swamps of Aguada and Aguadilla (Figure 2, insert B), 8 cores were taken for investigations on tsunami evidence. Aluminum pipelines 2.44 m (8 ft) long and 7.6 cm (3 in) in diameter were used to collect the cores. For the deepest sampling in Espinar, the services of a geotechnical company using portable drills were contracted. The materials were analyzed using a binocular 30X microscope. The generalized stratigraphy is presented in Figure 6.

### *7.2 Aguada Plain at Carrizales*

Two aluminum pipeline cores were taken in the area known as Carrizales (Figure 2 and 6) to a depth of almost 2.44 m deep. One was located about 50 meters from the actual beach and the other about 100 meters. The sites were named Aguada-Carrizales 1 (CA1) and Aguada-Carrizales 2 (CA2) respectively. CA1 presented strong difficulties because some layers were highly cemented, appearing as "beach rock" deposits, which made them very difficult to collect and analyze. However, two significant thin layers (0.02 cm) of fine sand with deposits of *Alemida Sp.* were observed, one at 0.20 m deep and the other at 1.85 m deep. The rest can only be described as very fine sand, silt and clay, sometime as hard as beach rock. The sand was mainly quartz and feldspars.

Due to its soft and clean sedimentological conditions, the most complete core was collected at CA-2. For CA2, from 0 to 0.15 m, the layer is composed of 40% coarse sandy deposits with quartz and feldspars and about 40% or more of calcareous sand, which appears mixed with recent soil. Also, white materials composed of pieces of broken shells are part of the layer. At this location there is a sudden change in grain size and composition of the materials at about 0.15 to 0.25 m depth. This change is characterized by a layer about 8 cm thick of white

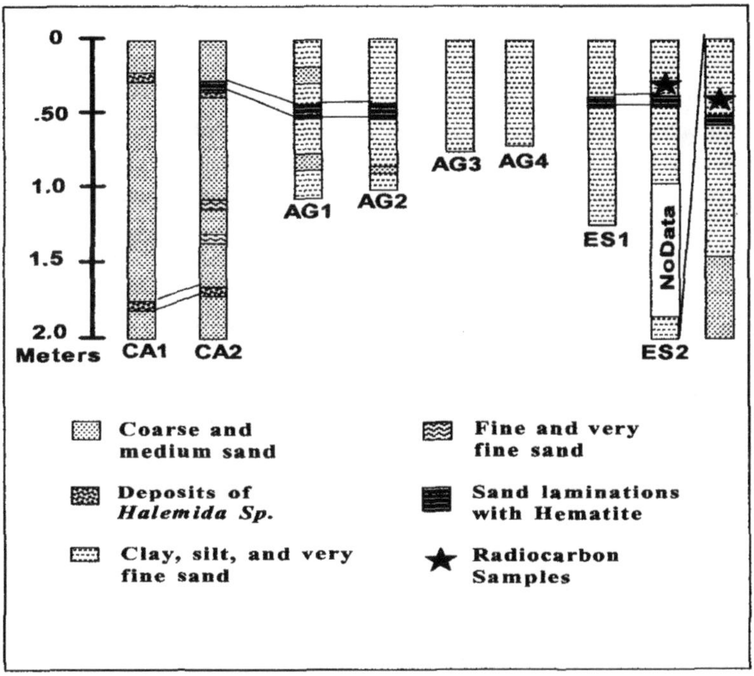


Figure 6. Stratigraphic record of the cores sampled in the swampy areas of Aguada and Aguadilla, Northwestern Puerto Rico. Triangles show locations of radiocarbon samples (Table I).

and gray, very well defined, fine calcareous sand and silt with abundance of pieces of *Halimeda sp.*, a marine algae which lives in the bottom of the shallow marine platform on the first few meters of depth. This layer appears in some areas as parallel cross-bedding laminations and resembles a clear depositional event, different from the massive deposition from the layer on top and underneath.

From 0.25 to 1.10 m the material is composed of massive white and gray calcareous fine sand. At about 1.10 m deep there appears a thin layer composed of fine quartz sand. This layer is only 0.08 - 0.10 m thick and suggests a single high-energy event. No marine organisms were observed in this layer. From about 1.20 to 1.30 m the layer is the same as from 0.25 to 1.10 m, basically massive white and gray calcareous sand. Again at about 1.30 to 1.35 m deep there is another layer similar to the one that appears from 1.20 to 1.30 m. This is a single layer of fine quartz sand about 4 to 5 cm thick. Layers at 1.10 m and 1.30 m deep are similar, presenting just fine quartz with no marine organisms. On these layers there is no evidence of marine organisms, so the possible source could be just the reworking of beach deposits.

From about 1.35 to 1.70 m deep, the layer is again massive as the others between 0.25 and 1.10 m and 1.20 to 1.30 m deep, suggesting similar coastal plain (beach) sequences interrupted by single events. At about 1.70 m deep, there is a layer of dense sand with some preserved pieces of *Halemida Sp.* In some pockets, deposits of Hematite are observed but without any defined sedimentary structure. The *Halemida Sp.* and the Hematite, which are at presently distributed on the shore, suggest that these deposits were carried from the beach about 100 meters to this place. Below this layer, the materials are composed mostly of massive coarse to medium gray calcareous sand and broken pieces of shells through 2.40 m, the depth of the end of the sample.

The presence of *Halimeda sp.* as a layer located at about 0.15 to 0.25 m deep is interpreted as the signature of a tsunami, as well as the layer at about 1.70 m deep containing Hematite. The changes in color and grain size are significant, and the presence of the *Halimeda sp.* suggests that the event was not related to river flooding.

### *7.3 Trenching at Carrizales*

One trench was also dug on the coastal plain of Aguada (Figure 2), searching for stratigraphic markers of the 1918 tsunami. The trench was almost 25 meters long and was oriented perpendicular to the beach in Barrio Carrizales, about 120 meters from the shoreline (Figure 2 insert B). According to the property owner, this swampy area has been used intensely for cattle in the last 50 years, reworking the first inches of the topsoil. This is the place pointed out by one of the survivors, who mentioned that the seawater was ponded for days in the area and that about 4 people were killed by the wave.

The stratigraphy found in the trench showed some differences with respect to some units observed in the cores (Figure 7). At the top of trench, as in log CA2, we observed a reworked soil with quartz and calcareous sand. Then, at about 0.25 m deep, a 0.12 m thick layer of gray calcareous coarse sand and shells defines the possible evidence of the tsunami observed in Core CA2. This layer presents some small well-preserved shells and oysters, which were deposited in this area. None of them were broken. Variations of different layers of sandy materials appear in the profile from about 0.35 m through about 1.80 m deep. At about this depth a gray layer of sand with *Halemida Sp.*, marine shells, and Hematite is evident. This layer presents a sharp change and is only 5 cm thick. The rest of the column is basically sand with different amounts of broken shells and sand with quartz and feldspars. Both layers, the one at 0.25 m and the

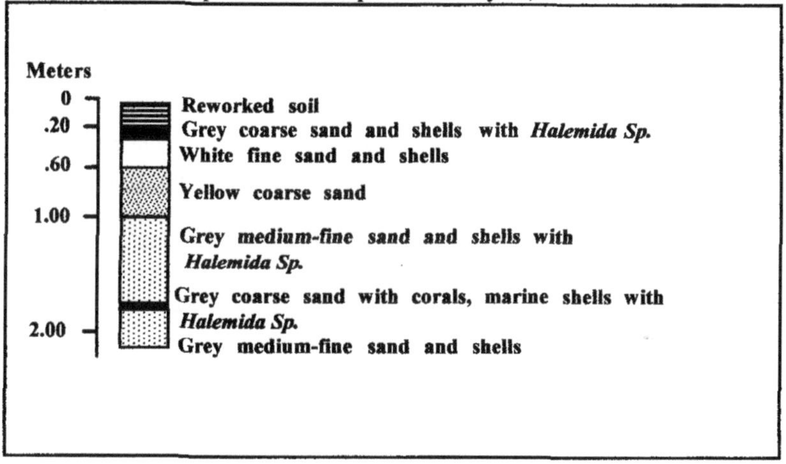


Figure 7. Generalized stratigraphic record observed in the trench opened in Barrio Carrizales in Aguada. This trench was opened to search for tsunami deposits.

other at about 1.80 m, present well-preserved pieces of *Halemida Sp.* This suggests a sudden sedimentation event which brought these materials from the sea to this swamp-type environment. Again, it is suggested here that the well-preserved layers of *Halemida Sp.* could be indicators of a tsunami. We have assumed that broken or reworked pieces of *Halemida Sp.*, or marine shells, would be associated with storms due to the continuous energy affecting the shorelines. The layer at 0.25 m has been interpreted as deposited by the 1918 tsunami. The differences between the information obtained from the logs and from the trench suggest that lateral variations in sedimentation have been present in the area.

### *7.4 Aguada Swamp at Espinar*

Three (3) cores were collected in the swamp of the area known as Espinar in Aguada, about 1200 meters along the coast to the east of the Culebrinas River. Espinar is another sector in Aguada where people said the houses of several fisherman where ripped out by the tsunami. In the Espinar area, according to local people, the swamp has receded about 20 to 50 meters since 1930. Also, about 50 meters of coastal erosion have caused the coast to retreat, so the swamp is now about 40 to 30 meters closer to the beach than in 1918. Here the sampling was difficult. One pipe got a good sample (ES1) and the other two had the problem of sand compression at the mouth of the cores. To resolve this problem, a geotechnical company (ES2) collected the samples using a vibrocore.

First, at the log obtained at ES1, from 0 to 0.25 m, a beige organic material appears covered by thin soil. This organic material is composed of silt and fine sand. From about 0.25 to 0.30 m deep there is a layer composed of 5 to 6 laminations of Hematite and gray and white medium to fine sand, which contrasts with the beige color of the organic material. Then, from about 0.30 to 1.2 m, the beige silty-sand materials of the swamp environment again appear (Figure 6).

At about 10 meters from the same place (the swampy area of Espinar) a geotechnical company took a 4.0 meters deep, 5 cm thick core with almost no perturbation (Figure 6). From 0 to about 0.50 m, the log showed that a fine silty sand gray-brown organic layer appears at the surface. At about 0.50 to 0.55 m deep there is layer of gray and black coarser sand with high content of Hematite. This layer is about 0.05 m thick and presents five oxidized laminations and very small broken pieces of shells and small pieces of preserved root plants. There is an important difference between the laminations of ES1 and ES2. In the laminations of ES1 no oxidation is observed; however, in ES2 the oxidation and the presence of small pieces of organic material, mainly roots, and less laminations suggest that the ES2 laminations may be different. A piece of organic material located in the laminations mentioned on ES2 was taken for radiocarbon dating.

Continuing with the ES2 description, from 0.55 to about 1.00 m it is again observed a fine sandy gray-brown organic layer, quite similar to the layer above the laminations. No data was collected by the firm in the log from about 1.0 to 1.8 m, due to sample liquefaction. From 1.8 to 2.43 meters deep in the core appears a gray-clear-beige silty-clay organic mud. Then from about 2.43 to 2.50 meters appears again a sandy-silt zone of oxidized laminations of medium sand with Hematite. This layer is about 0.04 m thick, similar to the one located at 0.50

m. These laminations contain important amounts of Hematite and small pieces of broken shells and tiny pieces of organic tissues such as root plants. There, another organic piece was taken for radiocarbon dating. The authors acknowledge that the sampling in the log could generate important errors in the interpretation. As observed, the lack of samples from 1.0 m to 1.8 m has made the data interpretation difficult.

### 7.5 Aguadilla Swamp at Parque de Aguadilla

In the Parque de Aguadilla (Figure 2) area four cores were taken in the mangrove along the channel that is located behind the park, where a few years ago an entertainment train used to be located. The samples were located on the tide channels and called AG1, AG2, AG3, and AG4 (Figure 6). On the AG3 and AG4 logs, all the material obtained from 0 to 0.80 m was beige and clear gray silty sand and clay deposits. On the AG1 and AG2 cores, the top of the samples is composed of beige and clear gray silty sand and clay deposits associated with this low energy environment. At about 0.39 and about 0.50 m deep, we observed in the two cores a set of about 6 to 7 laminations characterized by fine and medium gray and white sand layers stratified with fine sand composed of Hematite, and very scarce pieces of broken marine shells. This layer is similar to the layer found in Espinar, but no clear oxidation is observed. Here the laminations are more clear and spectacular. The lamination layers are common in both logs at almost the same depth and similar to the one in ES1.

In general, the main point in this analysis has been trying to define the signal for the tsunami record, for which it is suggested again that the presence of *Halemida Sp.* and Hematite distributed in laminations may be good indicators of tsunami records for the area. The thin laminations observed in the cores should represent the inputs of the marine sediments and heavy minerals located in the beach, which were brought to the swamp by the tsunami.

## 8. Evidence in the cliffs

### *8.1 Punta Agujereada*

The zone between Punta Borinquen and Punta Agujereada presented evidence of a run up with up to 6 m high (Figures 2 and 3) [*Reid and Taber,* 1919]. These authors mentioned that some rocks were ripped out from the sea, carried out by the seismic wave, and left on the beach at the moment of the event. Field evidence showed that along the coast there are boulders up to 10 meters in diameter located on the beach. The rocks are composed of limestones that have fallen out from the cliffs. Marine erosion has eroded the base of the cliffs and the rocks came down due to the gravity as rock falls.

The field evidence suggests that these rocks fell down from the cliffs to the sea and were later tossed out against the cliffs or were taken laterally along the coast until they reached the beach, as mentioned by a survivor in Aguadilla and as described by *Reid and Taber* [1919]. In the areas with high run-up, the wave brought boulders and corals from the sea to the beaches and the cliffs. Based on the interviews and descriptions of *Reid and Taber* [1919], some of these rocks were also tossed out against some of the beaches. *Taggart et al.*, [1993] reported for Isla de Mona similar boulders which were suggested to be associated to tsunami or hurricanes.

According to the Civil Defense Office in Aguada, during important storms or *"bravatas"*, storm surges in northwestern Puerto Rico have caused remotion of the sand in the beaches, so the sandy coast has retreated, some times up to 5 meters (Martin Concepción, personal communication). So, in the last 20 years shoreline erosion has been intense and beach retreat has taken the sand form the beach. In Punta Agujereada, just underneath the US Army Base on the beach, possible rock impacts of more than 1.0 m in diameter can be observed in the cliffs, in areas where boulders of this diameter are abundant along the coast. Figure 2 presents the area where the evidence of these possible impacts structures was observed. We believe that because the rocks are limestones, the lack of weathering, the presence of recent dissolution, and the formation of secondary carbonate in the exposed surface of a rock make the impact recognizable. This evidence consists of semi-circular broken surfaces or "chips" on the rocks, which follow the contour of the rock shape. *Bryant and Young* [1996] observed similar structures in the southern coast of Australia associated with tsunami. The height of the impact in the area varies in different locations.

A close-up of the areas of impact on the cliff surface showed that dissolution in the zone "chips" is less significant than the dissolution and weathering in the surrounding areas, as evidenced by change in color due to the lack of secondary carbonate in the limestone. In the same area, but on the beach, many rocks with diameters between 50 and 100 cm were found about 50 meters from the coast. However these rocks could have been carried by the sea wave or have fallen from the cliff.

### *8.2 Punta Borinquen*

The area of Punta Borinquen was also visited close to the old light house. There, a small creek comes to the beach. Located at about 70 meters from the beach inside the creek, there is a zone where we believe a group of rocks with diameters of about 0.50 to 1.00 cm where piled by the tsunami. These rocks appear isolated. Piled rocks like these have been observed in zones where tsunami have penetrated inland in zones that present steep rivers or creeks (*Bryant et al.*,1992; *Bryant and Young*, 1996). The presence of these rocks makes them very suspicious as indicators of tsunami penetration. The creek has not enough energy to carry out these rocks from the land to the sea because the water infiltrates in the karst.

### *8.3 La Bajura*

East of Punta Agujereada, in the area known as La Bajura (Figure 2 insert A), a coral of about 200 to 250 pounds and 1.00 to 1.20 m in diameter is observed behind the natural sand dune that protects the coastal plain (Figures 2 and 8). There are some indicators that in some areas the dunes, which are recent dunes, have been washed out in the past. *Minoura and Nayaka* [1991] have described this as a common process generated by tsunami. In La Bajura, this coral is located almost 50 meters from the beach behind a depression in the dune, at the front of a sandy fan-delta like feature. This coral may have been carried out from the sea by the tsunami. Here the dune was broken and a depression was formed, which we believe was indeed a fossil path of the tsunami. As discussed above, the extremely narrow shelf off this area makes storm surges very small, and never strong enough to remove or cross over the coastal dune in the area carrying big pieces of corals. People interviewed and living in this area, about 1-2 km east of the coral location, have not experienced a storm surge or inundation since 1918. This area appears in the historic records of the Civil Defense office as one of the areas where the penetration of the tsunami was more catastrophic, with a local fishing village destroyed by the 1918 tsunami.

Historical data about tsunami around the world show that boulders and corals are commonly carried out during these events. This is the case of the 1996 Indonesia tsunami, and the 1996 Nicaragua tsunami [NOAA, 1999], so the similarities of the coral in La Bajura and the boulders in Punta Agujereada and Punta Borinquen are not too different from the evidence found in other areas of the world. It is suggested here that many of the boulders and the coral in the beach were carried by 1918 tsunami.

Figure 8. Photo of the coral reef found behind the dunes in the beach known as La Bajura located on Figure 2. Insert A.

## 9. Radiocarbon Dating and Age of Events

Two radiocarbon dates were obtained from log ES2. It is assumed preliminarily that the laminations found are associated with tsunami deposition. It is also assumed that combined with log ES2, the logs of AG1, AG2, and ES1 have helped to define a possibly stratigraphic definition of tsunami in a swamp environment in northwestern Puerto Rico. Table I show the age of the radiocarbon samples.

It is preliminarily suggested that the first pre-historic record of tsunami in northwestern Puerto Rico has been observed in this core. After the calibration of the radiocarbon dates, the age of the upper event at Espinar 1

*Table I*

Radiocarbon dates obtained from the samples in the cores AGU-2 and AGU-3. Weighted averages calculated using program of Stuvier and Reimer (1993).

| Sample | material | ID | Mass (mg),d13C | C14 age (BP)$\pm$ 1s | 2s calendar age range |
|---|---|---|---|---|---|
| AA28490 | charcoal/sed | AGU-2 | 0.34-26.7 | 630-55 | 1270-1410 AD |
| AA28491 | charcoal | AGU-3 | 0.26-25.4 | 2,515-70 | 820-400 BC |

(ES2) was dated about 1270-1410 AD. However, the deeper event was dated about 820-400 BC. The events at AG1, AG21, and ES1 have been assumed to be associated with the 1918 tsunami.

It is important to mention that on other paleoseismic studies in western Puerto Rico, the first author found liquefaction features expressed as sand injections on an archeological site on Rio Yaguez (at the city of Mayagüez, just south of Rincón). The analysis of this site is out of the scope of this study, but two radiocarbon dates and paleoseismic interpretation of the site suggest that there was an important seismic event between 1300 and 1511 AD [Moya *et. al.*, in prep]. This is important because there is a correlation with the assumed tsunami deposits on ES2 in the upper laminations, which overlap the dates in the archeological site at Rio Yaguez. This data supports that these laminations on ES2 could be associated with a pre-historic seismic event occurring close to northwestern Puerto Rico.

## 10. Discussion

This study has presented some field investigations interpreted as associated with the 1918 tsunami and previous tsunami. The evidence suggested here to be associated with these events were compiled from historic, and field geologic and geomorphologic investigations, plus interviews from survivors and results from a numerical simulation.

We have to accept the limitation of the field techniques applied to the search of tsunami sedimentation in the swamps of Aguada and Aguadilla and this study should be considered preliminary. More detailed field sampling and investigations are needed. Despite the technical difficulties collecting the samples, there is a repetition of sand laminations in different levels on the logs, which suggest that at least for the samples in logs AG1, AG2, ES1, and ES2 these sand laminations may be the indicators of different tsunami waves, which brought inputs of sandy Hematite and *Halemida Sp.* to the low energy environment in the swamps. These sand inputs could be associated with seismic induced waves beyond the reach of storm surges.

The data from log CA2 was composed of calcareous sand and suggests that the sedimentation was different in this area. After comparing all the potential stratigraphic records of the 1918 tsunami, we suggest that there was not a unique stratigraphic record of this event. Local variations such as local sedimentary environment, sources of sediments, wave behavior, and geomorphic conditions in the local area, could change the stratigraphic record of tsunami. However, for the logs in the swamp of Aguadilla sand laminations were common.

The laminations on ES2 are sedimentologically similar to the others at AG1, AG2, and ES1. However, the oxidation in the layers on the Hematite suggests that the possible 1918 event was missed in the core and that the upper event, according to radiocarbon dating, corresponds to an event which occurred between 1270-1410 AD. An older event has been suggested here to have occurred about 820-400 BC (Figure 9), as suggested also by radiocarbon dating. Then, the events at AG1, AG21, and ES1 have been assumed as associated with the 1918 tsunami.

The date suggested by the radiocarbon date on the upper laminations on ES2 is very similar to the dates obtained from two other independent sources of liquefaction in western Puerto Rico (Moya, in prep). This suggests that an important seismic event may have occurred between 1300 and 1511 AC, as suggested as well by archeological interpretations on the Spanish occupation in western Puerto Rico [*Rivera and Silva,* 1997]. The overlap in the dates on the archeological site at Rio Yaguez [*Rivera and Silva,* 1997; and *Moya et al.,* in prep], and the date on ES2 are suggesting very strongly the existence of this important seismic event in western Puerto Rico between these dates.

The field investigations on the cliffs of Punta Agujereada and Punta Borinquen suggest that rock impacts observed on the cliffs may be associated with tsunami as well, mainly boulders located on the beach and far way from rock fall sources. The best evidence of boulders carried by the tsunami should be the coral observed in La Bajura. This data suggest that there are good conditions in northwestern Puerto Rico for rocks or corals that will be tossed by future tsunami against the coast.

### *10.1 Long-term geomorphologic changes and tectonic controls on the northwestern coast of Puerto Rico*

The morphology of the northwestern Puerto Rico resembles a big embayment, which is concave toward the sea where beaches and cliffs co-exist (Figure 5). in the marine canyon of Aguadilla [*McCann and Mercado, 1997*] (Figure 5). There, two active normal faults have contributed to the development of this important tectonic graben. These faults are the Borinquen and the Borinquen South Branch faults [*McCann and Mercado, 1997*]. Combined with the Mona Canyon or Mona Rift fault these tectonic features have contributed to the development of a natural depression in Quaternary times. This graben points directly to northwestern Puerto Rico, developing the conditions for a natural

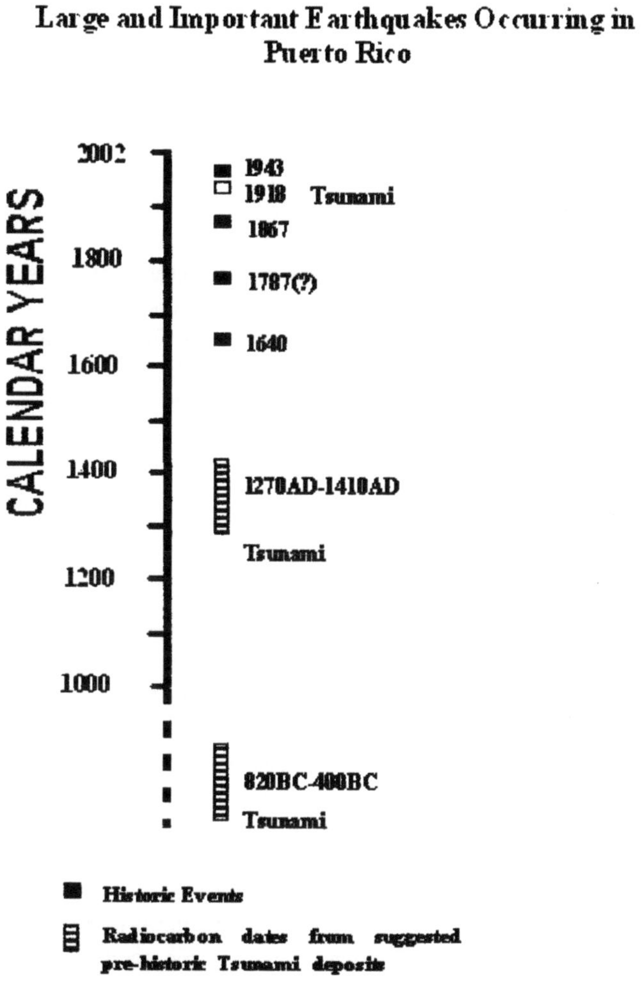


Figure 9. Historic and Pre-historic earthquakes as defined after this investigation

waveguide or natural funnel, which contributes to the fast arrival of tsunami, developing local conditions which are favorable for the impact of tsunami.

Observing the general geology and geomorphology of the marine platform north of Aguadilla and west of Rincon, and the active normal faults in the zone, it is here suggested that a minimum of 8 km of coastal retreat has occurred in the northwestern portion of Puerto Rico during the Quaternary. This is the result of tectonic subsidence forming the Aguadilla Canyon, which is connected with the Mona Canyon.

The second important condition is the presence of limestones exposed to dissolution. The karstic landscape in the area does not allow the rivers to mature and develop a good alluvial or coastal plain, so the contribution of sand along the beach is poor and limited. Only the Culebrinas River is the main source of sediments to the beach, but the concave morphology of the bay suggests that there is a negative sand balance which cannot sustain a sandy beach accreting in the area. Also, along the cliffs, the dissolution generated by the seawater at the base of the limestones is very important. This dissolution creates good conditions for the formation of rock falls (Figure *10*) in the cliffs of Aguadilla.

Field evidence showed that if an earthquake shakes the coast again, lateral forces will generate many rock falls and landslides again next to the cliffs. Combined with the rock fall activity and the lack of sand inputs from the rivers, we can expect an important geomorphic retreat in the long-term on the coast.

Finally, the last morphologic control of this coastal area is the tsunami itself. Accepting the limitations of only two radiocarbon dates to define three possible tsunami events in the last 2000 years, we can estimate preliminarily the long-term impact of tsunami in northwestern Puerto Rico. The data on historic and prehistoric tsunami events suggest one event in 1918 and other possible seismic event between 1300 and 1511 AC, and a third event about 820-400 BC. If we just consider the last two events, the 1918 and the previous event about 600 years ago, we have at least 2 events in the last thousand years. It is clear that there are other faults in the Mona Passage capable of generating tsunami [*McCann, 1998*]. For that reason, we should assume that the tsunami wave traveling in the Mona Passage will have important impacts in the northwestern portion of Puerto Rico, due to its local conditions. That makes the area very vulnerable to tsunami in the future, whatever should be the source in the region. As a matter of fact, numerical simulations of the potential tsunamis due to 30 active faults described by *McCann* (1998) show that the highest runups are located between Punta Agujereada and Punta Higuero *[Mercado, 2001*].

Figure 10. Photo of the rock falls along the cliffs conformed by limestone in northwestern Puerto Rico

Assuming this high risk, then it is reasonable to estimate that at least two large seismic tsunamigenic events can occur at least every one thousand years as response to the deformation in this active plate boundary zone and between *Puerto Rico and Eastern Hispaniola [Jansma et al., 2000]. Assuming the age of* the Aguadilla and the Mona Canyons as at least Quaternary features (at least formed in the last million years), two tsunami events per thousand years in the last million years, indicates great tsunami activity. If the conditions have been the same for the time suggested, even with just one event every thousand years means an important number of tsunami arriving to the coast.

The hypothesis is that the combination of tectonic subsidence in the Aguadilla Canyon, together with the lack of renewal in the sand sources in the beach and the rock fall activity in the cliffs, is reflected in cliff retreat. It is proposed here that this is the most important factor in the coastal evolution of northwestern Puerto Rico. Again, if the physiographic conditions have not changed in the last million of years it is logical to think that in the long-term the impact of tsunami has been very important, enough to make a significant change in the morphology of the coast.

## 11. Conclusions

Northwestern Puerto Rico has the conditions to be affected by future tsunami. The tectonic activity between Puerto Rico and Hispaniola in the plate boundary zone has generated a zone of extension occurring mainly on the Mona Passage, Eastern Hispaniola, and Western Puerto Rico. This means that potential seismic events associated with the faults in these areas could generate important earthquakes and tsunami. All the seismic sources located around Puerto Rico could induce important tsunami in northwestern Puerto Rico due to the submarine canyon coming from the Mona Rift to this area as it occurred in 1918. This is verified by numerical simulations using earthquake sources all around Puerto Rico [*Mercado and Justiniano, 2003*]. Also, any important submarine landslide on the slopes along the Puerto Rican Trench could induce and develop an important tsunami for northwestern Puerto Rico.

The evidence proposed here on past tsunami, suggests that these events have produced important geomorphologic changes to the coast of northwestern Puerto Rico, as well as modifications to the environments, and marine sand inputs to the swamps. A more detailed study should be conducted to define more specifically the geologic record of the tsunami in different environments in the island in order to develop a very realistic hazard plan against these events. This future study has to include the quantification of pre-historic tsunami to establish the recurrence time of these events in northwestern Puerto Rico.

In general, for western Puerto Rico the model of *McCann and Mercado [1997]* agrees with the description of *Reid and Taber* [1919], the interviews, and the field data. The areas to be impacted most by a future tsunami will be exactly the areas reported by these sources. Important attention should be focused on the Aguadilla and Aguada beaches, and mainly on La Bajura [*Mercado, 2001*].

It is not necessary to have another event of M7.3 as the event in 1918 to generate a catastrophic tsunami. As observed in other parts of the world, earthquakes with magnitudes of about M7 or less could be also catastrophic if submarine slides are generated. The conditions in the Mona Canyon and its secondary canyons present the conditions for the generation of submarine slides because of its steep slopes. The problem here is that historic information is too scarce to fully make the public aware of the potential hazard.

It is suggested here that another tsunami similar to the 1918 event occurred between 1270 and 1410 AD, and a previous event could have occurred about 820-400 BC. These data is not enough to define a recurrence time of tsunami, and has to be verified and validated with more field data based on detailed investigations. This work is only the beginning of this investigation. The authors accept that the data presented here is preliminary and has to be improved with better stratigraphy. The sand laminations need to be studied in more detail combined with other geologic indicators of tsunami sedimentation. Northwestern Puerto Rico is being controlled tectonically and has responded to the separation of Puerto Rico and Hispaniola, tsunami being one of the most important effects of this separation.

## 12. Acknowledgements

The Sea Grant Program in Puerto Rico funded part of this investigation. Important support was provided by Mr. Martin Concepción of the Civil defense Office in Aguada. Mr. Aibot Ramos kindly allowed trenching in his property. Mr. Pedro Quiñones analyzed the marine organisms. Dr. Carol Prentice made possible the two radiocarbon dates.

## 13. References

Acevedo, K.M., 1996, Reflexiones sobre la labor etnográfica: historias de vida acerca del maremoto del 1918, en la costa oeste de Puerto Rico. Class Report. University of Puerto Rico. Mayaguez Campus. 30 pp.

Asencio, E., 1980, *U.S. Open File Report.* 80-192, 135 pp.

Atwater, B. and Hemphill-Haley, E., 1997, Recurrence Intervals for great earthquakes of the past 3,500 years at northwestern Willapa Bay, Washington. U.S.G.S. Professional Paper 1576. pp.108.

Bryant, E.A., Young, R.W., and Price, D.M., 1992, Evidence of Tsunami sedimentation on the southeastern Coast of Australia. *Journal of Geology.* V.100:753-765.

Bryant, E.A., Young, R.W., and Price, D.M., 1996, Tsunami as a Major Control of Coastal Evolution, Southeastern Australia. *J. of Coastal Research,* 12, 4, 831-840.

Bryant, E.A., and R.W. Young., 1996, Bedrock-sculpturing by tsunami, south coast New South Wales, Australia. *J. of Geology,* 104, 565-582.

Byrne, D.B., G. Suárez, and W.R. McCann., 1985, Muertos Trough: subduction-microplate tectonics in the northern Caribbean? *Nature,* 317, 6036, 420-421.

Jansma, P. E., A. López, G. S. Mattioli, C. DeMets, T. H. Dixon, P. Mann, and E. Calais, 2000, Neotectonics of Puerto Rico and the Virgin Islands, northeastern Caribbean, from GPS geodesy, *Tectonics, 19*, 1021-1037.

Jolly, W.T., Lidiak, E.G., Schellekens, J.H., Santos, H., Lidiak, E.G. and Larue, D.K., 1998, Volcanism, tectonics, and stratigraphic correlations in Puerto Rico, Tectonics and geochemistry of the northeastern Caribbean. *Special Paper Geological Society of America, p. 1-34.*

McCann, W. R., 1985, On the Earthquake Hazards of Puerto Rico and the

Virgin Islands. Bull. Of Seis. Soc. Of Amer. 75(1): 251-262.
McCann, W.R., 1995, Analysis of the Tsunami potential in northwestern Puerto Rico. Earthquake Advisory Board, State Civil Defense Office of Puerto Rico. 40pp.
McCann, W. R., J.C. Moya, and C. Lithgow-Bertenolli, 1996, Lateral arc extension, accelerated tectonic erosion and platelet rotation in the eastern Greater Antilles: Oblique subduction of the Bahama Platform, *EOS*, 75, 44, San Francisco.
McCann, W. R., A. Mercado., 1997, Analysis of the tsunami potential of northwestern Puerto Rico, *Earthquake Advisory Board of State Civil Defense of Puerto Rico*, 134 pp.
McCann, W. R., 1998. Tsunami hazard of western Puerto Rico from local sources: characteristics of tsunamigenic faults. Tech. Report. Submitted to Sea Grant Program, University of Puerto Rico, 79 pp. [can be found at http://poseidon.uprm.edu]
Mercado, A., 1994. On the use of NOAA's storm surge model, SLOSH, in managing coastal hazards: the experience in Puerto Rico. *J. Nat. Hazards*, V. 10, 235-246.
Mercado, A. and W. McCann, 1998. Numerical simulation of the 1918 Puerto Rico tsunami. *J. Nat. Hazards*, V. 18, 57-76. [can be found at http://poseidon.uprm.edu]
Mercado, A., 2001. Determination of the tsunami hazard for western Puerto Rico from local sources. Tech. Rept. Submitted to Sea Grant Program, University of Puerto Rico, 207 pp. [can be found at http://poseidon.uprm.edu]
Mercado, A. and H. Justiniano, 2003. Puerto Rico Tsunami Warning and Mitigation Program, Task 1: Tsunami coastal flood mapping for Puerto Rico and adjacent islands. Tech. Rept. Submitted to the Federal Emergency Management Agency and the University of Puerto Rico, 100 pp. [can be found at http://poseidon.uprm.edu]
Minoura, K, and Nakaya S., 1991, Traces of tsunami preserved in inter-tidal lacustrine and marsh deposits: Some examples from northeast Japan. *J. of Geology*, 99, 265-287.
Minoura, K., Imamura, F., Takahashi, T., Shuto, N., 1997, Sequence of sedimentation process caused by the 1992 Flores tsunami: Evidence from Babi Island. *Geology*, V.25, No. 6, 523-526.
Moya, J.C., 1998, The Neotectonics of Western Puerto Rico. Ph.D. Dissertation. Department of Geological Sciences. University of Colorado at Boulder. 120 pp.
Moya, J.C., and W.R. McCann., 1991, Earthquake Vulnerability Study of Mayaguez, Western Puerto Rico, Cooperative Agreement, *Earthquake Safety Commission of Puerto Rico-Federal Emergency Management Agency, Internal Report 91-1: FEMAPR-0012*. 66 pp.

Moya, J.C., and W.R. McCann., 1996, Tsunami deposits in northwestern Puerto Rico: Evidence of earthquake hazards associated with rapid extension in Mona Passage, *Abstract with Programs, GSA Annual Meeting, A-283.*

Monroe, W., 1969, Geologic Map of the Aguadilla quadrangle, Puerto Rico, *U.S.G.S. Misc. Geol. Inv. Map I-569*, Scale 1:20 000.

Monroe, W., 1977, Las Divisiones Geomórficas de Puerto Rico, in *Geovisión de Puerto Rico, Aportaciones recientes al estudio de la geografía, edited by Maria Teresa B. de Caliñanes. Editorial Universitaria*, Universidad de Puerto Rico, pp. 1-43.

Monroe, W., 1980, Geology of the Middle Tertiary Formations of Puerto Rico, *U.S.G.S. Professional Paper 953*, 93 pp.

National Oceanic and Atmospheric Agency. Recent tsunami in the Pacific, *http//: www.NOAA.gov.*

Panagiotopoulos, D.G., 1995. Long-term earthquake prediction in Central America and Caribbean Sea based on the time- and magnitude-predictable model. *Bulletin of the Seismological Society of America, 85(4): 1190-1201.* Caribbean, Central-America, Lesser-Antilles, tectonic, kinematic, earthquake.

Reid , H. and S. Taber., 1919, The Porto Rico earthquake of 1918, with descriptions of earlier earthquakes (Report of the Earthquake Investigation Commission), *House of Rep. Doc. 269, Washington, D.C.*, 74p.

Rivera, J. and Silva, D., 1997, Proyecto Arqueológico Bo. Quemado, Mayaguez (Batey Delfín del Yaguez). *In: Ocho trabajos de Investigación Arqueológica en Puerto Rico. 2o. Encuentro de Investigadores. Instituto de Cultura Puertoriqueña.* p. 54-64.

Stuvier , M. and Reimer, P.J., 1993, Extended 14C data base and revised CALIB 3.0 14C age calibration program: *Radiocarbon,* v. 35, no. 1, p.215-230.

Taggart, B., Lundberg, J.I.,Carew, J.L., and Mylroie, J.E., 1993, Holocene reef-rock boulders on Isla de Mona, Puerto Rico: Transported by a hurricane or seismic sea wave. *Geological Society of America Annual Meeting Abstracts with Program* 25:61.

# NEW EVIDENCE AND DATINGS OF HOLOCENE PALEO-TSUNAMI EVENTS IN THE CARIBBEAN (BARBADOS, ST. MARTIN AND ANGUILLA)

ANJA SCHEFFERS and DIETER KELLETAT

*Department of Geography, University of Duisburg-Essen, Universitätsstr. 15*
*D-45117 Essen, Germany*

Recent sedimentologic research has documented new evidence of Younger Holocene tsunami in the Caribbean. The tsunami generating mechanisms have been attributed either to plate tectonic seismics, volcanic events, or submarine slides on the Caribbean plate itself. The tsunami evidence from Barbados, St. Martin and Anguilla, however, point to origins in the open Atlantic Ocean as the geomorphic and sedimentologic imprints are distributed along east-facing coastal sections. Here, large single boulders as well as boulder ridges are situated on top of high cliffs. Relative and absolute age dating reveal at least two tsunami events on Barbados and Anguilla, and one on St. Martin during the Younger Holocene.

## 1. Introduction

The scientific investigation of tsunami has been mainly the domain of geophysics and history within the last 100 years, and until hitherto inductive field analysis of tsunami signatures in the coastal environment has not found its adequate place in the tsunami research progress.

The tsunami history not only for the Caribbean, but also worldwide is predominantly based upon the seismic character of a certain region including volcanic activities and the thorough study of historical reports with respect to their validity (Lander and Whiteside, 1997; Lander, Whiteside, and Lockridge, 2002). The instrumental paleoseismic record for the Caribbean reaches back for more than hundred years, whereas the historical sources span the time period since the discovery of the New World by Columbus. Both databases have been used to evaluate the tsunami risk for the Caribbean (Zahibo *et al.*, 2003), but lack the verification of the suggested paleotsunami impacts within the sedimentologic or geologic record.

Yet, without the implication of this evidence, the seismic and historic arguments contain some pitfalls, in particular regarding tsunami energy and the time period on which the assessment is based. On the other hand, the usage of sedimentologic and geomorphologic field data represents a powerful tool to investigate the tsunami risk of a region. The geologic record of paleotsunami imprints may cover the Younger Holocene with 6000 years of a high sea level and give reliable values for run up, tsunami energy, the direction of wave attack, or an average repeating time of strong events, whereas the tsunami sources and generating mechanisms may remain unknown.

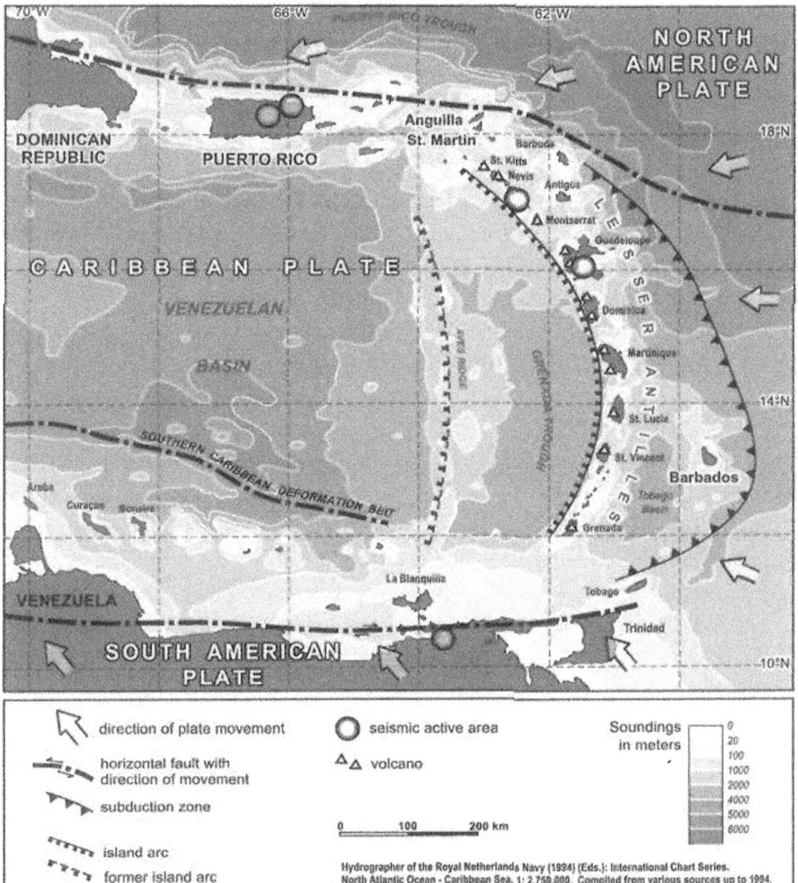


Figure 1. The Antillean Island Arc and the position of Barbados, St. Martin and Anguilla.

Within the last decade, several studies illustrated the widespread occurrence of Holocene paleotsunami deposits within the Caribbean Basin and in the wider vicinity of the Intra Americas Seas: e.g. Jones and Hunter (1992) on Grand Cayman, Mercado and McCann (1998) on Puerto Rico, Taggart *et al.* (1993) for Mona island west of Puerto Rico, Schubert (1994) at a location in northern

Venezuela, or Weiss (1979) on a cay north of that country. The most substantiate tsunami record of regional extension have been described by Scheffers (2002a, b and c) for Aruba, Curaçao and Bonaire (Netherlands Antilles). Relative and absolute age dating place them approximately in the periods 500 BP, *1500 BP and 3500 BP and the geographical distribution pinpoint to tsunami* waves approaching from the east. For the Late Pleistocene, Hearty (1997) and Hearty *et al.* (1998) have described evidence of tsunami with extreme magnitude. The extraordinary transport capability of the waves is documented by boulders with a weight of over 2000 t, which have been accumulated on the Bahaman island of Eleuthera. Moreover, chevron accumulations on the Bahamas and Exumas strengthen the assumption of Pleistocene tsunami impacts. Field surveys on the Antillean Island arc with the aim to clarify the source mechanism responsible for the tsunami sediments of the ABC–islands revealed evidence of Holocene as well as Pleistocene tsunami on the islands of St. Lucia, Grenada, Barbados, Guadeloupe, St. Martin/St. Maarten, Anguilla, Eleuthera, and Long Island of the Bahamas. In this paper, we will briefly present our field results from Barbados, St. Martin and Anguilla (Fig. 1).

## 2. Hurricane History

The Caribbean Basin is regularly affected by hurricanes, which move predominantly from the open Atlantic westward and curve to the north after passing the Antillean island arc. The islands in the southern Caribbean suffer less from these impacts, as they are situated at the southern fringe of the hurricane belt, but several strong hurricanes can hit Barbados within a century's time (Mah and Stearn, *1986*; Perry, *2001*, see also Fig. 2). The central and northern islands as well as the area north of the Larger Antilles are affected much more often, sometimes more than twice a year with hurricane categories of 3 and more (Fig. 3).

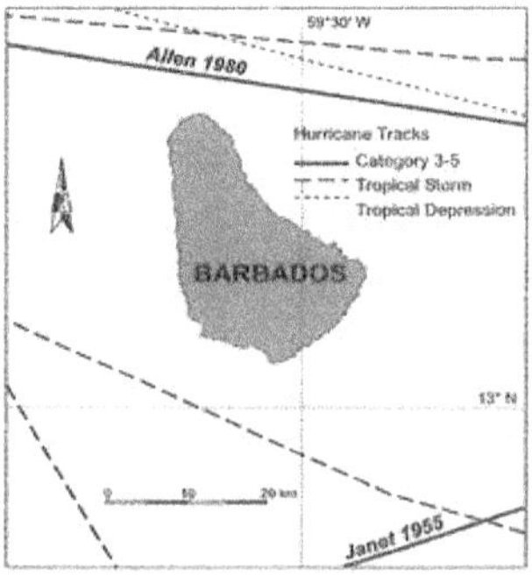


Figure 2. Tropical storms and hurricanes passing Barbados during the last 150 years (NOAA, 2001).

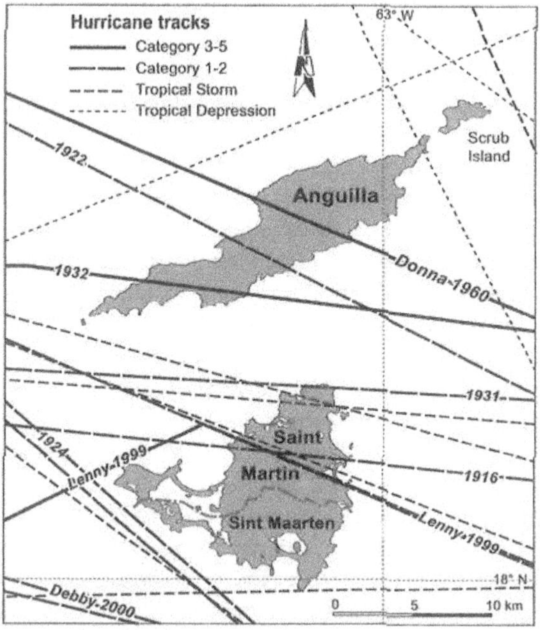


Figure 3. Tropical storms and hurricanes passing St. Martin and Anguilla during the last 150 years (NOAA, 2001).

## 3. Remarks on littoral boulder transport

The dislocation of littoral boulders due to extreme events is an example of how local legends and facts are often difficult to distinguish. Worldwide, the movement of boulders by hurricane wave has only rarely been observed directly.

According to the literature the most extreme example is a 200 t boulder, said to be transported in 1912 for about 50 m horizontally on a platform near Sydney (Süssmilch, 1912, see also Bryant, 2001). In contrast, that boulder did not move during the strongest storm (1974) during the last decades in that region. In congruence, the few empirical studies on boulder movement (Oak, 1984) confirm that the transport capacity of storm/hurricane waves is restricted to boulder weights of several 100 kg. That does not exclude that singular waves (e.g. freak waves) may transport or overturn occasionally larger boulders, but these scarce events are remarkably different compared to the sedimentary traces of tsunami. Here, the accumulation of extended boulder ridges, boulder assemblages or ramparts is a common geomorphic feature and usually occurring well above the range of even strong storm events. Moreover, the setting of the boulders into long and curved ridges, piled up with good imbrication is not characteristic for storm surf. The restricted transport capacity of storm waves is the reason why coastal engineers use artificial tetrapods and large natural blocks with weights of 4–6 tons as rip rap for coastal protection even in very stormy regions. In addition, the supratidal spray belt along rocky shorelines, which is only reached by strong waves during extreme storms mostly lack any coarse debris although thousands of rock pools constitute excellent sediment traps. Soil development, age of vegetation and weathering status often prove that no significant geomorphologic traces have been left by hurricanes along rocky shorelines for the last centuries and descriptions of hurricane events (compare Fletcher *et al.*,1995) do not mention any boulder deposits.

Nevertheless, tsunami researchers are frequently confronted with the question "storm or tsunami transport?" not only from the public, but also within the wider scientific community. This controversial view is illustrated by two studies describing extraordinary large boulders with weights exceeding 2000 t (at Tuamotu by Bourrouilh-Le-Jan and Talandier, 1985, or Eleuthera/Bahamas by Hearty, 1997), where the authors ascribe "superstorms" for the depositional process, although in both papers a tsunamigenic origin is discussed briefly. However, there are some objective criteria to distinguish storms and tsunami: Boulders >10 t dislocated against gravity have been reported worldwide only as extreme singularities, even after the passage of very strong storms and hurricanes.

During our own field research we only noticed two cases, where storm waves moved boulders with a weight of several tons for a short distance: Blocks of 2–3 t have been tumbled during a century storm with surf wave heights up to 9 m at Mallorca (Spain) in 2001 (Bartel and Kelletat, 2003) and a 7 t–boulder uplifted about 1 m and transported 3–5 m by "Lenny", a category 4 hurricane with surf wave heights of 6 m on Klein Bonaire on the Leeward Netherlands Antilles in 1999.

Besides, the argumentation based on field evidence is supported by theoretical equations calculating the necessary wave height to overturn a boulder of a specific shape and weight (see Nott, 1997, 2003 a,b). According to that, a joint bound cube with the dimensions 2 x 2 x 2 m (20 t) would need storm wave heights of about 16 m just to be overturned, cuboids require a wave height of 32 m, and as a platy-shaped boulder more than 40 m. These calculated wave heights will never occur in all littoral environments worldwide. Another very strong argument against a storm-induced deposition can be derived from the results of absolute age dating: The absolute age data show well defined clusters for the time period of the last 6000 years separated by gaps with no data over spanning 1000 years or more. This clearly pinpoints to a high-magnitude, but low-frequency event and to a tsunamig enic origin.

## 4. *How to date a tsunami event?*

Relative age indicators like the age of the vegetation on or close to tsunami deposits, soil development (in particular on bimodal or sandy deposits), the degree of weathering with karstification on carbonate particles, or the forming of eulittoral (notches) or supralitoral (rock pools) features by bioerosion may yield good approximations of the time of the tsunami impact. In particular, karst and bioerosive/bioconstructive processes are useful and easy to determine in the field using a meter, as the rate of littoral bioerosion by cyanophycae and chlorophycae and grazing gastropods like *Patella* sp. or *Littorina* sp. is mostly in the order of 1–2 mm/year in all coastal environments worldwide. In cases, historical or archaeological facts facilitate the dating.

Dating a tsunami event with absolute age determination techniques as radiocarbon ($^{14}C$) or Electron Spin Resonance (ESR) requires the presence of organic material (corals, bivalves, gastropods, barnacles, and serpulids) that was transported onshore by the tsunami from the foreshore area and ultimately being killed during this process.

However, there are some pitfalls associated with the sampling of adequate material: even very fresh looking organisms from the foreshore may have been buried under sediments and being protected from coating with calcareous algae or from boring organisms. Organic crusts like calcareous algae, vermetids, barnacles, oysters etc. attached to tsunami boulders are predominantly the result of a long lasting bioconstructive and bioerosive processes. Their dating may misrepresent the time of the impact by hundreds of years as it can never be excluded that a thin crust of these organisms have been lived and killed long before the transport event.

To overcome this problem sampling and dating of several organisms, preferably different species, attached to boulders or incorporated in a boulder deposit will give the most reliable absolute data of the tsunami impact.

## 5. Observations on Barbados

Barbados, an island of approximately 430 km², located at 13° 05'N and 59 30'W may be the best investigated island in the Caribbean due to the well developed staircase of uplifted coral reef terraces. The absolute dating of the fossil reef terraces has highlighted new insights into the sea level oscillations over a longer period of the Younger Pleistocene.

Barbados represents the only remnant on a now inactive fore arc east of the Caribbean plate, parallel to the active volcanic belt of the Lesser Antilles (Fig. 4). Its geology, geodynamic situation and physiography have recently been summarized in detail by Schellmann, Radtke and Whelan (2002; see also

Tomblin, 1975; Peter and Wertbrook, 1976; Fairbridge, 1975; Dengo and Gase, 1990; or Bouysse *et al.*, 1990). The Pleistocene reef terraces cover most of the island except for the Scotland District in the east, where disturbed Tertiary sandstones and claystones are exposed (Fig. 4). Along the coastline of the Scotland district, huge boulders with weights of over 1000 t are located. Their deposition has been attributed to sliding processes on the Tertiary clays (Schellmann, Radtke and Whelan 2002). The deep incised bioerosive notches confirm that they rest at their present position during a long time span of the Holocene.

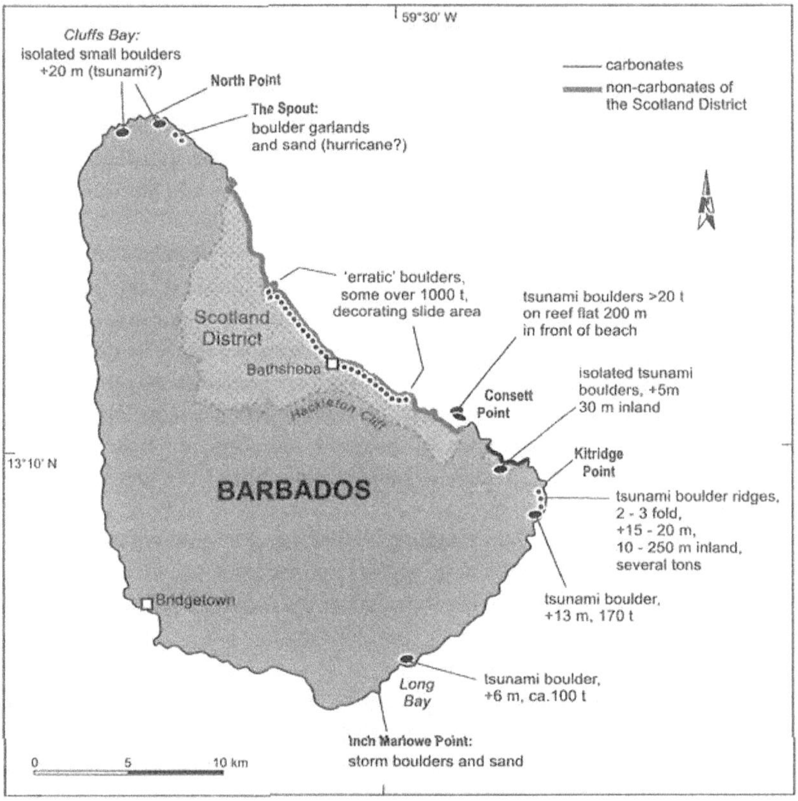


Figure 4. Barbados and the areas with significant boulder deposits along the coastline.

Coarse coastal sediments of unusual dimensions and at high elevations are situated at Palmetto Bay, and here, the depositional process can be attributed unambiguously to a tsunami impact approaching from the Atlantic Ocean. The most impressive sedimentary trace is a single large boulder of ~170 t (6 x 4 x 3 m,

Fig. 5) accompanied by some smaller ones. The boulders rest on the last interglacial uplifted reef terrace at 13 m asl and 30 m distant from the cliff edge. The growth form of coral in the boulder (*Acropora palmata, Siderastrea* sp. and others) indicates that the boulder has been overturned, when entrained in the wave. Moreover, the coral assemblage reveal that it has been broken off the cliff edge and due to the impact of the depositional process it was broken into five pieces. Relative age indicators confine the time period of the tsunami impact. Slow growing and several decade old shrubs have been settled on the boulder, indicating that the boulder has not been moved during this time span, which excludes all of the strong hurricanes of the 20th century. The flat upper surface of the block shows well developed signs of weathering by karst solution, whereas the boulder has protected the fossil reef surface from karst solution, so that at present a karst table with a height of 10 to 40 cm forms the foundation for the block. As the rate of terrestrial karst solution is in the order of approximately 0.1 mm/year (Pfeffer, 1978), it can be derived that the tsunami responsible for the deposition occurred several thousand years ago. Furthermore, the run up of the paleotsunami must have reached at least 16 m in height (which is the height of the modern cliff plus the thickness of the boulder ). Scattered smaller boulders are accumulated in the vegetation about 500 m to the NE, but more illustrative of the tsunami magnitude are two parallel boulder ridges 600 m in length (Figs. 6 and 7). The seaward ridge is located 40 m distant to the cliff edge at a height of 14 m asl, the inland ridge is deposited further inland (80 m) at 16 m asl. The majority of the boulders weight 2 to 5 t, but occasionally reaching 10 t. Seaward dipping imbrication are common as well as unstable positions. Smaller ridges with excellent imbrication run perpendicular to the coastline — they obviously have been piled up along a tsunami wave track.

Figure 5. A 170 t–tsunami boulder at 13 m asl, turned upside down and resting on a small karst pedestal.

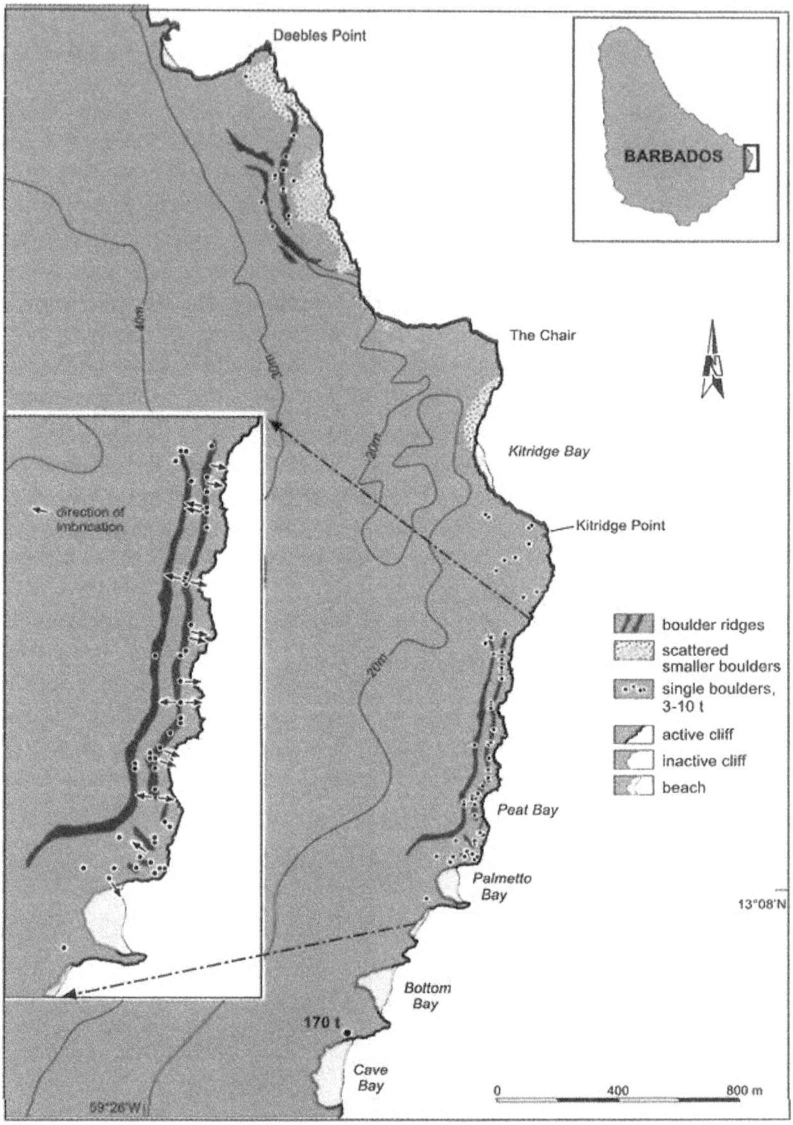


Figure 6. Sketch map of the tsunami boulder ridges along the east cape of Barbados.

Overall, it is difficult to differentiate the weathering status of the two ridges in order to decide whether they stem from one tsunami event (with two waves), or from two different impacts. The weathering of the inland ridge is taking place in a less salty, but more humid microclimate, so that it is difficult to determine their relative ages. More clarifying is the vegetation development. The seaward ridge acts as a windbreaker in this strong trade–wind exposed environment close to the surf zone, so that after its accumulation higher developed plants (shrubs, bushes) could settle, whereas the seaward front of the ridge is still bare and well exposed (Fig. 8). The composition of the boulders clearly indicate their former origin, as coral species like *Acropora palmata* is typical for the older reef crest, which today is exposed on top of the cliff at about 10 to 15 m asl.. Boulder ridges at this altitude and distance from the sea containing about 10,000 t of large boulders would require strong tsunami for their construction (compare also Nott, 2003), in particular since the Holocene reef flat extend several hundred meter into the sea and will greatly diminish the wave energy from the open ocean along this part of the coastline. At Deebles Point (see Fig. 6), wide-curved ridge forms are deposited at 150 m and 250 m from the cliff at heights of 19 m asl. These ridges are not as sharply marked as near Palmetto Bay, but they also point to at least two, maybe three, tsunami events. The tsunami evidence around the eastern Cape of Barbados point to a source in the open Atlantic ocean.

Figure 7. Oblique aerial photograph of two boulder ridges situated on top of a cliff about 15–16 m asl.

Further north near Consett Point, large boulders (>100 t) lie scattered over the outer reef flat, evidently being broken off the Holocene reef (Fig. 9). Undoubtedly, they have been tilted and dislocated. This event occurred at least several hundred years ago, as revealed by deep horizontal bioerosive notching of the largest of these boulders. The deep incised notches proof that they have not been moved or tilted by hurricanes during the last several hundreds of years, despite their rather open exposure close to the edge of the outer reef.

Figure 8. The seaward ridge north of Palmetto Bay with tsunami boulders of 1–10 tons.

Figure 9: Large tsunami boulders on the reef flat north of Consett Point.

Whether the reason for the Barbados tsunami were submarine slides along the continental slope outside the island arc or volcanic events on the other side of the Atlantic (Canary islands, Cape Verde islands) remains an open question, but it is highly unlikely, that geotectonic events in the deeper ocean or at the mid-Atlantic ridge are responsible, since seismic events during sea floor spreading are extremely rare and if, very small. Likewise sediment slides on the mid-Atlantic slopes is unlikely due to the lack of sediments.

Beta Analytics dated three samples of *Strombus* sp. Drom the seaward ridge, N of Palmetto Bay, with the radiocarbon method. Two samples date to 1550 ± 70 and 1360 ± 60 YBP, respectively (conventional ages — which is in agreement with the weathering state, vegetation, soil, and notch development, etc.), and one yield an age of 4500 ± 50 BP, which corresponds with the pedestal height of the 170 t boulder. Since these data indicate two different events, it is certain that the two parallel ridges near Palmetto Bay are not the result of two waves during the same tsunami, but derive from two separate tsunami occurring at different times during the Younger Holocene. Of course, three dated samples are not sufficient to understand the tsunami history of Barbados, but they are in good agreement with all field evidences.

## 6. Field evidence of tsunami impacts on St. Martin

At the northern end of the Antillean Island Arc (Fig. 1), St. Barthélemy and St. Martin/St. Maarten rise to several 100 m asl with a steep and hilly relief. The basement constitutes of volcanic rock of Younger Tertiary age. Whereas on St. Barthélemy no signs of tsunami boulders could be detected during an aerial survey, St. Maarten (Netherlands Antilles) exhibits large boulders on top of low cliffs around the westernmost peninsula (Terres Basses), which is built by carbonate rocks (Fig. 10). The size of the boulders at Pointe du Canonnier may reach about 10 t and their height above sea level up to 6 m.

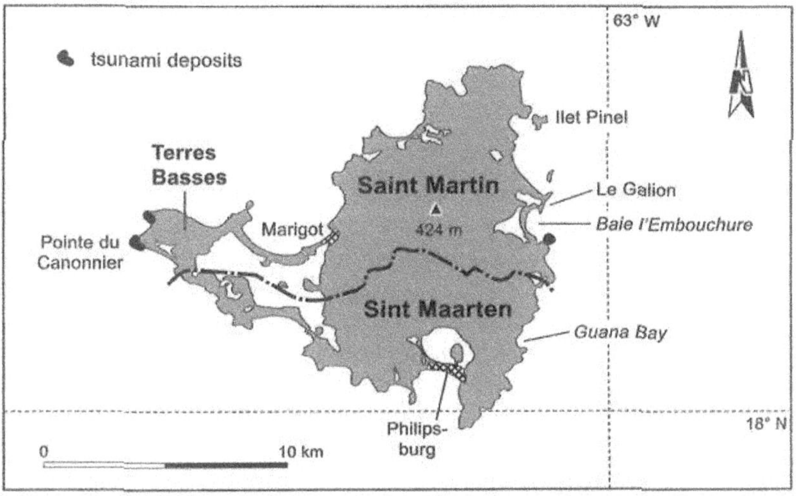


Figure 10. Saint Martin/Sint Maarten exhibiting only very few tsunami relics.

Nevertheless, compared with the geomorphic tsunami field evidence on other Caribbean islands, this may point to a tsunami impact, but the evidence is not undisputable. The coastline of the French part of the island (St. Martin) expose dark volcanic rocks and numerous woolsack boulders, which have been weathered and rounded by chemical processes *in situ*. However, incorporated carbonate boulders from the reefs along the coastlines were missing. The reason for the absence of Holocene tsunami boulders on non-carbonate islands (as on Grenada or St. Lucia in the south) remains an unsolved question.

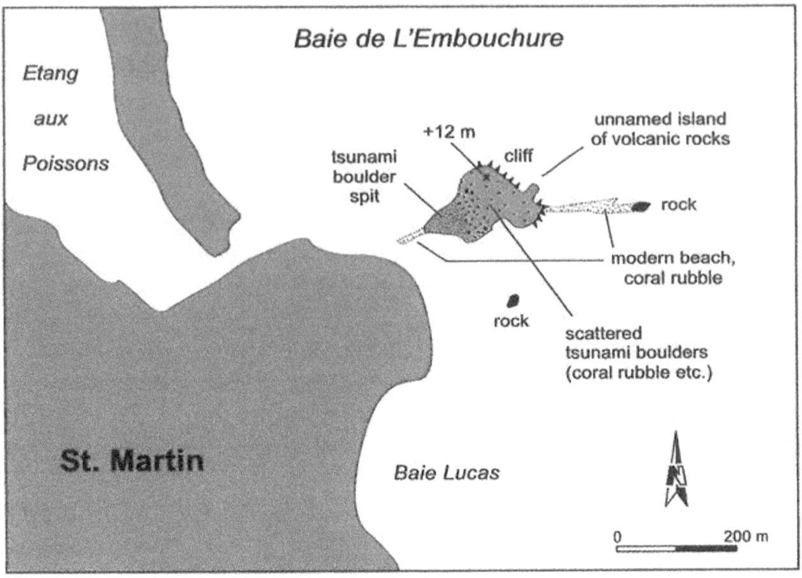


Figure 11. Tsunami deposits at the leeward side of a small island at eastern St. Martin.

Yet, extended coarse debris from a single and strong wave event is accumulated at the western spit of a small unnamed island south of Baie de l'Embouchure at the east facing coastline (Fig. 11). The deposit consists of coral fragments of in cases several 100 kg in weight and is distributed in form of two ridges, which converge at the western end and embanking a slight depression. The deposits reach 3 m asl at the southern flank and 2.7 m asl at the northern one. Strong storm events have reworked them on both sides and exposed the inner part of the deposit, which is characterized by coarse sand and broken shell under a 30–50 cm sand free upper layer. The spatial distribution suggests, that these coral debris (at least several 10,000 t in weight) were deposited by an event approaching from the east at the most leeward point of the islet in shelter of a steep, 12 m high and east facing cliff. Here, most of the debris accumulated at + 3 m asl (Fig. 12).

Evidently even the strongest past hurricanes did not impact this location, as no fresh coral rubble can be found on the dark weathered surface of these deposits. Moreover, old and now dead wind shaped trees had settled on the deposits indicating a higher age of the deposits. One sample of *Acropora palmata* (+7 m) were dated by 5 radiocarbon data and yielded a conventional age of about 500 ± 50 BP. Future tsunami research should focus on the lagoons of St. Martin, which certainly have preserved a longer and more differentiated tsunami history in their sedimentary archives.

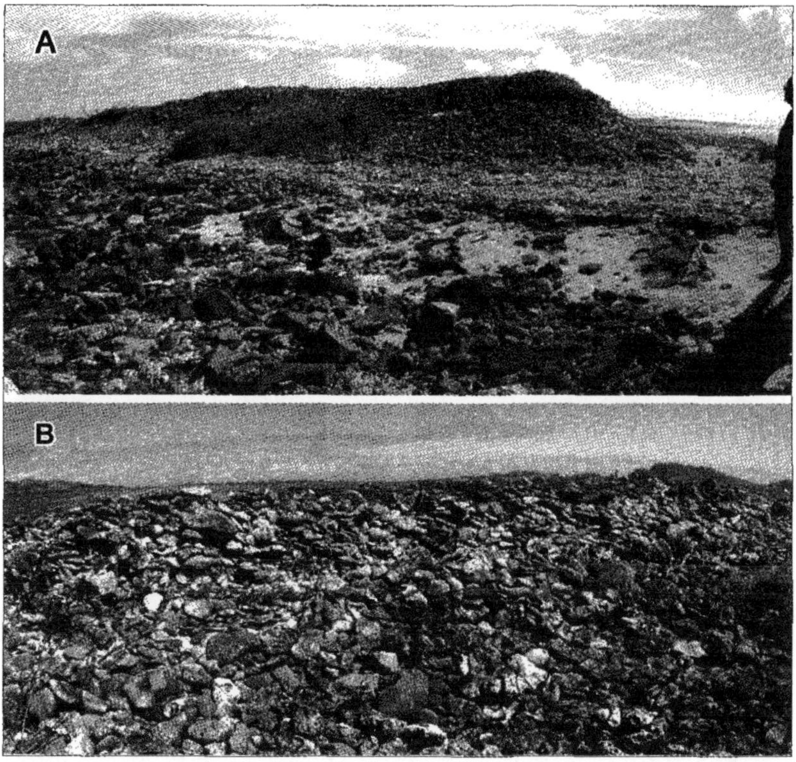


Figure 12. A: The spit like tsunami deposits at the westernmost point of the islet in Fig. 11 with dark weathered coral boulders up to +3 m asl. B: Closer view on the coral debris of the uppermost strata of the tsunami deposits.

## 7. Observations on Anguilla island

In a short distance to St. Martin, separated by a shallow channel (20–25 m), the Island of Anguilla extends elongated from WSW to ENE for about 22 km (Fig. 13). Its geology is characterized by rather flat lying carbonates from the Younger Tertiary with the greatest heights raising to + 60 m asl.. Occasionally a small belt of Younger Pleistocene eolianite may follow the contours of the coastline. In contrast to the non-carbonate islands in the south, Anguilla exhibits large tsunami boulders nearly all along its east facing shorelines, with a clear dominance in the eastern part of the island (Fig. 13). Impressive and nearly undisturbed tsunami deposits are located on Scrub island just east of Anguilla.

In accordance with findings from other islands in the IAS (St. Martin, Barbados, Guadeloupe, Bahamas), the distribution of the debris strongly suggests a tsunami approach from the open Atlantic. The north facing shorelines of East Anguilla are rather steep inclined with several meter high cliffs. Nevertheless, large boulders (20–50 t) are accumulated on top of them (Fig. 14). The boulder material consists of the Younger Tertiary limestone, which exhibits some gentle folding in this area. Structural weakness and joints within the rock strata has enabled the waves to break off large blocks, evidently from sublittoral positions up to at least 10 m asl, as can be derived by attached marine or littoral organisms. The lack of karren karstification on the lower slope sections is due to mechanical break off by huge waves. Imbrication as well as unstable setting of boulders is common (Fig. 15).

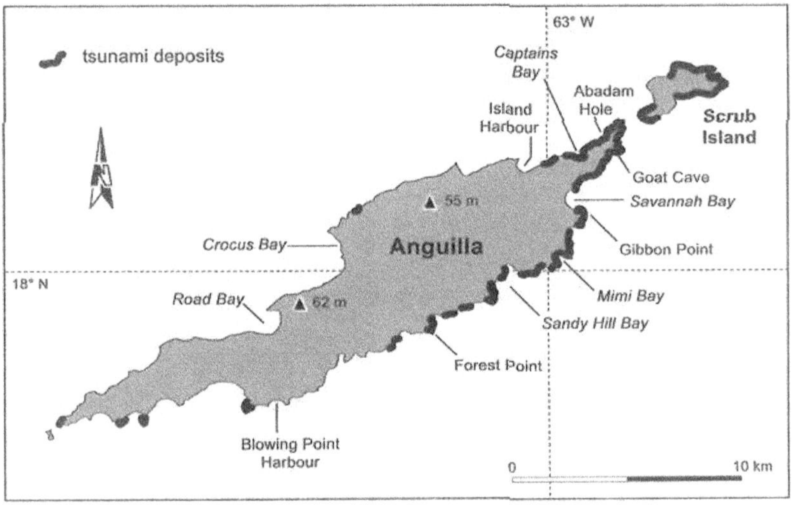


Figure 13. Anguilla and Scrub Island with extended coarse tsunami deposits.

Figure 14. Tsunami boulder fields at Abadam Hole on top of a terrace at about +12 m, separated from the cliff by a debris free belt.

Figure 15. Examples of unstable, vertical and imbricated boulders (each 5–20 t in weight) from the east-exposed southern shorelines of Anguilla.

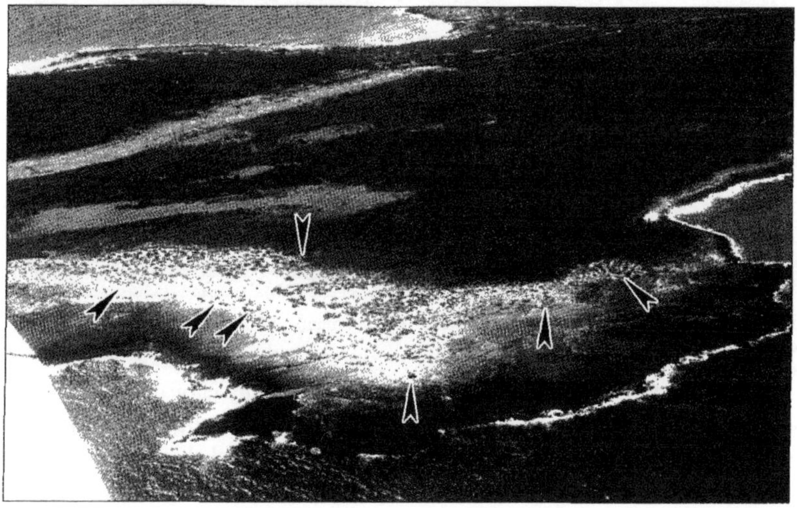

Figure 16. Oblique aerial photograph of the east coast of Scrub Island. Large boulders (>30 tons) are marked by arrows.

The size of the boulders is not restricted by wave energy, but evidently by rock structure. The Scrub Island (Fig. 16) coastal stretch from Windward Point to near Island Harbor inherits the most distinct remnants of the tsunami impact. In the western part of Abadam Hole, mining has exposed the lower sections of the tsunami deposits, which constitutes of sand, broken shell and numerous coral fragments to a height of + 15 m asl (Fig. 17). Here, the boulders become smaller in a denser vegetation. The foreshore area is deep (>12 m), so that strong waves are necessary to pick up sand or coral, which eliminates storms or hurricanes as a mechanism for boulder transport, as in congruence does the weight and height above sea level of the boulder deposits. Along the northern coast of Anguilla, the last significant boulder assemblage can be found in a small horseshoe embayment northeast of Crocus Bay.

**Figure 17. Bimodal tsunami deposits at Abadam Hole at around +17 m asl show boulders, sand and coral debris.**

The southern shores of Anguilla are lower with many curved beaches, barriers and lagoons, and the headlands are mostly gently submerging into the sea. Several headlands are exposed to the east or have no shelter from easterly waves, as the main island's direction is trending ENE. Therefore, the amount of large boulders accumulated in ridges or fields as well as single deposits is substantial. Overall, the landscape show a great resemblance between the headlands: The lower sections of the rocky coastal slopes exhibit a belt of bioerosive rock pools, either in the Younger Tertiary limestone or the superimposed Younger Pleistocene eolianites. This belt as well as a higher one, both 10 to 30 m wide and at the highest places around + 5 m asl, contain only smaller clasts in the deep rock pools, deposited evidently by the last stronger storms (hurricanes), but no significant large boulders. They have been deposited further inland, with considerable dimensions in front of the vegetation (shrubs, wind-shaped trees), but diminish in size with more distance to the coastline. In cases as e.g. Mimi Bay and Sandy Hill Bay, a field of coral clasts behind a first line of older vegetation is bare of any plants (Fig. 18).

**Figure 18. Headland south of Sandy Hill Bay with a debris ridge as well as large boulders in the vegetation at around +10–14 m asl.**

The largest boulders along this coastal section are nearly 100 tons in weight and were broken off from the eolianite near Gibbon Point. Only occasionally bimodal tsunami deposits with a higher amount of sand between the boulders are preserved (Fig. 17). Again imbrication is very typical as well as unstable or even vertical positions of boulders (Fig.15). Many boulders show tilted rock pools and on several blocks notches from the eulitoral are visible. Notably on these more gentle coastal slopes is a belt of dead shrubs and trees. Their stem diameter is considerable with regard to the salty environment and suggests old ages. The wind-shaped growth form is clearly perceptible due to the strong and steady trade wind impact. All plants are still rooted in fissures of the limestone (Fig. 19). They represent a time of mature and good growth conditions in front of the recent living and younger vegetation and seem to be killed by a strong event from the seaside at least decades ago. Only at one place (west of Sandy Hill Bay), a double boulder ridge may suggest more than one tsunami event, but may represent two waves of the same tsunami, as well. Here, within the supratidal as well as the subtidal fragments of *Acropora palmata* >1 ton is accumulated. Along the south facing shoreline of Anguilla (from Mimi Bay to Blowing Point Harbor) elongated bars consisting of a framework of very coarse *Acropora palmata* and other coral debris are situated. Their length is about 100 m with the maximum height asl of 2 m reached in the west of Blowing Point Harbor. These bars rest on the living reef flat and are separated from the coastline by 50 to 130 m of shallow water. Parts of them are cemented, and the highest parts show dark grey to black weathering and coating by Cyanophyceae and Chlorophyceae. We presume that they have been formed by a tsunami along the outer parts of the fringing reef, but have partly been transformed by subsequent hurricane waves.

Figure 19. Rooted, but dead trees and shrubs in front of tsunami boulders west of Mimi Bay.

Assuming a tsunami approach from the east, we can derive the run up by the depositional height of the boulders. At the southern coastline the highest largest boulders are deposited at +12 m asl and at about +15 m along the northern one. However, this may not represent the maximum run up due to the absence of higher relief. This also may limit the occurrence of backwash features. Therefore, we consider 16 m asl as an absolute minimum value for the tsunami run up at Anguilla.

East of Sandy Hill Bay (Fig.19), an interesting sequence of features may enable us to determine a more exact minimum age of the tsunami impact: Old shrub vegetation has settled on large tsunami boulders and subsequent to the deposition a belt of trees has developed seaward of the boulders, but today all plants are dead. The oldest trees have an estimated age of 200 years based on counting from tree rings. A sample from the youngest outer parts of the stem in Fig. 19 was dated with radiocarbon and yielded an age of 1860–1900 AD. Consequently the trees started to grow (some time after the tsunami event) at about 1700 AD or even earlier, and may have been killed by the Virgin Islands Tsunami in 1867. We can exclude any significant sea level oscillation was responsible as beachrocks, coral reefs as well as the rock pool belt and notches show no dislocations for a time span of many 100 years for Anguilla. However, for the determination of the tsunami age, we must extend at least the time period of growth and settlement, which will cover several decades or centuries.

Two radiocarbon ages from *Acropora palmata* samples collected at +8 m and + 10 m respectively at the northern and southern coast were dated to 500 ± 40 BP and 610 ± 50 BP and yield maximum ages for the tsunami event. These two data are corresponding with the absolute age determination for St. Martin (500 ± 50 BP). Two samples of organic material (tube worms and barnacles) directly attached to large tsunami boulders of the north and south coast have been dated also. They may provide rather exact data of their sudden death through their dislocation out of the littoral environment, but some pitfalls have to be considered. Kelletat and Schellmann (2002) pointed out that samples from the same crust at a boulder may exhibit age differences of centuries due to the fact that they are the result of longtime bioerosion and bioconstruction at the same spot. The barnacles at a large boulder from Abadam Hole at +11 m asl have been dated to a conventional age of 1880 ± 40 BP with AMS technique, and tube worms at a large boulder from the south coast at +10 m asl to 1600 ± 70 BP. Two other samples from large boulders (12 tons and 32 tons) in a vegetated area, dated with AMS technique, gave only modern ages because of contamination of the very thin tube worm sceletons, coated and bored by endolithic algae. It is highly likely that both islands (Anguilla and St. Martin) exhibit signs of the same younger tsunami, but that an even stronger impact has dislocated huge boulders more than 1000 years ago on Anguilla Island. Again, the many lagoons of Anguilla may hide a more complete tsunami history of this area at the northern end of the Antillean Island Arc.

## 8. Conclusions

Table 1 summarizes the Holocene paleotsunami impacts based on geomorphic and sedimentary evidences and their occurrence time revealed by radiocarbon and ESR–dating along a trajectory of more than 1500 km on the Antillean Island Arc to the northern Bahamas. But still, the spatial extent of each event may only come to light with the conduction of more intensified field surveys accompanied by dating. With the increase of absolute age data the tsunami history of the Caribbean may be understood in more detail, if e.g. more tsunami occurred during a certain time interval, which hitherto cannot be separated by the few data available. In particular, the event of 450–500 BP should be investigated thoroughly considering the implications of the reservoir correction with the radiocarbon method.

Since tsunami field research is somehow incidental (and restricted to disputable landscape marks e.g. boulder deposits or drill holes with remarkable stratigraphy), and tsunami imprints may be overseen, have been eroded, or tsunami never have left any depositional or geomorphologic relics, the number of them certainly is much higher than any catalogue will compile. But it has to be emphasized that tsunami impacts documented by field evidence show a significant greater run up and magnitude in terms of sediment transport or coastal destruction than formerly known from historical sources.

Table 1. Absolute age determination data of Holocene tsunami events in the wider Caribbean.

| **Region** | **No. of absolute data** [conventional $^{14}C$] | **Date of most probable tsunami event** [BP] |
|---|---|---|
| Aruba | 5<br>4<br>1 | 3500<br>1500<br>500 |
| Curaçao | 4<br>20<br>16 | 3500<br>1500<br>500 |
| Bonaire | 9 | 500 |
| Barbados | 1<br>2 | 4500<br>1400 |
| Guadeloupe | 5 | 2400 |
| St. Martin | 5 | 500 |
| Anguilla | 3<br>3 | 1500<br>500 |
| Eleuthera (Bahamas) | 4<br>4 | 3000<br>450-500 |
| Long Island (Bahamas) | 2<br>1<br>3 | 300<br>1400<br>450-500 |

It can be derived that risk analyses, and calculation of tsunami magnitude, inundation and run up values based solely upon historical sources underestimate the real tsunami risk of the region by a factor of five to ten and the so far neglected coasts facing the open Atlantic Ocean are particularly endangered. The run up values in Fig.20 represent conservative estimations: they mark the uppermost large boulders plus one third of the boulders height, which is the necessary water column to move the boulder in a tsunami wave. From modern tsunami events or the example of Cyprus (Kelletat and Schellmann, 2002) we have learnt, that deposits may only be accumulated in the lowermost part of a tsunami wave. So the real run up may be apparently higher and in the order of 20–30 m suggested by field data, which show minimum run ups of 12 m to 20 m. In comparison, with run up values described in historical catalogues we should be aware of the fact that the real tsunami risk in the Intra Americas Seas for the population and in particular for all infrastructure and investments in the local economies is certainly higher than hitherto noticed.

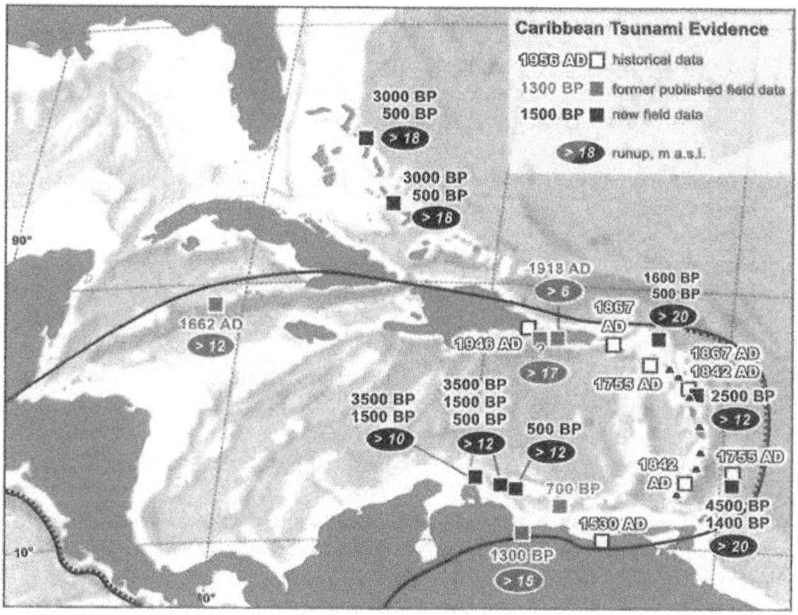


Figure 20. A compilation of absolute tsunami data for the wider Caribbean from historical sources, former published investigations and new field data.

More and systematic field investigations on sedimentologic and geomorphologic tsunami imprints and much more absolute data are needed to formulate a realistic scenario for the Caribbean and IAS tsunami risk. Investigations of run up values and the spatial extent of single strong tsunami events will allow to estimate tsunami energies much better. Moreover, both insights will enable us to indicate tsunami sources and origins along the Caribbean plate boundaries or even further more accurate. Finally, the sedimentary traces of tsunami impacts illustrate the destructive forces of tsunami very spectacular. We believe, that an assemblage of huge boulders dislocated by a single wave will convince the coastal population and decision makers and emphasize the need for warning programs more than geophysical modeling or theoretical source energy calculations.

**Acknowledgements**

Deutsche Forschungsgemeinschaft has supported our field work and dating with substantial grants. We also thank Sander, Hanni and Peter Scheffers for their assistance in the field as well as Beta Analytics for the prompt delivery of radiocarbon data.

**References**

1. Bartel, P. and D. Kelletat, Erster Nachweis holozäner Tsunamis im westlichen Mittelmeergebiet (Mallorca, Spanien) mit einem Vergleich von Tsunami- und Sturmwellenwirkungen auf Festgesteinsküsten, Berichte Forschungs- und Technologiezentrum Westküste der Universität Kiel, **28**, 93–107 (2003).
2. Bourrouilh-Le-Jan, F.G. and J. Talandier, Sédimentation et fracturation de haute énergie en milieu récifal, tsunamis, ouragans et cyclones et leurs effets sur la sédimentologie et la géomorphologie d'un atoll: Motu et Hoa, à Rangiroa, Tuamotu, Pacifique SE, Marine Geology, **67**, 263–333 (1985).
3. Bouysse, P., D. Westercamp and P. Andreieff, The Lesser Antilles Island Arc, in: Moore, J.C. and A. Mascle (eds.), Proceedings ODP, Scientific Results, **110**, 29–44 (1990).
4. Bryant, E., Tsunami. The Underrated Hazard, Cambridge, University Press, 320 pp (2001).
5. Dengo, G. and J.E. Gase (eds.), The Caribbean region, The Geology of North America, Vol. H, Geological Society of America, Boulder, Colorado (1990).
6. Fairbridge, R.W., Windward Islands, The Encyclopedia of World Regional Geology, Part I: Western Hemisphere. Stroudberg, Dowden, Hutchinson and Ross, 667 (1975).
7. Fletcher, C.H., B.M. Richmond, G.M. Barnes and T.A. Schroeder, Marine Flooding on the Coast of Kauai during Hurricane Iniki: Hindcasting Components and Delineating Washover, Journal of Coastal Research, **11**(1), 188–204 (1995).
8. Hearty, J.P., Boulder deposits from large waves during the last interglaciation in North Eleuthera Island, Bahamas, Quaternary Research, **48**(3), 326–338 (1997).
9. Hearty, J.P., A.C. Neumann and D.S. Kaufman, Chevron Ridges and Runup Deposits in the Bahamas from Storms late in Oxygen-Isotope-Substage 5e, Quaternary Research, **50**, 309–322 (1998).
10. Jones, B. and I.G. Hunter, Very large boulders on the coast of Grand Cayman, the effects of giant waves on rocky shorelines, Journal of Coastal Research, **8**, 763–774 (1992).
11. Kelletat, D. and G. Schellmann, Tsunamis in Cyprus: Field Evidences and $^{14}$C-Dating Results, Zeitschrift für Geomorphologie, NF **46**(1), 19–34 (2000).

12. Lander, J. and L.S. Whiteside, Caribbean tsunami, an initial history, Mayaguez Tsunami Workshop, June 11-13, Puerto Rico (1997).
13. Lander, J. F., L.S. Whiteside and P. Lockridge, A brief history of tsunamis in the Caribbean Sea, Science of Tsunami Hazards **20**(2), 57–94 (2002).
14. Mah, A.J. and C.W. Stearn, The effect of hurricane Allen on the Bellair fringing reef, Barbados, Coral Reefs, **4**, 169–176 (1986).
15. Mercado, A. and W. McCann, Numerical Simulation of the 1918 Puerto Rico Tsunami, Natural Hazards, **18**, 57–76 (1998).
16. NOAA (National Oceanic and Atmospheric Administration, Hurricane index, URL http://hurricane.csc.noaa.gov/hurricanes/index.htm (2001).
17. Nott, J., Extremely high wave deposits inside the Great Barrier Reef, Australia: determining the cause, tsunami or tropical cyclone, Marine Geology, **141**, 193–207 (1997).
18. Nott, J., Waves, coastal boulder deposits and the importance of the pre-transport setting, Earth and Planetary Science Letters, **210**, 269–276 (2003a).
19. Nott, J., Tsunami or Storm waves? Determining the Origin of a Spectacular Field of Wave emplaced Boulders using Numerical Storm Surge and Wave Models and Hydrodynamic Transport Equations, Journal of Coastal Research, **19**(2), 348–356 (2003b).
20. Oak, H.L, The boulder beach: A fundamentally distinct sedimentary assemblage, Association of American Geographers Annals, **74**, 71–82 (1984).
21. Perry, C.T., Storm induced coral rubble deposition: Pleistocene records of natural reef disturbance and community response, Coral Reefs, **20**, 171–183 (2001).
22. Peter, G. and G.K. Wertbrook, Tectonics of the southwestern North Atlantic and Barbados Ridge Complex, Bulletin Association of American Petroleum Geologists., **60**, 1078–1106 (1976).
23. Pfeffer, K.-H., Karstmorphologie, Erträge der Forschung, Darmstadt, Wissenschaftliche Buchgesellschaft (1978).
24. Scheffers, A., Paleo-Tsunamis in the Caribbean. Field Evidences and Datings from Aruba, Curaçao and Bonaire, Essener Geographische Arbeiten, **33**, 181 pp. (2002a)
25. Scheffers, A., Paleotsunami Evidences from Boulder Deposits on Aruba, Curaçao and Bonaire, Science of Tsunami Hazards, **20**(1), 26–37 (2002b).
26. Scheffers, A., Landschaftsspuren und Zeitstellung holozäner Tsunamis auf den Niederländischen Antillen (Aruba, Curaçao und Bonaire), Berichte Forschungs- und Technologiezentrum Westküste der Universität Kiel, Büsum, **28**, 75–92 (2002c).
27. Schellmann, G., U. Radtke and F. Whelan (Eds), The coral reef terraces of Barbados — a guide, Exkursionsführer for 4th Annual Meeting of IGCP Project 437 Coastal Environmental Change during Sea Level Highstands: A Global Synthesis with Implications for the Management of Future Coastal Change, International conference on Quaternary Sea Level Change, Barbados, W.I., 26. Oct.–2. Nov., Bamberg, 120 pp (2002).

28. Schubert, C., Tsunami in Venezuela. Some Observations on their Occurrence, in: Finkl., Ch. F. (ed.), Coastal Hazards. Perceptions, Susceptibility and Mitigation. Journal of Coastal Research, Special Issue, **12**, 189–195 (1994).
29. Sussmilch, C. A., Note on some recent erosion on Bondi Beach, Journal of the Proceedings of the Royal Society of New South Wales **46**, 155–158 (1912).
30. Taggart, B.E., J. Lundberg, J.L. Carew and J.E. Mylroie, Holocene reef-rock boulders on Isla de Mona, Puerto Rico, transported by a hurricane or seismic sea wave, Geological Society of America, Abstract with Programs, **25**(6), A-61 (1993).
31. Tomblin, J.F., The Lesser Antilles and Aves Ridge, in: Nairn, A.E.M. and F.G. Stehli (eds.), The Ocean Basins and Margins, Vol. 3: Gulf of Mexico and the Caribbean. Plenum Press, New York and London; 467–500 (1975).
32. Weiss, M.P., A saline lagoon on Cayo Sal, western Venezuela, Atoll Research Bulletin, **232**, 1–33 (1979).
33. Zahibo, N., E.N. Pelinovsky, A. Kurkin and A. Kozelkov, Estimation of far-field tsunami potential for the Caribbean Coast based on numerical simulation, Science of Tsunami Hazards **21** (4), 202–222 (2003).

# RECENT CARIBBEAN PROGRAMS AND STUDIES

# THE PUERTO RICO TSUNAMI WARNING AND MITIGATION PROGRAM

Aurelio Mercado-Irizarry
*Department of Marine Sciences*
*University of Puerto Rico, Mayaguez, P.R. 00681-9013*

Christa von Hillebrandt
*Puerto Rico Seismic Network*
*Geology Department*
*University of Puerto Rico, Mayaguez, P.R. 00680*

Victor Huerfano
*Puerto Rico Seismic Network*
*Geology Department*
*University of Puerto Rico, Mayaguez, P.R. 00680*

A recent compendium of the history of tsunamis in the Caribbean reports that since 1498 there have been 91 tsunamis in the Caribbean region. In fact, in the Caribbean Sea region we have all of the potential tsunami-generating sources: submarine earthquakes, subaerial or submarine landslides, and underwater explosions. Past and recent studies support the interpretation that Puerto Rico and the northern Virgin Islands sit within a broad zone of deformation between two larger plates: the North American to the north and the Caribbean to the south. Offshore faults along which major earthquakes can occur have been identified in the boundary zone of the North American and Caribbean plates. The landslide-tsunami threat for the island comes from the existence of the Puerto Rico Trench just north of the island, along which there is evidence of a large slump, with a potential volume of approximately 900-1500.

The Puerto Rico Tsunami Warning and Mitigation Program was funded by the USA Federal Emergency Management Agency and the University of Puerto Rico with the purpose of addressing the tsunami threat and concern in the island. It consisted of six tasks: 1) preparation of tsunami flood maps for the island; 2) education about this "forgotten hazard"; 3) local (Puerto Rico) and regional (Caribbean) seismic wave form analysis for rapid determination of earthquake source parameters; 4) development of tsunami warning and advisories protocols; 5) preparation of a PC-Windows based Atlantic and Caribbean Historical Tsunami Database; and 6) participation in the meetings of the USA National Tsunami Hazard Mitigation Program.

## 1. Introduction

Lander *et al.* (2002) reported that since 1498 there have been 91 “tsunamis” in the Caribbean region. Among them, Lander *et al.* have verified that twenty-seven are true, verifiable tsunamis and nine additional wave events are considered to be very likely true tsunamis. In fact, in the Caribbean Sea region we have all of the potential tsunami-generating sources: submarine earthquakes, subaerial or submarine landslides, and underwater explosions. According to

Jansma and Mattioli (2003), Puerto Rico and the northern Virgin Islands sit within a broad zone of deformation between two larger plates: the North American to the north and the Caribbean to the south. This microplate moves approximately 15 mm/yr relative to North America and 3 mm/yr relative to the Caribbean. Offshore faults along which major earthquakes can occur have been identified in the boundary zone of the North American and Caribbean plates. The landslide-tsunami threat for the island comes from the existence of the Puerto Rico Trench just north of the island, along which there is evidence of a large slump, with a potential volume of approximately 900-1500 $km^3$ (Schwab et al., 1993; Grindlay, 1998) and many other smaller ones (ten Brink and Smith, 2003).

It is not strange then that the northeastern corner of the Caribbean, the region encompassing the islands of Hispaniola, Puerto Rico, and the United States and British Virgin Islands, has experienced three destructive tsunamis in the last 136 years: 1867, 1918, and 1946. All three tsunamis affected the island of Puerto Rico, with the one in 1918 doing great damage along the west coast of the island (Reid and Taber, 1919; Mercado and McCann, 1998). This "forgotten hazard" has led state and federal agencies to sponsor a series of workshops and projects with the purpose of evaluating the magnitude of this threat. The University of Puerto Rico Sea Grant College Program and the Puerto Rico State Emergency Management Agency sponsored a Caribbean Tsunami Workshop in 1997 and a series of studies (McCann and Mercado, 1997; McCann, 1998; Grindlay, 1998; Mercado, 2001; Lynett and Liu, 2002; Mercado, 2002; Mendoza and Huérfano, this volume) that set the stage for the Puerto Rico Tsunami Warning and Mitigation Program (PRTWMP). The two years long PRTWMP was funded by the USA Federal Emergency Management Agency and the University of Puerto Rico. It consisted of six tasks: 1) preparation of tsunami flood maps for the whole island; 2) education about this "forgotten hazard" in the Caribbean (including a video, tsunami drills, workshops and the installation of tsunami warning signs along beaches); 3) local (Puerto Rico) and regional (Caribbean) seismic wave form analysis for rapid determination of earthquake source parameters; 4) development of a tsunami warning and advisories protocol for Puerto Rico and the Caribbean region; 5) preparation of a PC-Windows based Atlantic and Caribbean Historical Tsunami Database; and 6) participation in the meetings of the USA National Tsunami Hazard Mitigation Program.

It is the purpose of this presentation to summarize the most important results from each task and the impact it has had in raising the awareness about this latent coastal hazard.

## 2. Program Tasks

### *2.1 Task 1: Tsunami Flood Maps*

For the preparation of the tsunami flood maps we first had to characterize the systems of local faults which history has shown as the main threat for earthquake tsunamis. For this, data recorded by the Puerto Rico Seismic Network (PRSN) was used. This included obtaining the basic parameters that control the tectonic tsunami generation: quake moment (Mo), fault dimensions and rupture geometry. Significant progress in the understanding of the geologic and tectonic environment of the Puerto Rico region has been made during the past years. Details of the seismogenic zones are being identified and the concept of a block for Puerto Rico and the Virgin islands has been established. However, there remain a number of fundamental questions. The two most important ones are the relationship between the seismicity and the geologic structures and the stress regime in relation to the tectonic processes in the Local Zone of Puerto Rico (LZPR).

The northeastern Caribbean region, encompassing Hispaniola on the west and the Leeward Islands on the east (which defines our LZPR), was divided into 12 sub-regions according to major tectonic features, seismic distribution and stress pattern. These were labeled as: 19°N Fault Zone, Anegada Passage, Eastern Dominican Republic, Leeward Islands, Mona Passage (identified as McCann's faults), Mona Canyon, Muertos Trough, North Platform, North Puerto Rico Trench, Puerto Rico West to Southeast, Septentrional and Sombrero (see Figure 1). The seismicity associated with faults in the LZPR are currently located using Geiger's method (Lay and Wallace, 1995; Udías, 1999), that is a linearization of the travel time equation in a first order Taylor series that relates the difference between the observed and theoretical travel times to unknown adjustments in the hypocentral coordinates through the partial derivates of travel time with respect to unknowns. There are some techniques available to improve the quake locations with the goal of identifying structural details, such as the location of active planes, into well defined focus. But in general, all methods show a diffuse picture of seismicity. A very important observation is that the spatial distribution of smaller earthquakes delineates areas in which larger earthquakes are likely to occur (Kafka and Levin, 2000), but this requires a very reliable earthquake location method to delineate fault systems. Based in the hypothesis that any quake that occurs is due to the fracture along a specific fault, and that multiple and successive quakes delineate the fault plane, Huerfano (2003) proposed a complete set of faults using two methods. The first is based on the composite focal mechanism analysis while the second is based on geometrical analysis of hypocenters stratified by magnitude.

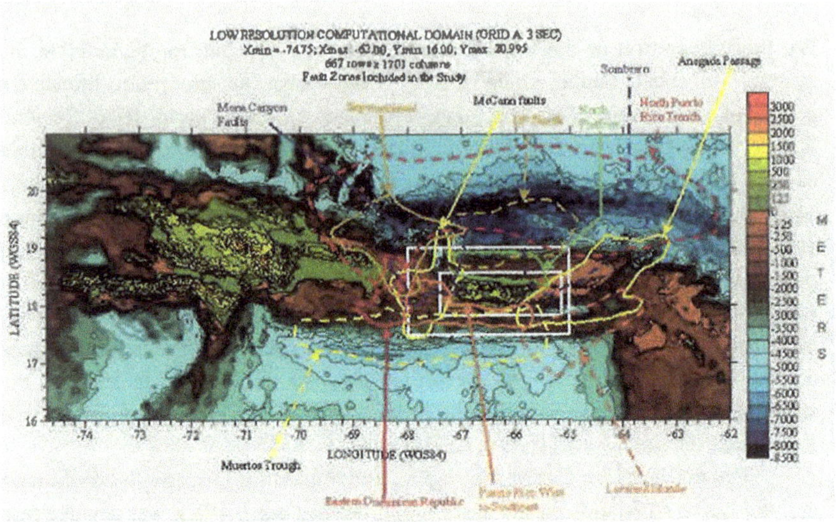


Figure 1. Computational nested grids and sub-regions used for location of hypothetical tsunamigenic earthquakes used in the tsunami flood mapping.

Earthquake location programs like PR-HYPO, the one used in the PRSN, need a realistic crustal model. Huerfano and Bataille (1994) proposed a flat homogeneous P-wave model for Puerto Rico and from Wadatti plots the Vp/Vs ratio (P wave to S wave velocity ratio) was determined. The complete PRSN seismic catalogue was recompiled with this model so that all solutions are based on the same methodology. A criterion used to determine the data quality is the network stability. Time history analysis indicates that from 1992 to present the data obtained by the PRSN remains mostly stable as evidenced by the number of readings and location errors. During the time period from January 1987 to June 2002, 4422 microseismic records from the PRSN catalogue with quality A, B and C (local scale) were selected to determine the seismic potential in the LZPR. Quality factors are based on the number of wave arrival readings, rms, hypocentral errors and azimuthal coverage (gap). These factors allow us to identify reliable solutions inside a specific region.

Based on the offshore seismic distribution of shallow earthquakes (depth less that 50 km) and mapping of known fault systems, the 12 sub-regions were proposed. Each one is defined by specific stress parameters and, in addition, each was previously analyzed by classic seismological methods like the Gutenberg and Richter b-value, Wells and Coppersmith relationships, composite stress patterns and the hypo-centering algorithm (Huerfano, 2003). The Puerto Rico Trench (PRT) was subdivided into eastern and western sections due to intrinsic differences in seismicity and tectonics. Finally, 238 spatially linear potential faults were characterized, each one generating their hypothetical earthquake. These sources were considered as local (relative to Puerto Rico) and it was found that the linear faults generate bigger tsunamis compared with the 102 spatially nonlinear or segmented faults which were also considered (Huerfano, 2003).

Based on the 340 potential faults determined under the methodology described above, a simulation was made of each one of them using the Japanese non-linear shallow water tsunami TIME model. Three nested grids were used, starting with the outer grid with a cell size of 27 arc seconds, followed by the intermediate grid with cell size of 9 arc seconds, and the inner grid with a resolution of 3 arc seconds. The region covered by Figure 1 shows the extent of the lower resolution grid, while the two rectangles show the location of the intermediate and higher resolution grids. It is in the higher resolution grid that the model is run in its non-linear mode, allowing for the computation of coastal flooding. The induced sea bottom deformation was determined for each one of the potential faults using the Mansinha and Smylie (1971) method. For nearshore bathymetry, recently acquired SHOALS bathymetry was used. For deeper shelf waters National Ocean Survey data was used, and for deep water use was made of the latest Sandwell and Smith (1997) bathymetry. For topography, use was made of the USGS Digital Elevation Model data for the island, with a resolution of 30 meters. The Japanese TIME tsunami model has been validated for Puerto Rico by comparison of runup model results with measured runup due to the 1918 Puerto Rico tsunami (Mercado and McCann, 1998). This reference can also be consulted for further details about the modeling aspects.

Adopting a standard hazard mitigation approach of preparing for the worst, the maximum flooding, irrespective of time of occurrence, at each computational cell for each scenario, was saved. Next, the maximum of the maximums (MOM) was obtained at each computational cell by going through each one of the 340 runs and picking out the maximum among all the runs. The maps prepared are based on the MOM. As in the simulation of the 1918 tsunami, a constant Manning coefficient of 0.025 was utilized. This approach is similar to the methodology followed in preparing storm surge coastal flood maps (Mercado, 1994).

The inland limits of the flood was overlaid on top of USGS quadrangle topographic maps, and recently acquired IKONOS 1-m resolution color photos, and made available through paper copies, CD's, and the Internet (http://poseidon.uprm.edu). From this site one can also download the final reports for this project. Figure 2 shows an example of the IKONOS photo for the Manati quadrangle (north coast) overlaid with the (blue) line showing the inland limit of the inundation.

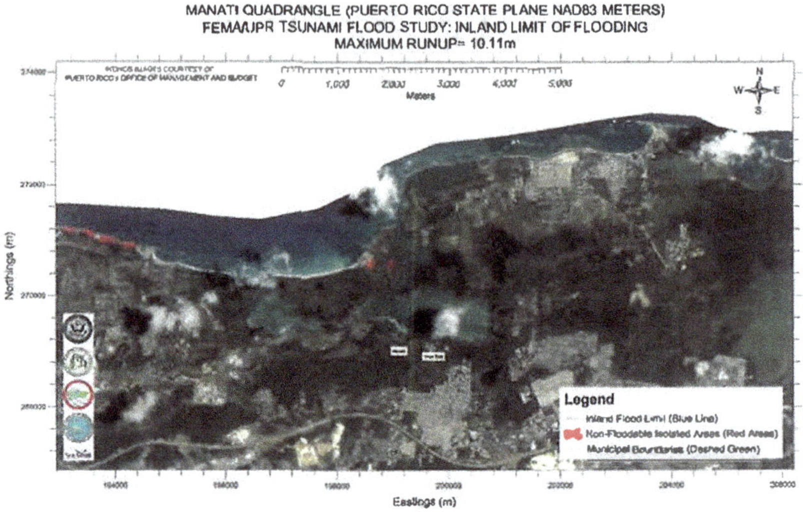


Figure 2. Sample tsunami flood map.

## *2.2 Task 2: Education/Outreach*

Together with the preparation of the tsunami flood maps (Task 1 above), the educational phase of the project is the most important one. Caribbean tsunamis (with the exception of the 1755 Lisbon earthquake) have all been local events. As such there is not much a tsunami warning system can do, and by having citizens knowledgeable about this hazard is the only way we can insure that the risk is minimized. For this purpose workshops were given throughout the island, in which representatives of various government agencies participated. This was accompanied by the distribution of a Spanish language tsunami video to all schools, government agencies, and interested citizens. Newspaper, radio and TV interviews were carried out in order to reach the widest audience possible. Tsunami drills have been carried out at public schools. Tsunami warning signs were prepared to be placed at all coastal areas which, according to the flood maps, are potentially vulnerable. For these, the then internationally accepted figure and message for tsunamis was used (see Figure 3). Finally, a WEB page was developed (http://poseidon. uprm.edu).

Figure 3. Tsunami warning sign installed at a beach.

## *2.3 Task 3: Local and Regional Seismic Waveform Analysis*

With the goal of establishing a Caribbean Tsunami Warning System for Puerto Rico, local and regional seismic wave form analysis algorithms were adapted to the Caribbean region. At-risk regions need real-time determination of earthquake source information to assess the nature of the hazard in order to optimize emergency response. In 1998 FEMA and the University of Puerto Rico funded a project to upgrade the instrumentation of the Puerto Rico Seismic Network (PRSN) of the Department of Geology of the University of Puerto Rico at Mayagüez. As part of this project, 9 digital telemetered broad-band stations were installed on the main island of Puerto Rico and offshore islands. In 2000, with funding from FEMA, the PRSN installed 4 additional broadband stations linked with TCP/IP to Mayagüez. The digital waveforms are recorded and saved at the facilities of the PRSN. Also, since 1998 the PRSN operates the data center for the Middle-America Seismograph Consortium (MIDAS). Currently available methodologies for the routine computation of earthquake source parameters were modified and adopted for use in the Caribbean region based on regionally and locally-recorded seismic waveforms.

An analysis of digital broadband waveforms from the Puerto Rico Seismic Network (PRSN) – for local analysis – and the Incorporated Research Institutions for Seismology (IRIS) Digital Management Information Center – for regional analysis - is carried out to provide earthquake-faulting information as well as an estimate of Mo in near-realtime. If possible, PRSN data will be supplemented with realtime broadband waveforms recorded by Global Seismograph Network (GSN) stations in the Caribbean and data submitted to the MIDAS Data Center.

For the regional analysis, the techniques developed by Ammon, Randall and Owens (Randall *et al.*, 1995) were implemented to perform a regional moment-tensor inversion using all three components (vertical, transverse, and radial) available from the digital seismograph stations the IU station SJG located in Cayey, PR.. An analogous methodology, now based on Zahradnik *et al.* (2001) was adapted for local analysis. In both methods, an inversion scheme is applied to identify the focal mechanism that best reproduces the seismic waveforms observed at regional and local distances (less than 12 degrees). For regional analysis theoretical waveforms are computed by applying a reflection-matrix technique that uses a prescribed crustal-velocity model. For local analysis, theoretical waveforms are computed by applying the Bouchon (1991) method that also uses a prescribed crustal-velocity model. Prior to inversion, the observed records are deconvolved to remove the instrument response and rotated to their R,T and Z components.

The application of these moment-tensor inversion methods in a semi-automatic manner should provide source parameters in near-realtime for earthquakes in Puerto Rico and the Caribbean.

We have also adapted to the Caribbean region the "Earlybird" system developed and used at the West Coast and Alaska Tsunami Warning Center (Whitmore and Sokolowski, 2002). This package provides rapid earthquake location and magnitude and will provide the moment-tensor solution. As part of this the PRSN has acquired a display system, composed of 9 PC monitors under one CPU, allowing the acquisition and display of waveforms, earthquake location on a local and regional basis and iterative waveform analysis, all on separate displays (see Figure 4). These activities are in harmony with the long term plans of NOAA to establish a Caribbean Tsunami Warning Center at Mayagüez.

Figure 4. Early Bird Warning System at the Puerto Rico Seismic Network.

### *2.4 Task 4: Warning Messages and Protocol*

Once a felt earthquake has been characterized in terms of magnitude, location and faulting mechanism and its tsunamigenic potential has been determined, this information needs to reach the affected communities as fast as possible. Warning protocols need to be developed at the municipal, state, and federal levels, and also within the rest of the Caribbean community. The key agencies in the Pacific states for warning messages and emergency response coordination are the National Oceanographic and Atmospheric Administration (Pacific Tsunami Warning Center, West Coast and Alaska Tsunami Warning Center and the National Weather Service Field Office, NWS), the USGS regional seismic networks, and state and local emergency response agencies. A similar setup will be established in Puerto Rico and the US Virgin Islands, and will include the Puerto Rico Seismic Network (PRSN) of the University of Puerto Rico, NWS, and the Puerto Rico State Emergency Response Agency. This will also involve the participation of Puerto Rico's Warning Coordination Meteorologist - WCM in the regular Tsunami Warning Coordination Meteorologists Training Workshops held on the West coast; the Pacific states WCM's have now included tsunamis as part of their coastal hazard community preparedness activities. The WCM's have been designated as the point of contact between the States and NOAA.

For this purpose, a "Tsunami Protocol Meeting" was held at the Mayagüez Campus of the University Puerto Rico during the 15-16th of January 2003. The objective of this meeting was to set the groundwork for a tsunami warning system capable of rapid dissemination of information throughout the Caribbean on potentially tsunamigenic events in Puerto Rico and the Virgin Islands. In addition, the Puerto Rico Tsunami Technical Review Committee was established. As a result of all this effort, as of this moment the Pacific Tsunami Warning Center is monitoring large earthquakes in the Atlantic and Caribbean basins and communicating to the PRSN any possible threat.

### *2.5 Task 5: Development of a Caribbean Historical Tsunami Database (CHTDB)*

A comprehensive earthquake and tsunami database, Historical Tsunami Database for the Atlantic, has been developed containing the written history of earthquakes and tsunamis that have occurred in the Atlantic Ocean, including the Caribbean Sea. The database includes all of the earthquake information available, while the tsunami information is based on the comprehensive compilation by Lander et al. (2002).

This database contains a PC-based Graphical User Interface for ease of visualization, and it was developed under contract at the Novosibirsk Tsunami Laboratory, Russia. The whole package is contained on CD, but it can also be run on-line at http://omzg.sscc. ru/tsulab/. Figure 5 shows a plot of the tsunami runup (yellow vertical bars) and earthquake locations (circles) available for the Caribbean region. A bar is shown at each location where runups have been measured and the bar height is proportional to the runup elevation. The same information can also be listed as a table for printing.

### *2.6 Task 6: Participation in the USA National Tsunami Warning and Mitigation Program*

At the national level there is a National Tsunami Hazard Mitigation Program (NTHMP), which presently includes the states of Washington, Alaska, Oregon, California, and Hawaii. This program is sponsored by FEMA, NOAA, USGS, and state co-sharing. During the PRTWMP we were able to attend various meetings and workshops sponsored by the NTHMP and we hope to be able to become a member in the near future. Attendance to some of these meetings will continue during extensions to the program.

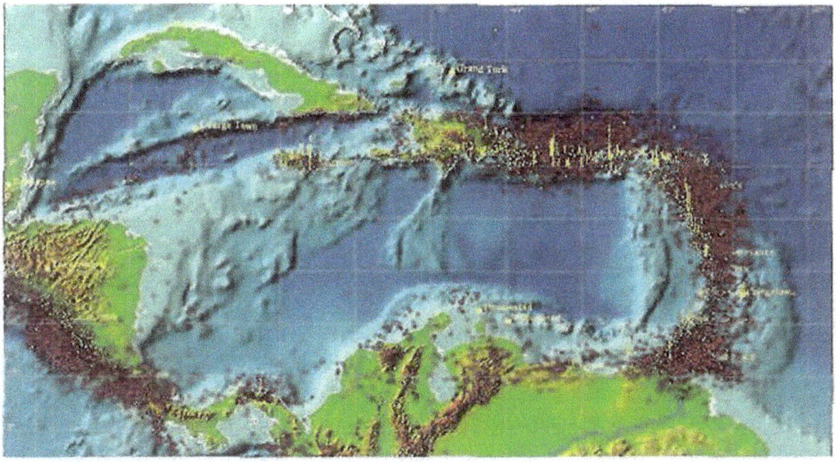

Figure 5 - Location of earthquake and tsunami historical data, based on CHTDB.

## 3. Conclusion

Based on the amount of feedback obtained from relevant government agencies (for example, Education Department, Puerto Rico Planning Board), private industries, civic organizations, and the public in general, we can conclude that we have reached our most important goal of raising the government and public awareness of this forgotten hazard. Another measure of how successful we have been in informing the general public has been the many radio, TV, and newspaper interviews made, including through the national press.

## *4. Acknowledgment*

This project was supported by the USA Federal Emergency Management Agency (through Commonwealth of Puerto Rico Governor's Authorized Representative of the Office of Management and Budget) and the University of Puerto Rico. The Sea Grant Program of the University of Puerto Rico and the Puerto Rico State Emergency Management Agency also contributed to the success of this program. The authors would like to acknowledge the significant contribution of Mr. Harry Justiniano (System Manager at the Physical Oceanography Laboratory) in all phases of this program. Other participants in the program were Dr. Carlos Mendoza (seismologist), Dr. Havidan Rodriguez (social scientist) and Ms. Maritza Pagan (Administrative Assistant).

**5. References**

Bouchon, M., 1981. A simple method to calculate Green's Functions for elastic layered media. *Bull. Seism. Soc. Am.*, 71, 959-971.

Grindlay, N., 1998. Volume and density approximations of material involved in a debris avalanche on the south slope of the Puerto Rico Trench. Technical Report. Sea Grant College Program, University of Puerto Rico, 13 p.

Huerfano, V., 2003, Susceptibility of Puerto Rico Due to Local Tsunami Effects. Mayaguez, Puerto Rico: University of Puerto Rico, Ph.D. thesis, 108p.

Huerfano. V., and bataille , K., 1994, Crustal structure and stress regime near Puerto Rico: preliminary locations of earthquakes recorded near Puerto Rico, Puerto Rico Seismic Network, Seismic Bulletin, January-December, Univ. of Puerto Rico, Mayaguez, Geology Department, 9 p.

Jansma, P. and mattiolI, G., 2003. Diffuse extension across an active faulting within the Puerto Rico and northern Virgin Islands microplate: GPS geodetic results from 1994-2002. Abstract. Seismological Soc. of America Annual Meeting, San Juan, P.R.

Kafka, A., and levin, S., 2000, Does the spatial distribution of smaller earthquakes delineate areas where larger earthquakes are likely to ocurrs?, *Bull. Seismol. Soc.* 724-738.

Lander, J. F., whiteside, L. S. and lockridge, P. A., 2002. A brief history of tsunamis in the Caribbean. *Science of Tsunami Hazards*, 20(2), 57-94.

Lay, T. and Wallace, T.C., 1995. Modern Global Seismology. California: Academic Press, 521 p.

Lynett, P. and Liu, P. L.-F., 2002. "A Numerical Study of Submarine Landslide Generated Waves and Runup," *Proc. Royal Society of London A.*, 458, 2885-2910.

Masinha, L., and Smylie, D. E., 1971. The displacement fields of inclined faults. *Bull. Seismol. Soc. Am.*, 61(5), 1433-1440.

McCann, W., and Mercado, A., 1997. Analysis of the tsunami hazard of northwestern Puerto Rico. Earthquake Advisory Board, Puerto Rico State Civil Defense. Earth Scientific Consultants, 38 p.

McCann W., 1998. Tsunami hazard of western Puerto Rico from local sources: characteristics of tsunamigenic faults. Technical Report. Sea Grant College Program, UPR, 88 p.

Mercado, A., 1994. On the use of NOAA's storm srige model – SLOSH – in managing coastal hazards: the experience in Puerto Rico. *J. Natural Hazards*, 10, 235-246.

Mercado, A. and McCann, W., 1998. Numerical simulation of the 1918 Puerto Rico tsunami. *J. Natural Hazards*, 18, 57-76.

Mercado, A., 2001. Determination of the tsunami hazard for western Puerto Rico from local sources. Technical Report. Sea Grant College Program. University of Puerto Rico, 154 p.

Mercado, A., 2002. Investigation of the potential tsunami hazard on the north coast of Puerto Rico due to submarine landslides along the Puerto Rico Trench. Technical Report, Puerto Rico State Emergency Management Agency and Sea Grant College Program, Uinversity of Puerto Rico, 245 p.

Randall, E., Ammon, C., and Owens, T., 1995, Moment tensor estimation using regional seismograms from a Tibetan Plateau portable network deployment, *Geophys. Res.Lett.*, 22, 1665-1668.

Sandwell, D. T. and Smith, W. H. F. (1997). Marine gravity anomaly from Geosat and ERS-1 satellite altimetry. *J. Geophys. Res.*, 102, 10039-10050.

Schwab, W. C., Danforth, W. W., Scanlon, K., and Masson, D., 1991. A giant submarine slope failure on the northern insular slope of Puerto Rico. *Marine Geology*, 96, 237-246.

Ten Brink, U. and Smith, S. , 2003. High-resolution bathymetric map of the Puerto Rico Trench: Implications for earthquake and tsunami hazards. Abstract. Seismological Soc. of America Annual Meeting, San Juan, P.R.

Whitmore, P. M. and Sokolowski, T. J., 2002, Automatic earthquake processing developments at the U.S. West Coast/Alaska Tsunami Warning Center. Transworld Research Network, 37/661.

Zahrandnk, J., Jansky, J. and Papatsimpa, K., 2001. Focal mechanisms of weak earthquakes from amplitude spectra and polarities. *Pure and Applied Geophys.*, 158, 647-665.

# LOCAL AND REGIONAL WAVEFORM ANALYSIS FOR RAPID MONITORING OF SHALLOW CARIBBEAN EARTHQUAKES

C. MENDOZA
*Centro de Geociencias UNAM*
*Campus Juriquilla, Qro., Mexico*

V. HUERFANO
*Puerto Rico Seismic Network*
*University of Puerto Rico*
*Mayaguez, Puerto Rico*

We examine the application of waveform analysis procedures for the rapid identification of shallow earthquake source parameters in the Caribbean using three-component digital broadband records recorded at local and regional distances. For locally-recorded events, a grid-search method is investigated that uses the entire wave train recorded at multiple stations. A comparative search is conducted between observed and theoretical amplitude spectra at frequencies up to the corner frequency of the recording instrument to identify the best-fitting strike, dip, rake, and focal depth. In the regional procedure, a formal moment-tensor inversion scheme is used with the source depth fixed at 10 km to derive the focal mechanism that best reproduces the waveforms recorded at a single station. The regional analysis provides a timely derivation of the source size and fault geometry that would be useful in the identification of potential tsunamigenic sources in the circum-Caribbean region if the procedure is implemented in an automated manner. The magnitude and mechanism can be obtained within minutes following the occurrence of the earthquake, well within the time period available for tsunami alerting. The local analysis, on the other hand, is too time-consuming to be useful in the timely derivation of source information for local tsunamigenic earthquakes. The method, however, could be beneficial for checking the accuracy and validity of source parameters derived using the regional procedure.

## 1. Introduction

The circum-Caribbean region is characterized by intense seismic activity resulting from a complex interaction of the Caribbean plate with surrounding tectonic plates (Figure 1). The Caribbean plate is bounded to the north and east by the North American plate and by the South American plate on the south. To the east, the Caribbean plate overrides the North American plate along the Lesser Antilles subduction zone. The northern boundary is more complex, with the northwestern segment accommodating strike-slip motion along the Cayman Trough and the northeastern portion corresponding to a combination of oblique subduction and underthrusting. The southern boundary of the Caribbean plate corresponds primarily to underthrusting beneath Panama and South America, although there is a strong component of right-lateral motion, for example along the Oca-El Pilar fault system in Venezuela, which accommodates a large portion of interplate motion. On the west, the Caribbean plate overrides the Cocos plate along the Middle America Trench off the southwest coast of Central America.

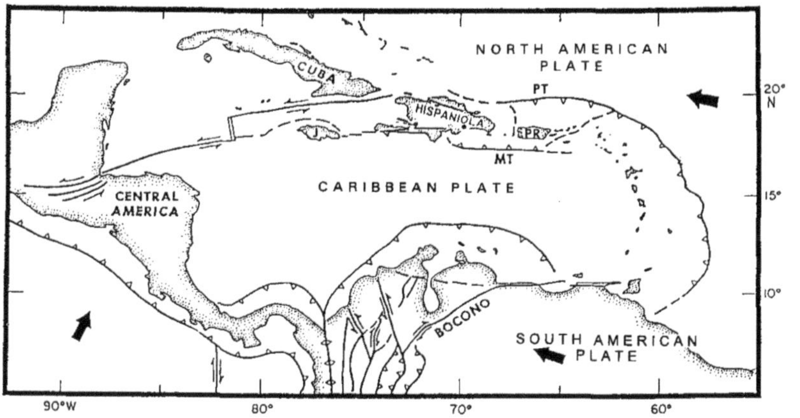


Figure 1. Plate-tectonic setting of the Caribbean region. PT: Puerto Rico Trench; PR: Puerto Rico; MT: Muertos Trough. Arrows show plate motion relative to the Caribbean Plate.

The variety and complexity in plate interaction in the circum-Caribbean region produces large-scale deformation that results in large damaging earthquakes. Because much of the deformation occurs at sea, these seismic disturbances can also generate large local and regional tsunamis. Lander et al. (2002) have identified more than 50 tsunamis in the circum-Caribbean region since 1530, including six that have resulted in fatalities. The implementation of a comprehensive earthquake monitoring system in the Caribbean would clearly benefit the identification of large potentially-tsunamigenic earthquake sources and also greatly aid the regional assessment of seismic hazard. The continuous, systematic monitoring of earthquake sources in the Caribbean region, however, has been difficult due to the limited land area available for the deployment of permanent seismograph stations. Although global seismic instruments currently exist that are used to locate moderate-to-large earthquakes worldwide, the station coverage is not sufficient to allow the routine and timely computation of earthquake source mechanisms and magnitudes for all events in the Caribbean region. Only moderate-size earthquakes (greater than about magnitude 5.0) can be recorded by a sufficient number of global stations to allow the routine retrieval of source properties. These computations are based primarily on teleseismic waveforms recorded beyond a distance of about 30 degrees from the source, thus requiring a relatively long time interval for both recording and analysis.

In a recent study, Mendoza and Malavé (2003) used a regional moment-tensor (RMT) inversion procedure to derive source mechanisms for nine shallow, moderate-magnitude earthquakes in the Caribbean region using three-component broadband waveforms recorded at the Global Seismograph Network station SJG located in Cayey, Puerto Rico. The procedure is based on the regional waveform analysis method developed by Randall et al. (1995) and uses a simplified, general velocity model to generate theoretical Green's functions. A 20-50 second passband filter is applied to both observed and synthetic waveforms to minimize the effects of any unmodeled details in the crustal structure. The focal depth of the earthquake is also fixed to reduce the time required for analysis, and the procedure is used to directly estimate the moment magnitude and the focal mechanism of crustal events.

We examine the feasibility of using the regional waveform inversion procedure described by Mendoza and Malavé (2003) in the development of a tsunami-warning system for the Caribbean. The procedure uses regional waveforms recorded within 1 to 12 degrees from the source, corresponding to a shorter recording time frame compared to teleseismic data and allowing a more timely derivation of the earthquake source properties. The use of the procedure in a routine manner using data recorded in real time, for example, could provide rapid seismic-monitoring capabilities useful for tsunami alerting.

We also investigate the use of a local moment-tensor (LMT) grid-search algorithm to identify earthquake source parameters using waveform data recorded at distances less than 1 degree from the source. These data are recorded before both regional and teleseismic data and can thus provide earthquake source information much more rapidly if analyzed in a timely manner. The estimation of seismic moment tensors at local and near-regional distances, however, is more challenging due to the strong effects of crustal and upper mantle heterogeneity on the recorded seismograms.

The LMT grid-search method examined here was developed by Zahradnik (2001) and uses the full waveform recorded at a selected set of broadband stations to derive the focal depth, seismic moment, and fault geometry of local events. All records are instrument corrected, band-pass filtered and rotated into radial, transverse and vertical components. The search procedure performs a comparison between observed and recorded waveforms in the frequency domain and includes frequencies up to the corner frequency of the recording instruments. Waveforms from multiple stations are used to perform the analysis, and theoretical Green's functions are generated for a point source using the discrete wave-number method of Bouchon (1981). The method additionally uses available P-wave polarities to identify the best-fitting source mechanism.

The acceptable LMT solutions form a compact family of nodal lines from which a single one is chosen using an L1-error function that is related to the earthquake focal depth and the crustal model. At each trial focal depth, the minimal error value is identified based on the strike-dip-rake combination that best fits the observed amplitude spectra and satisfies the P-wave polarities. The preferred solution is then determined from the minimum error for all depths used in the inversion.

## 2. Results For Caribbean Events

Huérfano (2002; 2003) applied the LMT methodology to three local earthquakes using velocity waveforms and polarities recorded by broadband stations of the Puerto Rico Seismic Network (PRSN). Seismograph stations that contribute to the PRSN database are listed in Table 1 and shown in Figure 2. Three-component broadband records with a good signal-to-noise ratio and located reliably by the PRSN were used over a 30-second time window starting at the earthquake origin time. Take-off angles of the first arrivals were calculated with the ray-method code of Jansky (2001). To avoid unrealistic take-off angles, a gradient velocity model was used based on the layered crustal structure of Huérfano and Bataille (1994), which is currently used to locate earthquakes in the Puerto Rico-Virgin Islands region.

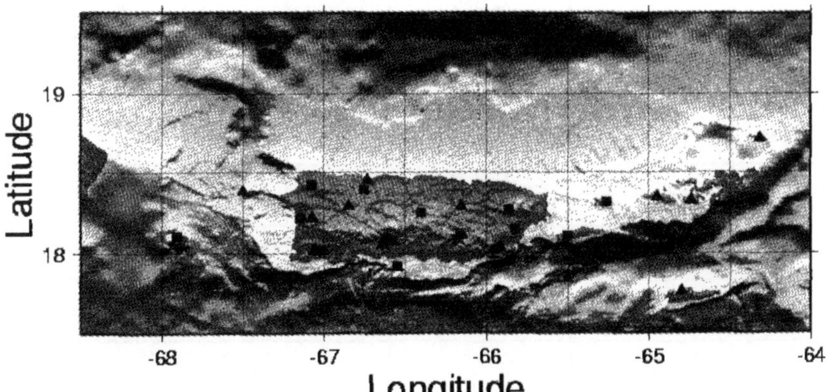


Figure 2. Geographic distribution of short-period stations (triangles) and broadband stations (squares) currently operated by the Puerto Rico Seismic Network. The star shows the broaband station SJG.

**Table 1.** Seismograph stations of the Puerto Rico Seismic Network. These include two U.S. Army Corps stations (marked with a ^ symbol) and the Global Seismic Network station SJG (marked with an asterisk). Stations not currently used in the earthquake location process are marked with a + symbol. PR=Puerto Rico; USVI=U.S. Virgin Islands; BVI=British Virgin Islands.

| Station | Site | Digitizer | Sensor | Lat (°N) | Lon (°W) | Elev (m) |
|---|---|---|---|---|---|---|
| +APR | Arecibo, PR | DSP-A64 | S-13 | 18.4440 | 66.7256 | 33 |
| CBYP | Cubuy, Canóvanas, PR | REF-TEK 72A-08 | CMG-40T | 18.2716 | 65.8566 | 607 |
| +CDPB | Cerro Punta, Jayuya, PR | REF-TEK 72A-08 | CMG-40T | 18.1732 | 66.5909 | 1285 |
| CPD | Yabucoa, PR | DSP-A64 | S-13 | 18.0369 | 65.9150 | 377 |
| | | REF-TEK 72A-07 | CMG-40T | 18.0368 | 65.9151 | 386 |
| CSB | Colonia Sábana, Bayamón, PR | DSP-A64 | S-13 | 18.2895 | 66.1576 | 447 |
| IIDE | Isla de Desecheo, PR | DSP-A64 | L-4 | 18.3852 | 67.4667 | 140 |
| +IMO | Isla de Mona, PR | REF-TEK 72A-08 | CMG-40T | 18.1094 | 67.9080 | 91 |
| LSP | Las Mesas, Mayagüez, PR | DSP-A64 | L-4 | 18.1757 | 67.0858 | 390 |
| LRS | Lares, PR | DSP-A64 | S-13 | 18.2914 | 66.8446 | 456 |
| MGP | Maguayo, Lajas, PR | DSP-A64 | S-13 | 18.0076 | 67.0891 | 35 |
| | | REF-TEK 72A-07 | CMG-40T | 18.0076 | 67.0891 | 35 |
| *SJG | Cayey, PR | DSP-A64 | L-4 | 18.1001 | 66.1502 | 457 |
| | | Q-680 | STS-1 | 18.1001 | 66.1502 | 457 |
| ^PORP | Ponce, PR | DSP-A64 | S-13 | 18.0629 | 66.6363 | 200 |
| ^CELP | Ponce, PR | DSP-A64 | S-13 | 18.0749 | 66.5792 | 196 |
| AGPR | Aguadilla, PR | REF-TEK C130 | CMG3ESP | 18.4675 | 67.1111 | 120 |
| ICM | Isla Caja de Muerto, PR | REF-TEK 72A-07 | CMG-40T | 17.8933 | 66.5210 | 77 |
| MTP | Vieques, PR | REF-TEK 72A-07 | CMG3ESP | 18.0972 | 65.5525 | 120 |
| +CULB | Culebra, PR | REF-TEK 72A-08 | CMG-40T | 18.3264 | 65.3006 | 50 |
| STVI | St. Thomas, USVI | DSP-A64 | L-4 | 18.3533 | 64.9622 | 300 |
| +SCVI | St. Croix, USVI | DSP-A64 | L-4 | 17.7317 | 64.7147 | 50 |
| +SJV | St. Johns, USVI | DSP-A64 | L-4 | 18.3456 | 64.7556 | 50 |
| ABV | Anegada, BVI | DSP-A64 | L-4 | 18.7386 | 64.3425 | 50 |
| MPR | Mayagüez, PR | NET-DAS | CMG-3T | 18.2117 | 67.1398 | 22 |
| HUMP | Humacao, PR | NET-DAS | CMG-3T | 18.1421 | 65.8489 | 79 |
| OBIP | Ponce, PR | NET-DAS | CMG-40T | 18.0428 | 66.6062 | 103 |
| AOPR | Arecibo, PR | NET-DAS | CMG-3T | 18.3466 | 66.7539 | 355 |

A frequency range from 0.1 to 0.6 Hz was selected to avoid the broadband signal-induced instabilities commonly known as the "mice" effect (Zahradnik and Tselentis, 1999). For each event, the focal depth was varied from 2 to 20 km, using increments of 2 km. A coarse grid search was then conducted for each trial depth that tested the strike from 10 to 360 degrees, the dip from 10 to 90 degrees, and the rake from 0 to 180 degrees, with increments of 10, 5 and 10 degrees respectively. A finer search was then performed using the focal depths with the least error to investigate the strike from 5 and 360 degrees, the dip from 5 to 90 degrees, and the rake from 0 to 180 degrees, in increments of 5 degrees.

The events studied by Huérfano (2002; 2003) are given in Table 2 together with the resulting source mechanisms. Also listed in Table 2 are the nine events and mechanisms obtained Mendoza and Malavé (2003) from the analysis of SJG regional waveforms using the RMT inversion procedure. The locations and mechanisms of all 12 events are shown graphically in Figure 3. Mendoza and Malavé (2003) additionally compared the RMT solutions with the source mechanisms obtained by the USGS National Earthquake Information Center (NEIC) and by Harvard University (HRV) and concluded that the regional analysis can provide an initial estimate of both the earthquake size and the type of faulting associated with an event. Moment magnitudes computed for the events are very similar to the NEIC and/or HRV magnitude estimates (Table 3), suggesting that a routine application of the method provides realistic estimates of the earthquake size that could prove useful in the comprehensive monitoring of Caribbean earthquake activity and the development of a tsunami alerting system.

## 3. Implications For The Rapid Derivation Of Source Parameters

The local moment-tensor (LMT) analysis procedure involves an iterative grid search over a broad parameter space that includes strike, dip, rake, and depth. This extensive search requires a significant time investment to identify the best-fitting earthquake source properties and magnitude. Approximately 10 minutes of CPU time are required for a 10-station grid search. This time interval is too long for the procedure to be useful in the timely identification of local potentially-tsunamigenic sources. However, the procedure could provide an independent check on the results of the regional moment-tensor (RMT) inversion procedure, which yields a more rapid determination of the moment magnitude and focal mechanism for moderate magnitude events.

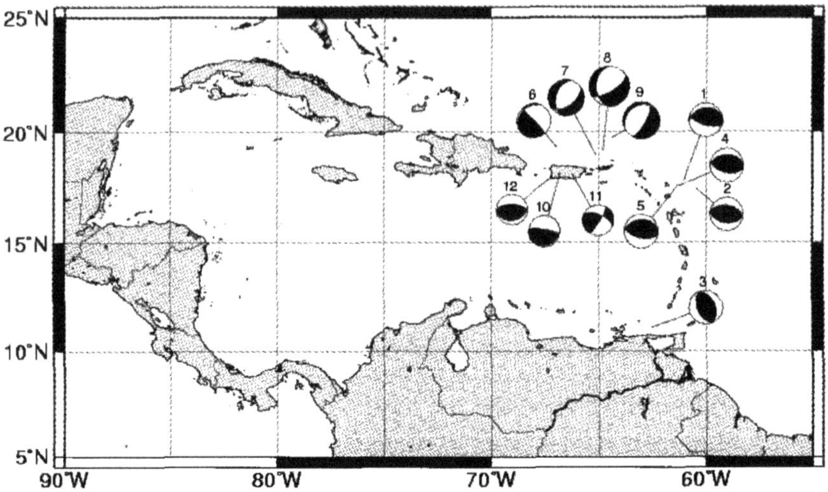


Figure 3. Geographic map showing the epicentral locations and focal mechanisms obtained by Mendoza and Malavé (2003) and by Huérfano (2002; 2003) using the local and regional three-component broadband data. Labels are keyed to the list given in Table 2.

We investigate the timeliness of the RMT procedure for deriving earthquake source parameters in the Caribbean region by examining the time requirements relative to the arrival of a regional tsunami. Figure 4 shows a sample set of three-component regional seismic waveforms recorded at station SJG for one of the events analyzed by Mendoza and Malavé (2003). The plot shows the P-wave arrival time relative to the origin time of the event and the waveform time interval required for the RMT inversion. We have tabulated these quantities in Table 4 for each of the events analyzed by Mendoza and Malavé (2003). Also listed in Table 4 is the computation time required for performing the RMT inversion on each of the events using a SUN workstation. Note that the computation time increases with increasing distance from the source due to the fact that a longer wavetrain is recorded at farther distances, requiring more computation time for the numerical inversion.

The sum of the P arrival time and the window length of the seismic record in Table 4 corresponds to the total time required for RMT analysis following the occurrence of each of the nine Caribbean events. We have plotted these RMT analysis times as a function of source-to-station distance in Figure 5 to examine the timeliness of the source derivation. Also shown in Figure 5 is the expected arrival time of a tsunami wave in the Caribbean for the same distance range. The expected tsunami arrival is based on a tsunami propagation velocity of 720 km/hr, corresponding to an ocean depth of about 4 km (Ward, 2001). The average ocean depth in the Caribbean basin is generally lower than 4 km,

**Table 2.** Source parameters of Caribbean earthquakes previously analyzed using local and regional seismic waveforms. Mechanisms (strike, dip, rake) for events 1 to 9 were derived by Mendoza and Malavé (2003) using the RMT procedure. For events 10 to 12, mechanisms were identified by Huérfano (2002; 2003) using the LMT grid-search methodology. Other source information is from the USGS National Earthquake Information Center for events 1 to 9 and from the Puerto Rico Seismic Network for events 10 to 12.

| Event | Date (mo/day/yr) | Origin Time (hr:min:sec) | Location (lat; lon) | Depth (km) | Magnitude | Strike (°) | Dip (°) | Rake (°) |
|---|---|---|---|---|---|---|---|---|
| 1 | 02/16/00 | 07:03:09.9 | 17.9°N;61.0°W | 33 | 5.1$m_b$ | 106 | 65 | 102 |
| 2 | 02/23/00 | 19:20:26.3 | 17.4°N;60.4°W | 33 | 4.8$m_b$; 5.1 $M_S$ | 96 | 45 | 87 |
| 3 | 10/04/00 | 14:37:44.7 | 11.1°N;62.5°W | 114 | 5.8$m_b$ | 159 | 48 | 91 |
| 4 | 10/27/00 | 19:02:53.3 | 17.6°N;61.2°W | 41 | 5.4$m_b$ | 268 | 37 | 78 |
| 5 | 10/30/00 | 03:07:10.2 | 17.6°N;61.3°W | 33 | 5.5$m_b$; 5.6 $M_S$ | 266 | 34 | 76 |
| 6 | 12/11/00 | 18:54:07.4 | 19.1°N;67.1°W | 53 | 5.1$m_b$; 4.8 $M_S$ | 318 | 88 | 78 |
| 7 | 10/16/01 | 15:27:27.7 | 19.3°N;64.9°W | 33 | 5.3$m_b$; 5.2$M_S$ | 55 | 53 | -65 |
| 8 | 10/17/01 | 11:29:09.9 | 19.3°N;64.9°W | 33 | 5.3$m_b$; 5.6$M_S$ | 57 | 68 | -61 |
| 9 | 10/24/01 | 09:51:50.9 | 19.3°N;65.0°W | 49 | 5.1$m_b$; 4.6$M_S$ | 28 | 72 | -98 |
| 10 | 09/05/02 | 02:19:48.72 | 18.1°N;66.8°W | 10 | 3.5Mpr | 60 | 10 | 50 |
| 11 | 09/07/02 | 04:01:06.02 | 18.1°N;66.2°W | 3 | 3.6Mpr | 30 | 80 | 30 |
| 12 | 03/22/03 | 18:25:42.41 | 18.0°N;67.1°W | 10 | 4.1Mpr | 80 | 30 | 80 |

which would suggest a tsunami propagation speed lower than 720 km/hr. Figure 5 thus identifies the minimum time interval available for tsunami alerting following the occurrence of an earthquake in the Caribbean if the RMT procedure were used to identify the magnitude and mechanism of the event. The

**plot indicates that at least 6 to 60 minutes would be available for tsunami alerting in the distance range of 1 to 8 degrees.**

**Table 3.** Moment magnitudes computed for events 1 to 9 listed in Table 1. RMT magnitudes are those computed by Mendoza and Malavé (2003) using the regional waveform data recorded at the single station SJG. HRV and NEIC magnitudes are those estimated by Harvard University and by the USGS National Earthquake Information Center, respectively.

| EVENT | MOMENT MAGNITUDE | | |
|---|---|---|---|
| | RMT | HRV | NEIC |
| 1 | 5.3 | 5.3 | - |
| 2 | 5.5 | 5.4 | - |
| 3 | 6.2 | 6.2 | 6.1 |
| 4 | 5.8 | 5.6 | 5.6 |
| 5 | 6.0 | 5.8 | 5.8 |
| 6 | 5.6 | 5.5 | - |
| 7 | 5.8 | 5.6 | 5.7 |
| 8 | 6.1 | 5.8 | 6.0 |
| 9 | 5.3 | 5.0 | - |

**Table 4.** Observed P-wave arrival relative to the earthquake origin time for events 1 to 9 from Table 1. Also listed are the record lengths and time intervals required for RMT waveform inversion for each earthquake. The sum of these three quantities denotes the total time required for RMT analysis following the occurrence of an earthquake at the specified distance from the recording site.

| Event | Distance to SJG (°) | P-wave Arrival (sec) | Record Length (sec) | Computation Time (sec) | Sum (sec) |
|---|---|---|---|---|---|
| 1 | 5.0 | 73 | 175 | 28 | 276 |
| 2 | 5.1 | 77 | 175 | 28 | 280 |
| 3 | 7.8 | 112 | 200 | 76 | 388 |
| 4 | 4.7 | 70 | 175 | 28 | 273 |
| 5 | 4.7 | 70 | 175 | 26 | 271 |
| 6 | 1.3 | 22 | 140 | 8 | 170 |
| 7 | 1.7 | 28 | 145 | 11 | 184 |
| 8 | 1.7 | 29 | 145 | 10 | 184 |
| 9 | 1.6 | 26 | 145 | 14 | 185 |

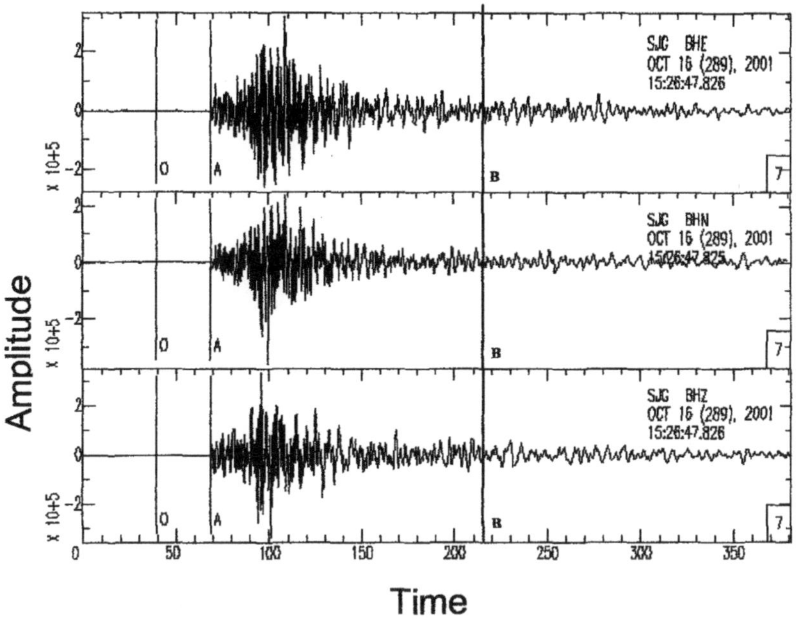


Figure 4. Three-component broadband regional waveforms (BHE, BHN, and BHZ) recorded at station SJG for event 7 listed in Table 1. Markers O and A indicate the origin time of the earthquake and the P-wave arrival, respectively. B marks the end of the record window required for RMT inversion for an event recorded at this distance.

## 4. Conclusions and Discussion

We have examined the time required for the application of a fixed-depth regional moment-tensor (RMT) inversion procedure following the occurrence of moderate-magnitude earthquakes in the Caribbean. The procedure inverts three-component broadband seismic waveforms using a general crustal velocity model to generate synthetic vertical, radial, and transverse Green's functions in the time domain. Because the analysis is based on the inversion of regional waveforms, a relatively rapid estimate of earthquake source parameters can be made compared to methodologies that use globally-recorded data.

We have also conducted a similar examination of a local moment-tensor (LMT) grid-search procedure applied to smaller magnitude events. The LMT search procedure, however, is too time-consuming to allow an effective examination of source parameters prior to the potential arrival of any tsunami

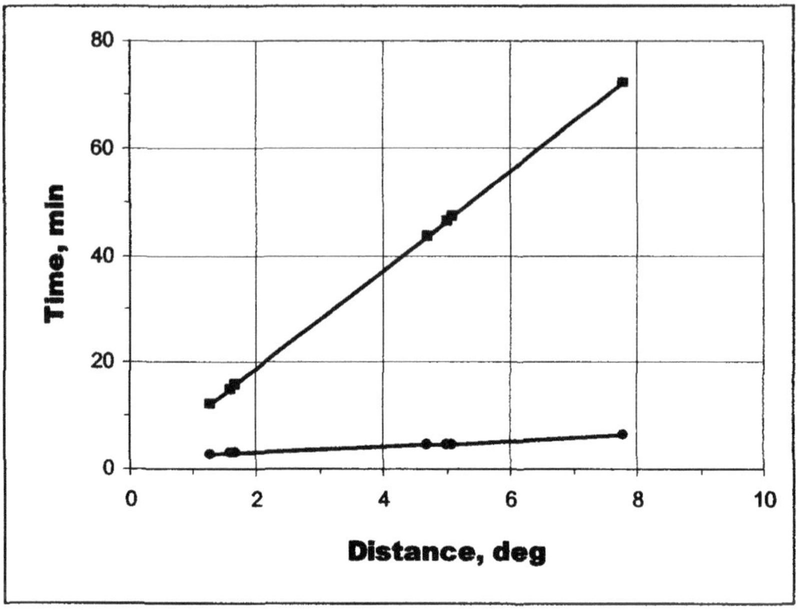


Figure 5. Time-distance plot showing the total time intervals (solid circles) in minutes required for the RMT analysis following the occurrence of each of the events listed in Table 1. Also shown are the tsunami arrival times (squares) expected at the same distances for an ocean depth of about 4 km. The plot shows the minimum time interval available for tsunami alerting within the distance range of about 1 to 8 degrees from the recording site.

generated within 1 degree of the recording stations. Nonetheless, the procedure could be useful in the testing of the RMT procedure as it could provide an independent check on the magnitude and fault geometry derived using the regional waveforms.

Our tabulation of the time required for the application of the RMT inversion procedure to events located in the Caribbean region and recorded by the single station SJG indicates that the total time required for an evaluation of the source parameters would vary from about 3 to 7 minutes following the occurrence of an earthquake located between 1 and 8 degrees from the recording site. These time intervals are well within the time period available for tsunami alerting. For a tsunami traveling at a speed of 720 km/hr, for example, the initial tsunami wave would arrive about 9 to 70 minutes following the occurrence of the event for the same distance range. This would allow time intervals of at least 6 to 60 minutes for issuing tsunami warning messages for earthquakes located 1 to 8 degrees away.

These results indicate that a magnitude and mechanism can be obtained within minutes of the occurrence of an earthquake if the regional procedure is implemented in an automated manner. Information on the earthquake fault mechanism may be useful for discriminating between tsunamigenic and non-tsunamigenic events. The timely estimate of the earthquake size is especially important for tsunami alerting since generally only earthquakes with magnitudes greater than about 7 have the potential for generating destructive tsunamis. Events analyzed to date using the RMT procedure have been of moderate size ($M_W \sim 6$), and it will be important to evaluate the performance of the procedure for earthquakes of much greater magnitude. The effects of source finiteness on the derivation of moment magnitude, for example, may need to be evaluated. Source finiteness effects for earthquakes of magnitude about 8 and greater can be pronounced in the period range used in the RMT procedure. The effects, however, may be less of a concern if the only criterion used for issuing a tsunami warning is that the magnitude be greater than 7.

Another consideration for tsunami alerting is the possibility of issuing a false alarm in the case where focal-depth information is not immediately available and the event is actually too deep to be tsunamigenic. In this case, a secondary check on the generation of an actual tsunami should be used to discontinue an alert. Such secondary monitoring, using tide gauges for example, is common in current programs of tsunami alerting in other coastal regions.

## 5. Acknowledgements

This work was conducted under the Puerto Rico Tsunami Warning and Mitigation Program sponsored by the Federal Emergency Management Agency (FEMA, PR-0077) and the Puerto Rico Sea Grant Program (R-122-2-04).

## 6. References

Bouchon, M., (1981). A simple method to calculate Green's functions for elastic layered media, *Bull. Seism. Soc. America*, vol. 71, pp: 959-971.

Huérfano. V. and K. Bataille (1994). Crustal structure and stress regime near Puerto Rico, *Seismic Bulletin: preliminary locations of earthquakes recorded near Puerto Rico*, January-December, Univ. of Puerto Rico-Mayaguez.

Huérfano, V., (2002). Determinación de mecanismos focales para temblores pequeños o moderados: región del Caribe nororiental, *Geofisica Colombiana*, 6, p. 40-42.

Huérfano, V., (2003). Susceptibilidad de Puerto Rico ante el efecto de maremotos locales, *PhD Thesis*, Univ. of Puerto Rico-Mayaguez.

Jansky, J., (2001) Ray-method calculations of the travel times and take-off angles in gradient models, program ANGGRA. *Res. Report, Faculty of Math. And Phys.*, Charles University, Prague.

Lander, J. F., L. S. Whiteside and P. A. Lockridge, 2002. A brief history of tsunamis in the Caribbean, Science of Tsunami Hazards, 20, p. 57-94.

Mann, P. and K. Burke (1984). Neotectonics of the Caribbean, Rev. *Geophys. Space Phys.*, 22, p. 309-362.

Mendoza, C. and G. Malavé (2003). Earthquake source parameters using a single regional broadband station, Proceedings, First Congress of the Americas on Environmental Geophysics, Special Issue, *Revista Geofisica*, in press.

Randall, G.E., C.J. Ammon, and T.J. Owens (1995). Moment tensor estimation using regional seismograms from a Tibetan Plateau portable network deployment, *Geophys. Res. Lett.*, 22, p. 1665-1668.

Ward, S.N. (2001). "Tsunamis", in *The Encyclopedia of Physical Science and Technology*, Third Edition, R.A. Meyers (ed.), Academic Press, 15,453 pp.

Zahradník, J. and G. Tselentis (1999), Broadband Stations in Western Corinth Gulf, *Orfeus Newsletter 1*, No 3.

Zahradník, J., J. Jansky, and K. Papatsimpa (2001). Focal mechanisms of weak earthquakes from amplitude spectra and polarities, *Pure and Appl. Geophys.*, 158, 647-665.

# EMERGENT TSUNAMI WARNING SYSTEM FOR PUERTO RICO AND THE VIRGIN ISLANDS*

CHRISTA G. VON HILLEBRANDT-ANDRADE

*Puerto Rico Seismic Network, University of Puerto Rico, PO Box 9017 Mayagüez, PR 00681-9017, USA*

VÍCTOR HUÉRFANO MORENO

*Puerto Rico Seismic Network, University of Puerto Rico, PO Box 9017 Mayagüez, PR 00681-9017, USA*

A tsunami warning system has been under development for the Puerto Rico/Virgin Islands (PRVI) region since 2000 as part of the Puerto Rico Tsunami Warning and Mitigation Program (PRTWMP). This system is in response to the historical tsunamis which have impacted Puerto Rico and the Virgin Islands and the potential for tsunamis to affect the region in the future. This emergent warning system has five thrust areas: definition of the tsunami scenarios, tsunami detection, tsunami protocol, communication, dissemination and education. The detection capabilities of the system are based on the detection of potentially tsunamigenic earthquakes by the Puerto Rico Seismic Network (PRSN) and the Pacific Tsunami Warning Center (PTWC). A protocol to respond to a potential tsunami has been developed. The messages would be broadcast through the Civil Emergency Alert System of the San Juan Field Office of the U. S. National Weather Service in coordination with the Puerto Rico Seismic Network and the Puerto Rico State Emergency Management Agency. Educational initiatives have been taken so that the threatened population is aware of the hazards and can respond effectively in case of a tsunami.

## 1. Introduction

Historical records, paleotsunami studies, modeling and research efforts suggest that low lying coastal areas of the Puerto Rico and Virgin Islands region can be impacted by tsunamis. The main sources of tsunamis for the region are earthquakes and submarine landslides. The hazard associated with volcanic activity is low given the distance of the volcanoes.

Faults around the Puerto Rico region have the potential of generating local tsunamis. Historically there have been three earthquakes which generated destructive tsunamis in the northeastern Caribbean (Figure 1).

The tsunami of November 18, 1867 was triggered by a Ms 7.3 (Pacheco and Sykes, 1992) earthquake in the Virgin Island Basin. Within 5 minutes waves of up to 6 meters in Charlotte Amalie, St. Thomas and 9.0 meters in Fredricksted,

* This is a draft of a paper which is to be included in the proceedings of the Caribbean Tsunami Workshop which was sponsored by NSF, PR Emergency Maangement, Sea Grant and the UPR on March 30 and 31, 2004.

St. Croix were flooding low lying downtown areas. The tsunami had a runup of 6 m in southeastern Puerto Rico and almost 20 m in Guadeloupe

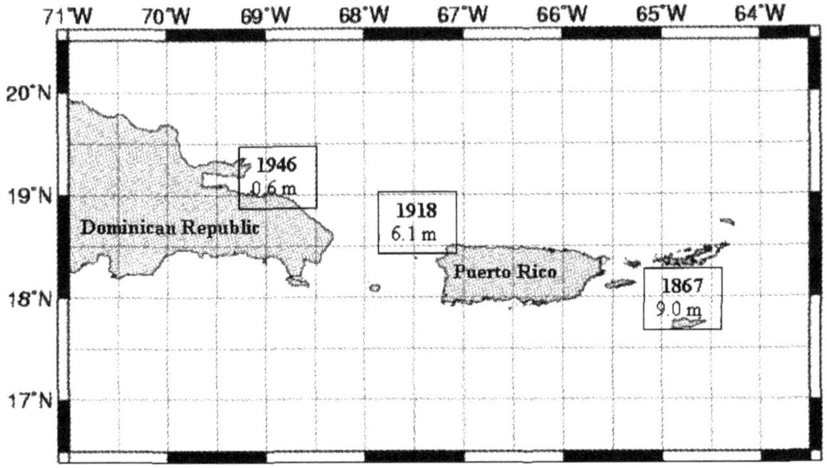


Figure1. Schematic location of the 1867, 1918 and 1946 earthquakes and the maximum runups recorded in the Virgin Islands (1918) and Puerto Rico (1918 and 1946).

(Lander et al, 2002). In the Virgin Islands 17 people died as a result of the tsunami, while 23 deaths were reported in Guadeloupe; no one was reported to have died in Puerto Rico as a result of the earthquake.

*On October 11, 1918 a Ms 7.3 earthquake (Pacheco and Sykes, 1992) in the* Mona Canyon, 40 km WNW of Aguadilla, Puerto Rico generated a tsunami. According to Reid and Taber (1919) the water was observed to have receded within a minute of the earthquake itself along the north westernmost point of Puerto Rico. Wave heights of up to 6 meters were observed in western Puerto Rico where 40 people lost their lives and the greatest damage was observed.

Moya and Mercado (2004) documented 3 tsunami deposits in northwestern Puerto Rico, one of which corresponded to the 1918 event and the other 2 were dated at 1270-1410 AD and 820 – 400 BC. The source of these tsunami deposits is uncertain but reinforces the idea that tsunamis have been a recurrent phenomenon.

Besides these historical and prehistoric tsunamis in the PRVI, Huérfano (2003) and Mercado and Justiniano (2003) modeled 269 earthquake scenarios associated with faults in the region, determined the initial tsunami conditions and associated potential flooding. According to his results, the hazard is greatest in the western part of the island and then, then in order of decreasing exposure in the north, south and east coasts. These local tsunamis are characterized by very short travel times (can affect areas within a minute of the earthquake) and potentially high waves, 20 – 30 feet, close to the epicenter which decay quickly with distance.

Earthquakes of magnitude greater than 7.5 which occur beyond the Puerto Rico/Virgin Island region waters also have the potential of generating tsunamis which could reach Puerto Rico. On August 4, 1946 a Ms 7.6 (Pacheco and Sykes, 1992) earthquake along the north shore of eastern Dominican Republic triggered a tsunami. This tsunami, which caused much damage and fatalities in the Dominican Republic, was observed and recorded (0.6 m) in San Juan, Puerto Rico (Lander and Lockridge, 1989), but caused no damage in Puerto Rico. According to their catalogue, a strong aftershock of this earthquake on August 8 caused the sea to retreat 24 m at Aguadilla, 76 m at Mayagüez and was also recorded in San Juan (.6 m). Other sources for such large earthquakes and tsunamis could be the Lesser Antilles subduction zone, the faults located offshore of Venezuela and northern Colombia, on the Caribbean side of Central America and the North America – Caribbean plate boundary to the north. No modeling has been performed on these sources, but according to a tsunami travel time program developed by Gusiakov (2001) tsunamis generated beyond PRVI, but within the Caribbean could take from 30 minutes to four hours to reach the PR/VI platform.

The PRVI region can also be affected by distant tsunamis. These tsunamis travel more than 1000 km. On November 1, 1755 an earthquake with an estimated magnitude of 9 was generated offshore Portugal. This earthquake, which caused great destruction itself, generated a tsunami which impacted Portugal, England and the coast of southern Europe and Africa. It reached the Lesser Antilles 7 – 8 hours later where waves of 7 meters were observed at Saba, 4.5 m at St. Martin and 3.6 m at Antigua and Dominica (Mader, 2001; Lander, Whiteside and Lockridge, 2002). There is no record that it affected Puerto Rico.

Another important source for a tsunami are submarine landslides. The deposits of submarine landslides have been mapped offshore Puerto Rico in the Puerto Rico Trench and the Mona Canyon (ten Brink, 2004). Mercado et al (2003) modeled the flooding that could have been caused by a landslide which was mapped in the Puerto Rico Trench. Although the lateral extension of these slides is less than those associated with earthquakes, the height of these waves can be substantially larger.

There are several active volcanoes in the eastern Caribbean. Gisler et al (2004) simulated catastrophic eruptions from the submarine volcanoe Kick 'em Jenny and concluded that it is unlikely to pose significant danger to Puerto Rico or the Virgin Islands. Others have studied the tsunami hazard associated with subaeral volcanoes, specifically with Soufriere Hills on Montserrat, both from partial collapse of the edifices as well as flows entering the surrounding waters. The results predict that the tsunami hazard associated with both of these phenomena is only significant for the surrounding and nearby islands.

Another source for tsunamis could be the impact of an object from space but, given the low frequency and probability of such events it is not taken into consideration for tsunami warning purposes.

## 2. Puerto Rico and Virgin Islands Tsunami Warning System

Since 2000 the Puerto Rico and Virgin Islands Tsunami Warning System (PRVITWS) has focused on tsunamis which can be originated from earthquakes. Other scenarios have not been given priority because of the difficulty of detection (landslide generation), low hazard (volcanic eruptions) and low probability (impact from outer space). Up to the present the system has focused on the detection and location of tsunamigenic earthquakes, tsunami protocol, dissemination and response.

### *2.1 Tsunami Scenarios*

There are three types of earthquake tsunami scenarios: local tsunamis, regional tsunamis and distant tsunamis.

#### *2.1.1 Local Tsunamis*

Curtis and Pelinovsky (1999) defined local tsunamis as those that have up 24 minutes of travel time. These tsunamis could be generated by a local earthquake within the PRVI. The PRVI is defined as the region between latitudes $17^{\circ}$ and $20\,^{\circ}$ N and longitudes $63.5^{\circ}$ and $69\,^{\circ}$ W. The 1867 and 1918 earthquakes are examples of historical local tsunamis.

#### *2.1.2 Regional Tsunamis*

Regional tsunamis are considered as those that would take between 24 minutes and 2 hours. For Puerto Rico and the Virgin Islands earthquakes generated within the Caribbean or just beyond the Caribbean and North America/South America plate boundary. A historical example of such a tsunami was the one generated by the August 4, 1946 earthquake.

#### *2.1.3 Distant tsunamis*

Distant tsunamis are those that would have travel times of over two hours to Puerto Rico and the Virgin Islands. For the PRVITWS, the source of these tsunamis would be in the Atlantic Ocean and Western Caribbean. The 1755 Lisbon earthquake generated such a tsunami.

### *2.2 Detection and Location of Tsunamigenic Earthquakes*

The Puerto Rico Seismic Network (PRSN) was established by the USGS for the Puerto Rico Electrical Power Authority in 1974. In 1987 this short period network was transferred to the Geology Departmentt. of the University of Puerto Rico at Mayagüez. It presently consists of almost 30 seismic stations installed from Mona Island to the west thorough Anegada Island in the British Virgin Islands to the east (Figure 2). The PRSN consists of short and broad band stations, data of which are routinely analyzed. Strong motion data of the Puerto Rico Strong Motion Program of the University of Puerto Rico at Mayagüez are also available in the case of larger earthquakes.

In January 2003 the automatic earthquake processing system developed by the West Coast and Alaska Tsunami Warning Center (ATWC), known as Early Bird (EB), was installed in the PRSN. In Puerto Rico, this system has been

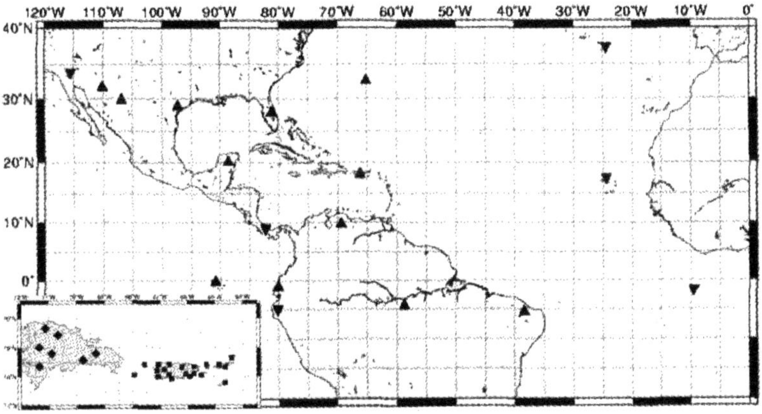


Figure 2. Seismic stations used by the Early Bird System running in the Puerto Rico Seismic Network for the detection of potentially tsunamigenic earthquakes. The insert corresponds to stations of the Puerto Rico Seismic Network (squares) and the "Instituto Sismolögico Universitario" (ISU) of the Autonomous University of Santo Domingo (UASD) in the Dominican Republic. Stations of the IRIS/IDA (inverted triangles) and IRIS/USGS (triangles) networks are also used.

tailored to detect and locate earthquakes in the Puerto Rico, Intra America Seas and the Atlantic Ocean. It monitors the data stream of the stations operated by the PRSN, the "Instituto Sismolögico Universitario" (ISU) of the Autonomous University of Santo Domingo (UASD) and the "Pontificia Universidad Católica Madre y Maestra" in the Dominican Republic as well as 17 stations of the Global Seismographic Network in the mid latitudes of the western hemisphere that are available in real time via the Internet (Figure 2).

Between February 2004, when a significant number of global stations in the Mid Atlantic region were added to the system, and September 2004, EB detected and located 292 earthquakes. Of these events, 115 were located in the Puerto Rico region (17 – 20 N and 63.5 – 69 W), their magnitudes ranged from 2.1 to 5.3. The other events corresponded to regional and distant earthquakes, many from the Pacific Ocean. Although it is still in trial stage, for most local and regional events, Early Bird has been providing, in general, faster solutions than those available from the National Earthquake Information Center (NEIC), the Pacific Tsunami Warning Center (PTWC) and the West Coast and Alaska Tsunami Warning Center (WCATWC). The information has also been just as accurate, for tsunami warning purposes, in terms of location and magnitude. As an example of the timeliness and accuracy of the locations and magnitudes of the events located by the EB system at PRSN, Table 1 compares the location of a major earthquake which occurred November 15, 2004 offshore western Colombia.

Table 1. Comparison of the location, magnitudes and timeliness of the solution of a 7.0 magnitude earthquake which occurred at 9:06 UTC on November 15, 2004 off western Columbia.

| Source | Lat. | Long. | Depth (km) | Magnitude | Time (UTC) Solution Received at PRSN |
|---|---|---|---|---|---|
| INGEOMINAS (National Seismic Network of Colombia | 4.8 N | 77.8 W | < 30 km | 6.7 | Accessed through INGEOMINAS Home page 12:00 |
| EB-PRSN (first solution) | 5.1 N | 78.25 W | 40 | 7.5 | 09:13 |
| PTWC | 4.8 N | 77.4 W | | 7.2 | 09:20 |
| WCATWC | 4.8 N | 78.25 W | | 7.2 | 09:22 |
| EB (last solution) | 5.11 N | 78.25 W | 40 | 6.9 MS | 09:37 |
| USGS/NEIC | 4.61 N | 77.54 W | 10 | 7.0 | 10:11 |

Once Early Bird detects an event the information is distributed automatically over the Internet. Outside working hours, the personnel of the PRSN receive the information on pagers and cell phones. The PRSN presently only distributes earthquake information which has been reviewed by an analyst. As more confidence in the system is acquired, the automatic information could also be distributed to first responders and the NWS.

### *2.3 Tsunami Protocol*

There are several earthquakes and tsunami warning centers, emergency management agencies and weather offices that could be involved in the case of a significant earthquake or tsunami (Figure 4). For Puerto Rico the four lead agencies are: PRSN, the Puerto Rico State Emergency Management and Disaster Administration Agency (PRSEMA), the National Weather Service Weather Forecasting Office in San Juan and the Pacific Tsunami Warning Center (PTWC). The PRSN would also be in contact with the U. S. Virgin Islands Territorial Emergency Management Agency (VITEMA) and the Department of Disaster Management of the British Virgin Islands (DDM). The National Earthquake Information Center (NEIC) of the USGS could also

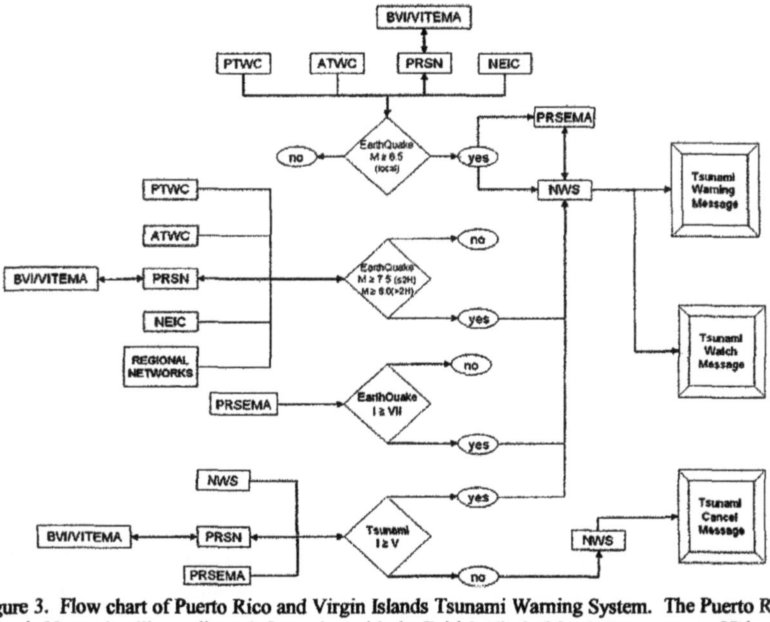


**Figure 3. Flow chart of Puerto Rico and Virgin Islands Tsunami Warning System. The Puerto Rico Seismic Network will coordinate information with the British Virgin Islands Department of Disaster Management.**

provide critical earthquake information. For regional and distant earthquakes other regional and international seismic networks and centers and emergency management agencies could be contacted in case of a significant earthquake or tsunami.

At the moment only PRSEMA, NWS and PTWC are manned 24 hours, 7 days a week. The established protocol has taken this into consideration.

The PRSN which is the authoritative source for earthquake information for the PRVI may take between 5 minutes (during working hours) and 2 hours (outside working hours) to respond with reviewed locations. The PTWC could provide information of earthquakes greater than 6.5 to PRSN and NWS within 20 minutes. The response time for NEIC is usually 20 minutes or more. The protocol includes 4 different messages: felt earthquake, tsunami warning, tsunami watch and "all clear". At the moment the criteria that are used to issue the messages are earthquake magnitude and location, earthquake intensity and/or reliable reports. The criteria of intensity is very important for local earthquakes given that the tsunami can reach the shores of the region within a minute of the earthquake and the time taken to get parametric information (magnitude, location, including depth) can take much longer. At present there are no sea level data that can be used for tsunami purposes in Puerto Rico.

**Felt earthquake messages** are issued for all felt events which meet one or more of the following criteria and for which there is no potential for a tsunami:

1. Magnitude of less than 6.5 in the PRVI/local region, 7.5 in the region or 8.0 at greater distances.
2. Earthquakes with a focal depth greater than 60 kilometers.
3. The earthquake is generated under land.
4. The event is felt with a intensity less than VII

Between 1994 and 2004 the PRSN released between 4 and 20 felt earthquake messages a year.

A **tsunami warning** would call for an evacuation of all the low lying coastal areas vulnerable to tsunamis. The PRSN will recommend a tsunami warning if there was the possibility that a tsunami could affect the Puerto Rico and Virgin Islands within 2 hours. For this, one or more of the following conditions would have to be met:

1. An earthquake of intensity VII or greater is felt in Puerto Rico or the Virgin Islands.
2. An earthquake of magnitude 6.5 or greater and shallower than 60 km is detected by the PRSN or the PTWC offshore in the PRVI. A threshold of 6.5 is used because of the potential for seismically induced submarine landslides.
3. An earthquake of 7.5 or greater is detected by the PRSN or PTWC beyond the PRVI region, but within a two hour travel time as calculated with the Tsunami Travel Time program of Gusiakov (2000). In the Caribbean Sea, this corresponds to all regions roughly East of Longitude 80° W and in the Atlantic Ocean between Longitudes 35 ° and 75 ° W and Latitudes 10 ° and 35 ° N (Figure 4). This area corresponds roughly to southern Cuba, Jamaica, Haiti, Dominican Republic, the Lesser Antilles and northern Venezuela.

4. Reliable reports were received that a tsunami has been observed locally, in the Eastern Caribbean or the western mid Atlantic Ocean as defined in 3.

It is important to point out that according to catalogues available through USGS (http://neic.usgs.gov) between 1900 and present, only 8 earthquakes met these criteria in 1900 (2), 1906, 1918, 1943, 1946 (2) and 1974. Of these 8 earthquakes, 1 (1918) generated a destructive tsunami in the PRVI and 2 (both in 1946) generated tsunamis which were detected, but caused no damage in the region. Therefore, there is the possibility that warnings could be issued and no tsunami be generated, but we feel the level is acceptable. If local and regional sea level monitoring data existed and were incorporated into the system, unnecessary evacuations could be reduced. A **tsunami watch** would instruct the population to make preparations for a possible evacuation because of a potential tsunami. The PRSN will recommend a Tsunami Watch under the following conditions:

1. An earthquake of magnitude 8.0 or greater at a depth shallower than 60 km is generated in the Caribbean Sea roughly west of Longitude 80° W, in the Gulf of Mexico or in the Atlantic Ocean beyond the longitudes 35 ° and 75 ° W and Latitudes 10 ° and 35 ° N.
2. Reliable reports are received that a tsunami has been observed which has the potential of reaching the Caribbean.

Between 1900 and the present there were only four earthquakes that met these criteria in 1929 (Sandwich Islands) and 1941, 1969 and 1975 (Azores Fracture Zone).

A **cancellation, "all clear"** of the tsunami watch or warning would indicate that the region is no longer at imminent risk from a tsunami. The PRSN would recommend a cancellation under the following conditions:

1. The PRSN determines that the conditions to issue the warning or watch were not met.
2. Within an hour of the conditions being met no reports are received that a tsunami has been generated.
3. The behavior of the sea has returned to normal levels according to reliable reports.

### *2.4 Communication and Dissemination*

There are several levels of communications for the system to be effective, between the key responders and then to the general public. At the petition of PRSEMA and/or upon the recommendation of the PRSN, the NWS/WFO in San Juan would activate the Emergency Alert System (EAS) issuing a Civil Emergency Message with tsunami information (felt earthquake, warning, watch and all clear/cancellation). These messages would be broadcast in English and Spanish over the radio and TV stations as well as through the NOAA weather radio. The NWS has developed a protocol to respond to the earthquake and tsunami scenarios. The NWS can also issue information messages based on intensity of the felt earthquake or other information it may receive from reliable sources, but will not issue a tsunami warning or watch unless the PRSEMA or PRSN requests them to do so. The NWS/WFO would only issues messages for Puerto Rico and the US Virgin Islands.

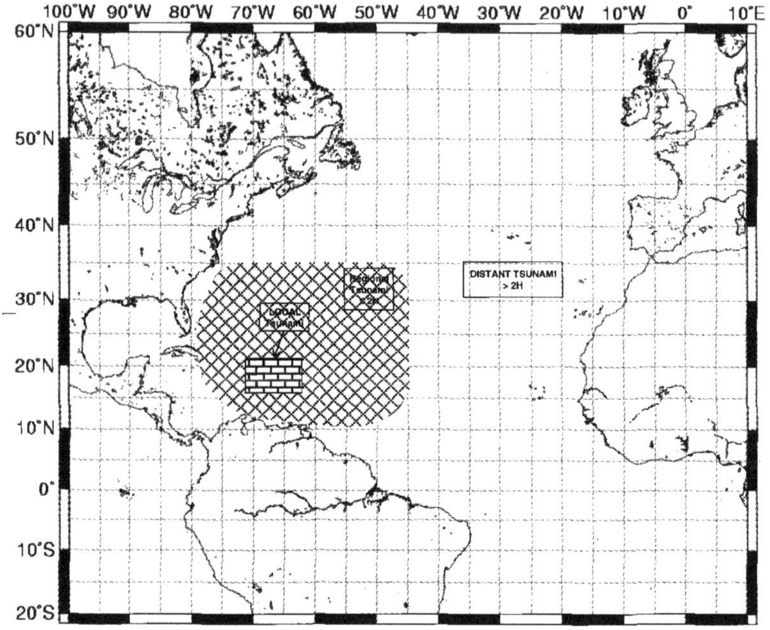


Figure 4. Schematic of areas for which tsunami warnings (Local or regional, less than 2 hours) or watches (distant, greater than 2 hours) would be issued.

In case of emergencies the main means of communication between the PRSN and the other agencies (PRSEMA, NWS, PTWC, VITEMA and DDM) is voice commercial land and cellular telephones. There is a PRSEMA radio base station at the PRSN which is also used for communication between the two agencies. FAX and Internet are also used between the agencies to share information, as well as with the media and other responders.

At present the only way to massively disseminate the information is through the EAS of the NWS. There are no alarms or sirens installed in any of the coastal communities at risk.

### *2.5 Education*

To be able to seek an appropriate response of the public it is necessary that all those potentially at risk are conscious of the tsunami threat and know how to respond in the event of a tsunami emergency.. State, regional and local emergency management officials and the general population have been oriented to:

1. Review the tsunami inundation maps that have been prepared for Puerto Rico to become familiar with the extent of the potential flooding.
2. Learn to recognize the natural signs of an impending tsunami emergency.
    a. An earthquake of intensity VII or greater
    b. Water along the coast sea recedes or advances in an uncharacteristic fashion.

    Under either of these circumstance people should immediately move inland, towards higher ground or up a building or object.
3. Evacuate or prepare to evacuate if a tsunami warning or watch is issued.

To reach the public, the following initiatives have been carried out by the Puerto Rico Tsunami Warning and Mitigation Program.

1. Production and distribution of the documentary "Tsunami in Puerto Rico: The Forgotten Danger". As of 2005 this video is available in VHS and DVD format, in both English and Spanish with subtitles and sign language.
2. 250 tsunami warning signs were prepared and installed throughout much of Puerto Rico's coasts.
3. The website http://poseidon.uprm.edu is continuously updated.
4. Workshops have been held for emergency response personnel for tsunami issues
5. Four tsunami evacuation exercises have been held in schools located in tsunami flood zones. Personnel from other schools and emergency management agencies have participated so that they can carry out other drills.
6. Preparation and distribution of a tsunami fact sheet.
7. Newspapers, TV and radio stations have covered tsunamis.

In 2004 the Puerto Rico Tsunami Technical Review Committee was established to review and advise on tsunami issues. Representatives of 15 private and public institutions meet regularly to discuss the program and the protocol which has been developed for the region.

### 3. Conclusions and Recommendations

Since 2000 Puerto Rico has been developing a tsunami warning system that responds to the tsunami hazard in the Puerto Rico and Virgin Islands region. It is very important that the research, monitoring, protocols, communication, dissemination and education initiatives be strengthened. The priorities for the development of the system include: the establishment of a 24 X 7 operation to monitor earthquakes, the establishment of a network of tsunami ready tide gauges throughout Puerto Rico, the installation of the Tsunami (DART) buoys developed by NOAA to detect regional and distant tsunamis, strengthen the communication systems between the lead agencies and continue to educate the public on this phenomena. The possibility of extending these operations to other parts of the Caribbean has also been contemplated.

### 4. Acknowledgments

The Puerto Rico Tsunami Warning and Mitigation Program has been supported by the University of Puerto Rico, University of Puerto Rico at Mayagüez, the Federal Emergency Management Agency, Puerto Rico State Emergency Management and Disaster Administration Agency and the Puerto Rico Sea Grant Program. The development of the system has greatly benefited from discussions with Dr. Charles McCreery (Pacific Tsunami Warning Center), Dr. Laura Kong (International Tsunami Information Center) and personnel from the National Weather Service Forecasting Office in San Juan and Southeast Region of the National Weather Service.

NOTE: In the aftermath of the Tsunami of December 26, 2004 significant new resources are expected to be made available to consolidate the tsunami warning system for Puerto Rico and the Virgin Islands and the Caribbean in general.

### 5. References

Curtis, G. D., and Pelinovsky, E. N., 1999. Evaluation of tsunami risk for mitigation and warning. Science of Tsunami Hazards 17(3): 187-192.

Gusiakov, V., 2000. Tsunami Travel Time Calculation Program for the Caribbean Region, Version 2.4. Tsunami Laboratory, Institute of Computational Mathematics and Mathematical Geophysics, Siberian Division of the Russian Academy of Sciences, Novosibirsk, CD.

Gusiakov, V., 2001. Historical Tsunami Data Base for the Atlantic. CD. Intergovernmental Oceanographic Commission, US National Weather Service Pacific Region, Novosibirsk Tsunami Laboratory, Center for Research and Development, UPR.

Huérfano, V., (2003). Susceptibilidad de Puerto Rico ante el efecto de maremotos locales, PhD Thesis, 114 pp, Univ. of Puerto Rico-Mayaguez.

Lander, J. F. and Lockridge, P. A., 1989. United States Tsunamis (Including United States Possessions) 1690-1988. National Geophysical Data Center, Boulder.

Lander, J. F., Whiteside, L. S. and Lockridge, P. A., 2002. A Brief History of Tsunamis in the Caribbean Sea, Science of Tsunami Hazards, Vol. 20, No. 1, pp. 57-94.

Mader , 2001. Modeling the 1755 Lisbon tsunami, Science of Tsunami Hazards, Vol. 9, No. 2, 2001

Mercado, A. and Justiniano, H., 2003. Tsunami coastal flood mapping for Puerto Rico and adjacent islands, Final report for Task 1, Puerto Rico Tsunami Warning and Mitigation Program (FEMA-PR-0077), http://poseidon.uprm.edu, 100 pp.

Mercado, A. and McCann, W., 1998. Numerical simulation of the 1918 Puerto Rico Tsunami, Natural Hazards 18, Kluwer Academic Publishers, pp. 57-76.

Mercado Irizarry, Grindlay, N., Lynet, P. and Liu, P., 2003. Investigation of the Potential Tsunami hazards on the North Coast of Puerto Rico due to submarine landslides along the Puerto Rico Trench, http://poseidon.uprm.edu.

Moya, J. C. and Mercado, A., 2004. Geomorphologic and stratigraphic investigations on historic and prehistoric tsunami in Northwestern Puerto Rico: Implications for long term coastal evolution. Abstract presented at NFS Caribbean Tsunami Workhsop, http://nsfctw.uprm.edu.

Pacheco, J. F. and Sykes, L. R., 1992. Seismic moment catalog of large shallow earthquakes, 1990 to 1989, Bulletin of the Seismological Society of America, Vol. 82, No. 3, pp. 1306 -1349.

Reid, H. F. and Taber, S., 1919. The Porto Rico Earthquake of 1918 with Descriptions of Earlier Earthquakes. Report of the Earthquake Investigation Committee, U. S. House of Representatives, 66th Congress, 1st Session, Document No. 269. Government Printing Office, Washington, 74pp. Also available at http://poseidon.uprm.edu.

Sokolowski, Thomas, J., 2002. Automatic Earthquake Processing at the US West Coast/Alaska Tsunami Warning Center, Abstract, Cordilleran Section, AGU.

Ten Brink, U., 2004. New seafloor map of the Puerto Rico Trench helps assess earthquake and tsunami hazards.

# TSUNAMIS IN THE LESSER ANTILLES

**Narcisso Zahibo**

*Département de Physique, Laboratoire de Physique Atmosphérique et Tropicale, Université Antilles Guyane, Pointe-à-Pitre, Guadeloupe (F.W.I.)*

**Efim Pelinovsky**

*Laboratory of Hydrophysics and Nonlinear Acoustics, Institute of Applied Physics, Nizhny Novgorod, Russia*

The problem of tsunami risk for the Lesser Antilles is analyzed. A total of 19 events can be considered as true tsunamis over the 1690-2004 period. Ten of them have seismic origin, and nine – volcanic origin.

## 1. Introduction

A tsunami catalogue has been recently created for the Caribbean Sea (Lander et al, 2002; HTDB/ATL, 2002; Loughlin and Lander, 2003). In the past 500 years this region has had devastating tsunamis causing damage in many states of the Caribbean Basin (Figure 1). The last event occurred in Montserrat, Lesser Antilles, on 12-13 July 2003, when a large pyroclastic flow entered the sea has generated a tsunami, which was recorded on Montserrat and Guadeloupe (Pelinovsky et al, 2004). The present paper analyzes the state of the art of tsunami study for the Lesser Antilles.

## 2. Historical Data

Historical material of tsunamis in the Lesser Antilles contained in various literature sources and datasets has been examined during the last 5 years (Shepard, 2001; Zahibo and Pelinovsky, 2001; Lander et al, 2002; O'Loughlin and Lander, 2003; Pelinovsky et al, 2004). A total 17 events can be identified as true or very probable tsunami events; they are listed below.

**1690, April 16:** An earthquake with magnitudes $Ms > 8$ (intensity 9 at Antigua) occurred in the Leeward Islands, and generated waves after substantial recession of the sea at many locations. Eyewitnesses in Charlotte Amalie, St. Thomas reported an earthquake around 4 pm, which lasted one fourth to one-half hour and caused the sea to recede so that it was possible to walk out 18 meters and pick up fishes. At St. Kitts, large earth cracks opened. The earthquake caused

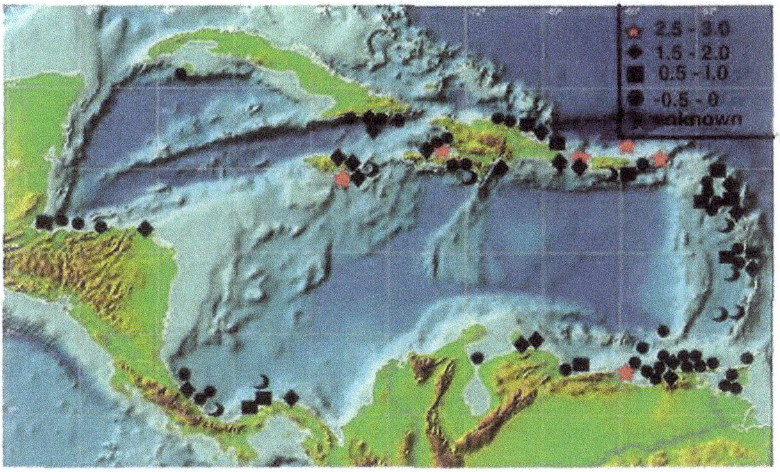


Figure 1. Historical tsunami distribution in the Caribbean Sea (numbers – tsunami intensity)

the collapse of the Jesuit College and all other stone buildings at Nevis, where landslides generated on volcanic Nevis Peak caused the sea to withdraw 201 m from Charleston, before returning in two minutes. Guadeloupe also incurred a lot of damage.

**1755, November 01:** A teletsunami was generated by a strong earthquake in Lisbon, Portugal. This tsunami reached Antigua about 9.3 hours later. Waves with estimated runup heights of 7 m were observed at Saba, Netherlands Antilles. At St. Martin, the runup was 4.5 m. The full height of the tsunami could have been as high as ten meters. Antigua and Dominica each had runups of 3.6 m. At Barbados, the waves were 1.5 - 1.8 m, and were reported to have a very short period (only 5 minutes). The water looked as black as ink (perhaps due to a local landslide). Waves were also reported at Samana Bay, Dominica. At Martinique, the water was reported to have withdrawn 1.6 km and returned to inundate the upper floors of houses. Lowlands on most of the other French islands were inundated.

**1761, March 31:** An earthquake near Lisbon, Portugal generated the second teletsunami and caused a 1.2 m wave at Barbados.

**1767, April 24:** Shocks have been reported at Martinique and Barbados. "An agitated sea ebbed and flowed in an unusual way at Martinique and Barbados".

**1823, November 30:** At 2:45 LT a strong earthquake was followed by a tsunami at 3:10 LT that caused damage in St. Pierre Harbor, Martinique.

**1843, February 8:** A disastrous earthquake (Mw=8.3) occurred at Pointe-a-Pitre, Guadeloupe. It was felt at Antigua, St. Lucia, St. Kitts, Montserrat, Martinique, and other islands. At Antigua, the sea rose 1.2 m and sank again immediately.

**1867, November 18:** An earthquake occurred in the Anegada Passage between St. Croix and St. Thomas, U.S. Virgin Islands and generated a tsunami with waves that reached the shore about 15 minutes later. The waves observed from Puerto Rico to Grenada, possibly reached the northern coast of South America. At Deshaies, Guadeloupe, according to published letters, the sea receded 100 m and returned as an 18.3 m wave about 5 km broad, damaging dwellings and carrying all floatable objects away; at Sainte-Rose, the wave height was 10 m. These data were examined and the wave heights in both locations were re-estimated as less than 10 m (Zahibo and Pelinovsky, 2001). At Basse-Terre, the height was 1.0 m, and the sea retreated far from the coast. At Isles des Saintes, there was a slight swell, and at Fond-du-Cure, houses were inundated with 1 m depth water. At Pointe-a-Pitre, there was a slight swell. At St. Christopher, waves were also observed. At St. Martin and St. Barthelemy, damage was also reported. At St. Johns, Antigua, the waves had a height of 2.4 to 3.0 m. The wave observed at Martinique, and St. Vincent had unusually high water. The wave height was 3 m at Grenada, and Gouyave (Charlotte Town) and 1.5 m at St. Georges. Waves were 1.8 m at Bequia, in the Grenadines.

**1902, May 8**: There was a devastating eruption of Mont Pelee, Martinique, killing in St. Pierre about 3,000 inhabitants. Fort-de-France was covered with ashes, stones were falling, and a tidal wave which was flying to the hills added to the terror of the population.

**1902, May 20:** Continuing eruptions of Mont Pelee, Martinique caused disturbances of coastal waters. The heavy fall of volcanic matter compelled the crews to seek shelter, and the tidal waves recurred rapidly, causing great danger. At noon the sea began to recede (at Fort-de-France) with a heavy ground swell tossing the shipping so several vessels broke from their moorings. Then a long, rolling wave spread over the sea front, but it did little damage, and the sea again receded and left a considerable area of the shore permanently uncovered. The sea itself seems troubled. It has invaded Le Precheur, undermining several houses, and adding the ravages of inundation to those of fire." A severe inundation at Basse Pointe, on the northeast coast of this island swept away twenty houses. A tidal wave has destroyed a portion of the village of Le Carbet.

**1902, August 30:** At 1 p.m. LT a great volcanic cloud flowed from NW to SW from the crater of Mont Pelée, Martinique to about half the distance to Fort-de-France. A violent eruption at 9 p.m., comparable to the May 1902 eruption, advanced almost to Fort-de-France with a light fall of ashes and small stones. The sea retreated at 9:25 p.m., followed by a rapid rise of about 1 m, which covered the quays and came to the border of the grassland area.

**1911, November 3:** A volcano-related tsunami produced extraordinary waves at Trinidad, following an explosion of a mud volcano island.

**1939, July 24:** The eruption of Kick 'em Jenny submarine volcano generated a tsunami which had an amplitude of 1-2 meters in northern Grenada, the southern Grenadines and Barbados.

**1965 October 30:** The eruption of Kick 'em Jenny submarine volcano generated the minor tsunami.

**1969, December 25:** A magnitude 7.6 earthquake was felt in Guadeloupe, Dominica, Martinique, St. Vincent, Antigua and Barbados. A wave was recorded at Barbados, Antigua, and Dominica, with maximal amplitude of 46 cm at Barbados. Tide-gauge records of this tsunami are presented in Figure 2.

**1985 March 16:** A moderate earthquake ($Mw = 6.3$) caused damage and injuries to six people in Guadeloupe and minor damage at Montserrat. It was also felt at Antigua and St. Kitts. A several-centimeter tsunami was recorded at Basse-Terre, Guadeloupe.

**1990 March 26:** The eruption of Kick 'em Jenny submarine volcano resulted to turbulent water.

**1997, July 9:** A $Ms = 6.8$ earthquake occurred off the coast of Venezuela, near Isla de Margarita, causing extensive damage and landslides in the Cariaco-

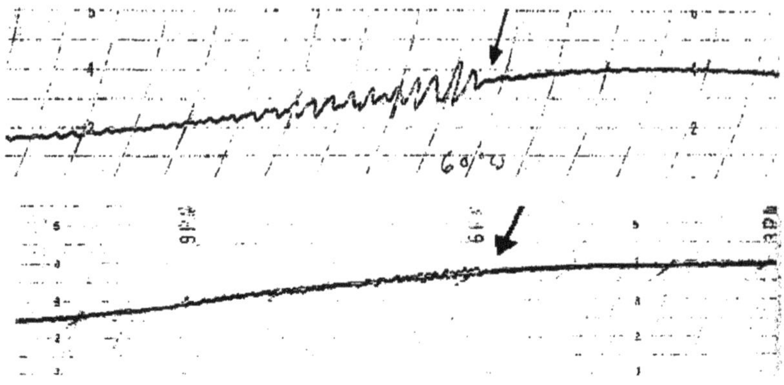

Figure 2. Tsunami registration at Bridgetown, Barbados (upper) and Dominica (below) (Shepard, 2001)

Cumana region. At least 76 people perished and 500 were left homeless. It was reported from Trinidad that a wave come ashore then recede on the south coast of Tobago, few minutes after the earthquake.

**1997, December 26:** A volcanic debris slide of about 60 million cubic meters occurred in the White River Valley, Montserrat. Pyroclastic flows entered the sea at the mouth of the White River. There were reports of a wave inundating the Old Road Bay area, 10 km from the landslide site. The wave was estimated to have been about 1 m higher than the road which lies 2-m above sea level, and to have moved inland a maximum distance of 80 m. A variety of objects, including a small wooden boat, a shelter's roof, and a stone table were displaced several meters inland and a large log was carried even further by the wave. Impact marks up to 1 m were also on the side of palm trees facing the sea. The grass was oriented in such a way as to indicate the retreat of the wave. An observer reported that the sea move out and then back in.

**2003, July 13:** The climax of the large dome collapse at the Soufrière Hills Volcano, Montserrat occurred 03:35 GMT on 13 July. The impact of a large pyroclastic flow on the sea (Tar River Valley) generated a tsunami, which was recorded at Montserrat, 2-4 km from generating area and Guadeloupe, 50 km from Montserrat. The wave height at Montserrat may have been about 4 m according to the location of the strandline of charred trees and other floating objects at Spanish Point (Figure 3). The wave height in Guadeloupe according to "direct" witnesses was about 0.5-1 m at Deshaies and near Plage de la Perle. The

Figure 3. Strandline of charred trees and other floating objects on the coast at Farm Bay (Spanish Point)

tsunami at Deshaies scattered boats, as confirmed by fishermen and local authorities.

Geographical locations of all tsunamis in the Lesser Antilles are presented in Figure 4.

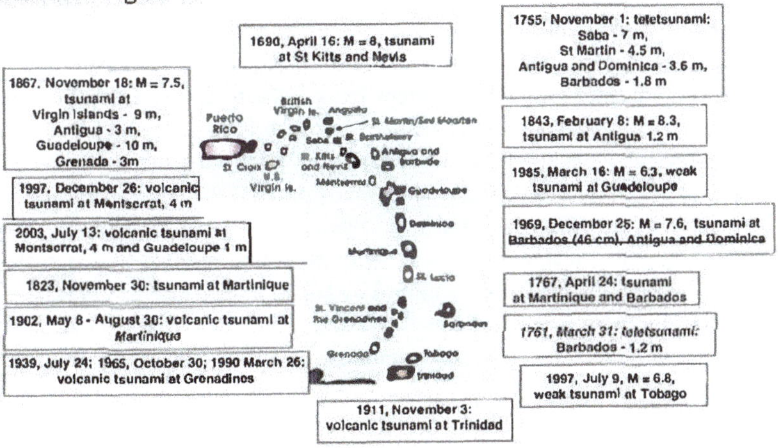


Figure 4. Information on the true tsunamis in the Lesser Antilles from 1690 to 2004

**3. Analysis**

At least nineteen tsunamis occurred in the Lesser Antilles from 1690 to 2004, or one event – every 16.5 year. Therefore, tsunami can be considered as frequent phenomenon for states of the Lesser Antilles. Most of reported tsunamis occurred in the last $20^{th}$ century (11 events). The new millennium began with the tsunami on 12-13 July 2004 at Montserrat and Guadeloupe after the volcanic eruption of the Soufrière Hills Volcano, Montserrat. Therefore the probability of new tsunami in the Lesser Antilles in a near future is high.

Almost a half of tsunamis has seismic origin (10 events). According to given information, the distant tsunamis were more dangerous. For instance, tsunami waves generated by the Lisbon earthquake in 1755 crossed the Atlantic Ocean and induced a 7 m wave at Saba, Netherlands Antilles. The wave height of the 1867 tsunami, after the strong earthquake at the Virgin Islands, was about 10 m at Deshaies, Guadeloupe (maximum recorded value of tsunami height in the Lesser Antilles). Both catastrophic events have been modelled (Mader, 2001; Zahibo et al, 2003a), and the numerical results are in good agreement with observed data. Recently Zahibo et al (2003b) studied the far-field tsunami potential of the Caribbean Sea based on numerical simulation of probable events. According to the calculations, the wave height attenuates with distance with exponent value (–2/3 and –1, Figure 5) and, therefore, the underwater earthquakes with magnitudes less 8 cannot induce strong tsunamis on distances that exceeds 3000 km, except for some coastal locations with "amplified" relief. Detection of such locations is a very important task for tsunami researchers.

For the Lesser Antilles the probability of tsunamis from volcanic origin is very high. For approximately 100 years (1902-2004) nine events have volcanic origin. The volcanic eruptions of Mont Pelée, Martinique and Soufrière Hills Volcano, Montserrat induced 5 tsunami events with maximum runup height of 4 m. Volcanic tsunamis have very local character, but during the last eruption on Montserrat (12-13 July 2004) tsunami waves reached the coast of Guadeloupe and scattered the boats at Deshaies. Possible scenario of tsunami waves generated by the volcanic eruption in Montserrat that resulted in a pyroclastic flow entering the sea has been developed by Heinrich et al (1998, 1999), and the last event in 2004 confirmed the theoretical scenario (Zahibo et al, 2004). Deplus et al (2001) found many submarine evidence for large-scale debris avalanches in the Lesser Antilles (Figure 6), and, therefore, the volcanic origin of tsunamis should be taken into account in the tsunami warning system.

Geographically, tsunamis were recorded frequently in Martinique (5 events), Barbados and Antigua (4), Guadeloupe (3). Unfortunately, the

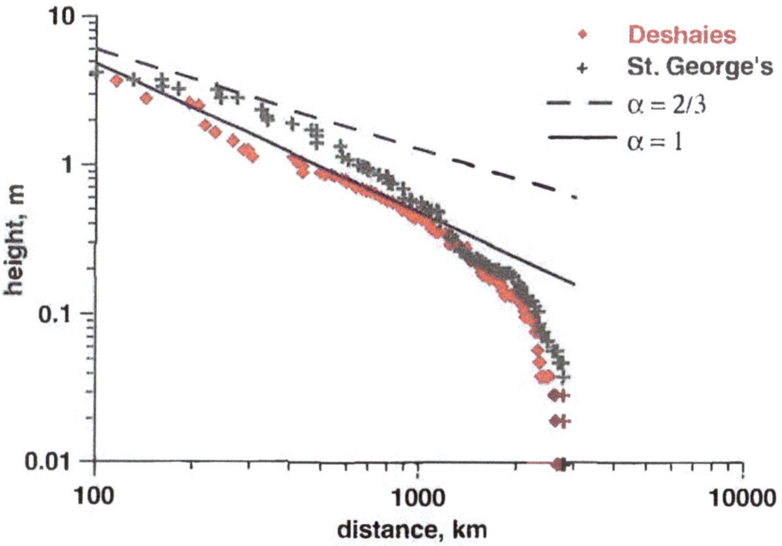


Figure 5. Computed tsunami height at Deshaies (Guadeloupe) and St. George's (Grenada) versus the distance to the source (Zahibo et al, 2003b)

quantitative information for each coastal location is poor: three values of the wave heights quantitative information forfor Barbados and Antigua only. Meanwhile, we plot the cumulative frequency for both islands (Figure 7). It confirms that tsunami wave with height of 1 m can be expected for 100 years, but the accuracy of these estimates is very low.

## 4. Conclusion

At least, 19 events can be considered as true tsunamis in the Lesser Antilles from 1690 to 2004. Therefore, tsunami is a frequent phenomenon for the Lesser Antilles. The sources of tsunami are: underwater earthquakes (10 events), debris avalanche from the volcano (6 events), and underwater volcano eruption (3 events). Maximum wave height was recorded at Guadeloupe (Deshies), it was about 10 m during the 1867 tsunami. Mathematical modelling of historical and possible tsunami occurred in the Lesser Antilles has been conducted for seismic tsunami as well as volcanic tsunami. Next and important step is to evaluate the tsunami risk for each coastal location in the Lesser Antilles using all collected information.

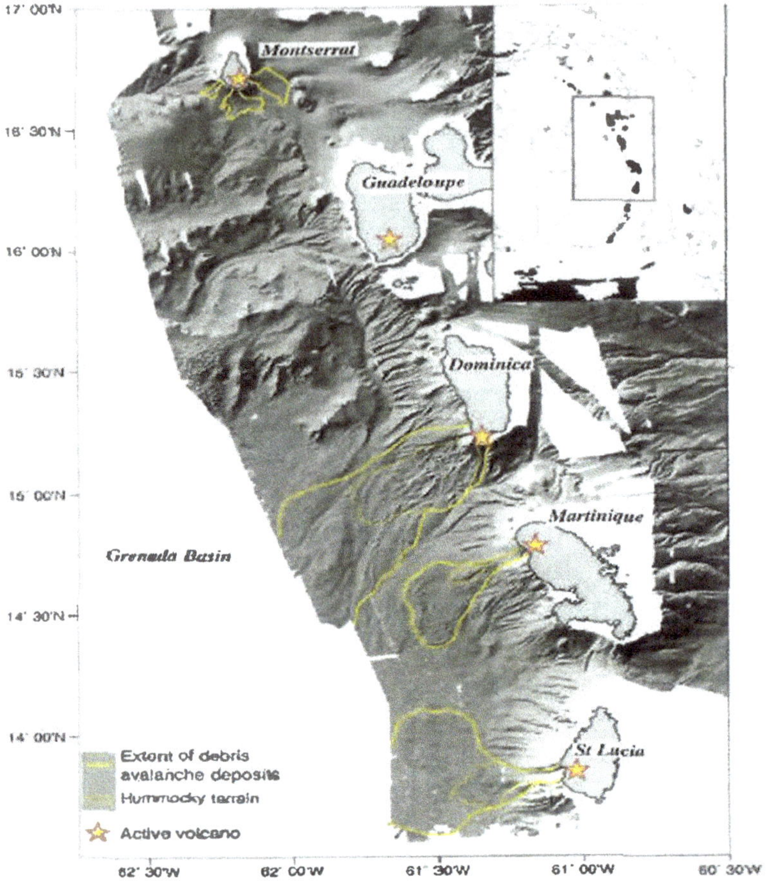


Figure 6. Submarine evidence for large-scale debris avalanches in the Lesser Antilles (Deplus et al, 2001)

## 5. Acknowledgements

This study is supported by the grants from EGIDE (04500YH) and INTAS (01-2156).

## 6. References

Deplus, C., Le Friant, A., Boudon, G., Komorowski, J.C., Villemant, B., Harford, C., Segoufin, J., and Cheminee, J.K. Submarine evidence for large- scale debris avalanches in the Lesser Antilles Arc. *Earth and*

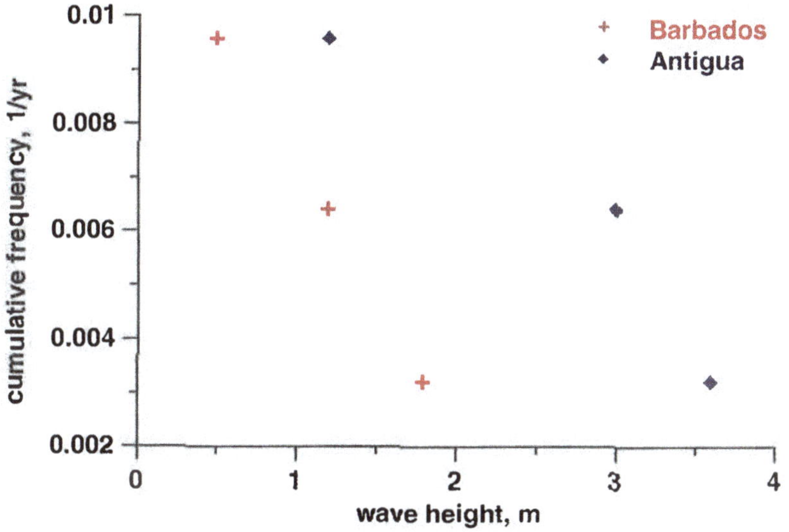


**Figure.** 7. Cumulative frequency versus tsunami wave height


*Planetary Science Letters*, 2001, vol. 192, 145-157.

Heinrich, F., Guibourg, S., Mangeney, A., and Roche, R. Numerical modelling of a landslide-generated tsunami following a potential explosion of the Montserrat Volcano. *Phys. Chem. Earth*, 1999, vol. A24, 163-168.

Heinrich, F., Mangeney, A., Guibourg, S., and Roche, R. Simulation of water waves generated by a potential debris avalanche in Montserrat, Lesser Antilles. *Geophys. Research Letters*, 1998, vol. 25, 3697-3700.

HTDB/ATL *Expert Tsunami Database for the Atlantics*. Version 3.6 of March 15, 2002. Tsunami Laboratory, Novosibirsk, Russia, 2002.

Lander, J.F., Whiteside, L.S., and Lockridge, P.A. A brief history of tsunami in the Caribbean Sea. *Science of Tsunami Hazards*, 2002, vol. 20, 57-94.

Mader, C.L. Modeling the 1755 Lisbon tsunami. *Science of Tsunami Hazards*, 2001, vol. 9, 93-98.

O'Loughlin, K.F., and Lander, J.F. *Caribbean Tsunamis: A 500-Year History from 1498–1998*. Advances in Natural and Technological Hazards Research, v. 20, Kluwer, 2003.

Shepherd, J.B. Tsunami Hazard in the Eastern Caribbean. *Workshop on volcanic and seismic hazards in the Eastern Caribbean* (May 28th - June 1st, 2001).

Zahibo, N., and Pelinovsky, E. Evaluation of tsunami risk in the Lesser Antilles . *Natural Hazard and Earth Sciences*, 2001, vol. 3, 221-231.

Zahibo, N., Pelinovsky, E., Yalciner, A., Kurkin, A., Koselkov, A., and Zaitsev, A. The 1867 Virgin Island Tsunami: observations and modeling. *Oceanologica Acta*, 2003a, vol. 26, No. 6, 609 – 621.

# AN 1867-CLASS TSUNAMI: POTENTIAL DEVASTATION IN THE US VIRGIN ISLANDS

ROY A. WATLINGTON
*University of the Virgin Islands*

This paper describes the potential devastation that might result in the United States Virgin Islands if a tsunami identical to the Virgin Islands tsunami of November 18, 1867 were to occur today. Comparisons are made based on the reported runup in 1867 and also on the changes in population, coastal infrastructure, shipping activity, and in warning and mitigation systems for natural hazards. To find examples and make comparisons to modern demographics and infrastructure, the impacts of recent damaging hurricanes are presented. Recommendations are made on preparing for and mitigating the effects of a local tsunami.

## 1. Background

The 1867 Virgin Islands' tsunami was initiated by an earthquake along a fault beneath the Virgin Islands Basin. This deep body of water is defined by the islands forming its topographic barriers. St. Thomas, St. John and the British Virgin Islands separate it from the Atlantic; St. Croix separates it from the Caribbean Sea, while Vieques and Culebra define its western boundary. To the east the Basin opens to the Atlantic through the Anegada Passage. It connects to the Caribbean's Venezuela Basin through the Jungfern Passage and Grappler Channel. Interpretation of historic data and the accounts of witnesses have placed the earthquake's epicenter at a location approximately equidistant from St. Croix, St. Thomas and Vieques.

At approximately 3:00 p.m. on November 18, 1867, inhabitants of the islands today known as the United States Virgin Islands (known before 1917 as the Danish West Indies) were jolted by a severe earthquake. Church steeples were toppled and many buildings were either damaged or destroyed. Residents rushed out of doors fearing for their lives. Shocks were also felt by sailors on boats in harbors of the towns of Charlotte Amalie on St. Thomas, Christiansted and Fredericksted on St. Croix (van Housel, 1868). Evaluation of damage done to structures and of eyewitness reports from 1867 has allowed approximation of the earthquake's intensity from 7.5 on the Richter scale (McCann, 1984). Within approximately five minutes of the earthquake, tsunamis propagated outward from the earthquake epicenter, where basin depth exceeds 4000 meters, to the gradual island slope south of St. Thomas and to the steeper north and west coasts of St. Croix. On all shores ringing the basin, witnesses report the sea retreating from the shorelines then advancing in waves of heights up to seven meters. Approaching Christiansted harbor, the wave lost some of its energy as it encountered fringing reef and islet. Fredericksted had no such protection. The town was severely damaged by the waves. Vessels were thrown onto its shores. Most noteworthy of these, the USS Monongahela was stranded on the

Figure 1. Naval Historical Archive's 1868 photograph of the USS Monogahela on the shore in Fredericksted, St. Croix.

Fredericksted shore as shown in Figure 1. In Charlotte Amalie vessels were torn from their moorings, lifted onto rocks, sunken or otherwise damaged. With a runup of approximately 2.4 meters, the waves pushed into the commercial and residential community leaving shops and warehouses filled with water. Tsunamis from this earthquake also caused destruction on St. John, Tortola, the British Virgin Islands, and Puerto Rico. Wave heights on Guadeloupe were reported to be even greater than in the islands nearer to the epicenter (Zahibo, 2004). Vessels were damaged as far away as in the harbor of St. Georges, Grenada. Small shoreline homes were swept into the sea wherever the tsunamis impacted. Many people were injured but the loss of life was limited. Although more would die from disease that would follow the earthquakes, tsunamis and the deadly hurricane of two weeks earlier, the death toll attributed to the tsunamis, was approximated at 16, with most fatalities resulting from drowning (Watlington *et al.*, 1997). Estimating the damage done to the economies of these Caribbean communities by the tsunamis of 1867 is made difficult because the waves had followed the earthquake but also the devastating hurricane of October 29, 1867. This sudden storm killed between 500 and 600 persons in the British Virgin Islands, St. John, St. Thomas and Puerto Rico and caused damage to and destruction of vessels and property.

It is known however that, in addition to disabling many vessels, the 1867 tsunamis swept away or ruined goods that had been stored in the many low lying shops and warehouses. In the aftermath, moisture extended the ruination. Subsequent to the destruction the onset of yellow fever then claimed the lives of some survivors (Watlington *et al.*, 1997). For the Danish West Indies as a whole, the destructiveness of the earthquake and tsunamis added to the hurricane's effects in disrupting treaty negotiations to transfer ownership of St. John and St. Thomas from Danish to American rule. Following these disasters, the treaty championed by U.S. Secretary of State William A. Seward was ridiculed as "one for the annexation of hurricanes and earthquakes..." (Tansill, 1932). Fifty years would pass before St. Croix, St. John and St. Thomas would become the United States Virgin Islands.

## 2. Comparing Disasters

Occurring in the present or near future, tsunamis initiated in the same manner as the 1867 event can be expected to visit the same communities as well as many new shoreline settlements and to have significantly greater impacts. Although these communities grew very slowly or shrank somewhat in the last part of the 19th Century and into the first half of the 20th, the blossoming of Caribbean communities and the growth of Caribbean tourism that has continued since the 1960s has created an entirely new profile of vulnerability.

Today these areas are heavily populated and have become repositories of great volumes of material goods, including very highly valued luxury items assembled for sale to tourists. In the US Virgin Islands, duty free status, guaranteed by the terms of the treaty, has resulted in the concentration of a treasure of expensive jewelry, perfumes, cameras, fine clothing and other luxury items. Both retail and warehousing occur within 400 meters of the water's edge and less than 2.4 meters above sea level. Busy container ports, cruise ship docks, schools and an oil refinery that ranks second in supplying the US market are among other activities that concentrate wealth, important infrastructure, and large gatherings of people. In 1867, neither population density, wealth nor infrastructure was comparable to what is found today within striking range of an 1867-class tsunami.

To anticipate the kinds of loss that today's communities might suffer, we consider the impacts of the destructive hurricanes Hugo and Marilyn. Both cyclones affected all of the Virgin Islands, with Hugo devastating St. Croix the most severely in 1989 and Marilyn doing the same for St. Thomas in 1995. These storms intruded into demographic profiles of population, wealth and infrastructure that more closely resemble today's profile in the Virgin Islands than does the profile of 1867.

On September 15, 1989, after not experiencing the direct impact of a hurricane for many decades, the Virgin Islands experienced Hurricane Hugo. The storm crossed both St. Croix and St. Thomas before moving on to cause destruction in Culebra, Puerto Rico and in the Carolinas (USA). Roads were made impassible, electric power was disrupted for periods ranging from weeks to months, businesses were shut down for weeks, both major airports were rendered inoperative for days, and many vessels in port were sunken, beached or severely damaged. Although a hurricane warning time of >24 hours allowed major vessels to avoid the storm, damage to port facilities and to general infrastructure caused cruise ship arrivals to be temporarily halted, not returning to previous levels for two years (Bureau of Economic Research, 2002). On St. Croix Hugo damaged 90% of the buildings, seriously damaging 80% of the homes. Some businesses never recovered. Oil refining on St. Croix, estimated at >100,000 barrels per day was disrupted for many days. The loss of infrastructure and of confidence in local emergency mitigation contributed to the long-term loss of St. Croix's market share in tourism.

Six years later on September 17$^{th}$, 1995, Hurricane Marilyn returned destruction to the Virgin Islands. It had been preceded by a glancing blow from Hurricane Luis two weeks earlier. This time, the greater share of the destructiveness was focused on St. Thomas. Once again, airports were closed, electric power was cut, many buildings were seriously damaged (80% on St. Thomas) and small boats were damaged or lost. Of 635 vessels surveyed, 251 (39.5%) were either destroyed, aground or beached, sunk, dismasted or otherwise damaged (COBEX, 1995). Once again, even though cruise ships were well and safely away from the storm, damage to on-shore infrastructure interfered with cruise ship tourism. As the hurricane approached, Hess Oil Virgin Islands had shut down its 545,000 barrels per day operations. It would resume production 12 days later (Oil and Gas Journal, 1995). Overall estimates of damages from both hurricanes amounted to approximately $2.1 billion, costing taxpayers $450 million while insurance companies paid out $750 million (Witt, 1998).

Cruise ship calls reflected the disruption caused. Comparing the fourth quarters of 1995 and 1994 for all three islands, one finds that 18.1% fewer ships called with 9.4% fewer passengers, even though St. Croix was able to accommodate ships that stopped going to St. Thomas. In addition, civilian employment was down by 9.9% and declined further the next year. The number of visitors arriving by air declined by 54.9% and declined further in 1996. The value of construction permits was down by 74.6% and did not recover to the 1994 value for several years afterwards (Bureau of Economic Research, 2003). According to this source, hotel occupancy rates were up 2.2% while the actual number of nights of hotel occupancy dropped by 27.4%, reflecting the reduced availability of units even as FEMA and off-island construction workers moved to the Virgin Islands to help with the recovery.

Taken together, these storms killed fewer than two dozen people in the Virgin Islands but they cost in the range of $3 billion to $4 billion in expenditures for repairs, rebuilding and recovery of homes, businesses and infrastructure (Office of Territories and Internal Affairs, 1999). Further, they lessened the attractiveness of the US Virgin Islands as a fourth quarter destination. In these ways, both storms hurt the economy of these islands measurably. Figure 2 illustrates the 8.1% loss in revenues that resulted from Hurricanes Marilyn and Bertha, which impacted the islands 10 months after

Marilyn (Virgin Islands Department of Finance, 2004). Fiscal Year 1996 represented a $30 million lowering of general fund revenues.

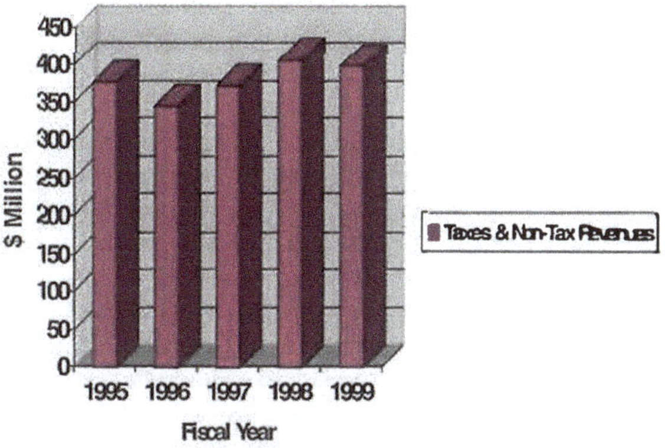


Figure 2. General Fund Statement of Revenues, FY 1995-FY 1999. Department of Finance, Virgin Islands Government.

There are limitations of how hurricanes can serve as surrogates for an 1867-class tsunami in modern times. Besides the greater concentration of wealth and infrastructure and the larger population in the modern communities, another significant difference between the occurrences of these recent hurricanes and that of the 1867 earthquake, tsunami (and hurricane) is the immensely greater warning time for hurricanes today. Without radio, the most rudimentary of instant communication devices, no significant warning times or warning procedures were possible for any of these hazards in 1867. That makes a great difference for hurricanes. However, for the "local tsunami" -- one that is initiated less than one-wavelength (typically ~100 km) away -- these advances have limited value.

### 3. Profile of the Today's Virgin Islands

Today the US Virgin Islands is an unincorporated territory of the United States. It elects its own government and sends a delegate to the United States House of Representatives. Its gross domestic product is reported variously as between $ 2.6 million and $3 million. Although the surface area of the Territory is only approximately 352 $km^2$ (136 square miles), the population has reached approximately 124,800. The coastal population density, especially during the day when businesses are operating and tourists are shopping, is much higher than it was in 1867 when approximately 37,000 people populated the Danish West Indies (Dookhan, 1974). It is reasonable to assume that, during the day, many of these people were working in the plantations and estates, most of which are located at elevations safe from tsunamis.

The Territory's main industry is tourism and the oil refinery represents the second most significant employer. In 2001, 1.8 million cruise ship passengers visited the islands spending $489.9M during their visits. The Hess oil refinery on St. Croix (which has changed corporate structure and is presently known as Hovensa) is the largest single exporter of petroleum to the United States. In 2001 it produced and exported to the United States approximately 236,000 barrels per day.

Today, within the likely runup and penetration of an 1867-class tsunami, one finds preferred recreational activities for visiting tourists – shopping, beach bathing, sailing, coastal sightseeing. The oil refinery lies within 400 meters of the sea as do many Virgin Islands Government offices, several schools, both airports and the electricity and desalination plants of the Virgin Islands Water and Power Authority. On many days in Charlotte Amalie harbor three or more ships may be found at the West Indian Company's Havensight dock, at the Crown Bay dock or anchored in the harbor. The largest cruise ships in the world visit this harbor. These vessels often carry 2,600 to 2,800 passengers attended by approximately 1,000 staff and crew members. Figure 3 shows several vessels anchored in the harbor and docked at Havensight. More than 12,000 persons (passengers, crew, and local providers of tourism services) may assemble around these vessels during a four-cruise ship day. This estimate is based on a typical occupancy of four of the average-sized vessels that visit and on the number of merchants, taxicab drivers and port officials working in the area.

Figure 3. The Charlotte Amalie Harbor, St. Thomas, U.S. Virgin Islands, on a normal day during the tourist season. Three cruise ships at the West Indian Company's Havensight dock. (St. Thomas – St. John Chamber of Commerce photograph).

## 4. Potential Tsunami Threats

For the Caribbean as for most other regions, the most likely causes of tsunamis from local sources are earthquakes, landslides, volcanic explosions and submarine avalanches resulting from slope failure. Faults in the Virgin Islands Basin, Anegada Passage, Puerto Rico Trench are active and occasionally release stresses with earthquakes. Slope failure along the submerged north coast of Puerto Rico, such as in the Loiza amphitheater, has been listed as a potential mechanism to trigger large tsunamis (ten Brink *et al.*, 2004). In the last few decades, small tsunamis have been observed to result from the volcanic activity of Montserrat's Soufriere Hills volcano and from Kick `em Jenny submarine volcano, located near Grenada. Tele-tsunamis remain a threat for the entire

Atlantic basin. In addition, there is a certain amount of concern about the eventual likelihood of a repeat of the seismic events that caused the 1755 Lisbon earthquake and about submarine debris avalanches that could be triggered by earthquakes or volcanic activity in the Canary Islands (Bryant, 2001).

## 5. Focus on the Cruise Industry

All sectors of Virgin Islands society and economy are susceptible to significant disruption by the natural hazards mentioned above. However, because of its unique vulnerability to tsunamis, the cruise ship industry today can be used to demonstrate potential impacts of a tsunami resembling that of 1867. As mentioned before, cruise ships are generally free to avoid hurricanes. Within an hour most modern vessels can easily move out of port. Capable of speed >20 knots, most can outrun or outmaneuver most big storms. And although their passengers on shore could be at considerable risk, cruise ships themselves are not highly vulnerable to significant damage from earthquakes. Even for tele-tsunamis, with adequate reporting of distant events and a minimal escape plan, cruise ships can cast lines and get to the safety of deep water when a tele-tsunami is expected. Even from as close as the location of Kick `em Jenny, the Virgin Islands, if adequately warned, could have as much as 80-90 minutes to prepare for the onslaught (Smith, *et al.*, 1994). It is generally understood that tsunami wavelengths are so long in deep water that vessels at sea are not likely to even perceive the passage of a tsunami that may go on to wreak disaster on some shoreline community. Adequately safe, deep water can be reached quickly from Virgin Islands' ports.

For the typical large cruise ship, many existing safety measures can minimize the destructive consequences of an encounter with a tsunami in shallow water. Unlike the sailing vessels and coal-fired steamers of 1867, they have powerful engines that can be brought to bear very quickly. Given reasonable warning they can maneuver to take the wave on bow or astern. These vessels have communication equipment that can receive warning messages under the most unfavorable circumstances. Many are designed to roll to angles as great as 40° without capsizing (Sabino, 2004). In their interior spaces, heavy object are fastened securely, hatches and doors are managed consistent with a safety plan.

But as was the case in 1867, the local tsunami that may originate in any of the locations listed above would have only the shortest travel and allow very little warning time. Even without fear of capsizing or sinking, anticipating the impacts of a local tsunami on a cruise ship in shallow water leaves much for worry. Docked end-to-end as three are shown in Figure 3, vessels may ram each other. And, instead of capsizing, the vessel only has to roll moderately for passengers to tumble and receive injury as they collide with interior barriers. With a sudden moderate pitch of the vessel, chairs, plates, serving dishes and other items can also become projectiles. Rather than sink, the vessel may merely be sufficiently damaged that it has to remain restricted to port for official inquiry and for repairs.

The case of the fire aboard the cruise ship M/V Angelina Lauro at the West Indian Company's dock in 1979 demonstrates this point. A fire that had started in the galley of the vessel resulted in its "total constructive loss". Fortunately no lives were lost. However, the vessel took on enough water to rest on the bottom, tilted over at 25° (US Coast Guard, 1981). There it remained for four months, preventing revenue-generating use of its part of the dock and presenting an unwelcome eyesore for visitors on the vessels that continued to use the harbor (Esannason, 1979). Filling the hold with enough water to cause the vessel to rest on the bottom contributed to the Angelina Lauro's extended stay at the site.

## 6. The Danger on Shore

The fortunate aspect of the Virgin Islands' situation with respect to tsunamis is the closeness of hills and headlands to heavily populated low-lying areas. In all of the major towns, the alerted pedestrian may only have to run 200-300 meters before reaching an elevation high enough to provide safety from the waters of an 1867-class tsunami. However, given the demographic profile of the Virgin Islands today, it may be most devastating if the 1867-class tsunami occurred in the day rather than at night. In contract to the situation in 1867, most Virgin Islands residents live at moderate to great elevations above sea level. Many hotel buildings are elevated above the reach of a 2.4 meter runup. And while there are many shoreline residents in Virgin Islands' communities, a greater portion of the population resides at elevations that would allow them to sleep safe from tsunamis.

During the day, however, all of the coastally related tourism recreation takes place near sea level. While thousands of residents pour into town to work in the low-lying towns, schools located in flood zones are also occupied mostly in the day.

Further, in crossing the island in east-west directions for business or pleasure, resident drivers spend a considerable time confined in cars just a few feet above the normal level of the sea. Figure 4 shows a junior high school located within 200 meters of the sea and on filled land. Possible safety routes for students and staff are interrupted by the road in the foreground along which heavy flows of cars speed.

Figure 5 is a photograph of one of the several quaint alleys that are usually filled with tourists and residents during a busy day in "high season" in Charlotte Amalie. These alleys end at the point where the ocean stopped before the area was filled and the shoreline highway was constructed. In the aftermath of the 1867 tsunami, these same passages were filled with water.

Receding waters up to 1.5 meters (4-5 feet) in depth prolonged the hazards caused by the advancing sea (Watlington *et al.*, 1997).

## 7. Planning for an 1867-class Tsunami

Today's demographic circumstances offer many advantages in preparing to mitigate the effects of an 1867-class tsunami. However, in spite of the relatively

Figure 4. (above) Addelita Cancryn Junior High School in Charlotte Amalie, St. Thomas, is situated just 2 meters above sea level within 400 meters of the sea and is constructed on filled land.

lower hazard that tsunamis represent after-hours in these unique Virgin Islands circumstances, there still are Virgin Islands communities of people sleeping within the tsunami runup range in Coral bay, Cruz Bay, Gallows Bay, Fredericksted and Charlotte Amalie.

Given that travel times for local tsunamis are on the order of a few minutes, an effective tsunami emergency response plan should place much focus on citizen education and awareness. Residents and visitors alike should know to "duck, cover and hold" for earthquakes and then immediately after a strong tremor to look to the sea and run to the nearest hill if they see anything suspicious. Common wisdom should be developed so that everyone would know not to approach or stay near the sea if the water recedes from the shore. Residents and tourists renting cars as well should be advised that if they are driving along the shore and an imminent tsunami is announced or if they can see disturbed water heading their way, they should abandon their cars rather than try to escape in them and move hastily to high ground. An official emergency plan might stop all shoreline traffic forcibly so that the lives of drivers can be saved and so that these cars do not inflict injury on pedestrians running for the nearest safe elevation.

Although it may be likely that the oil refinery, the schools and cruise ship docking facilities all have emergency plans that deal with greatly disturbed seas, sound emergency plans should require excellent communication among local police, emergency service providers, dock masters and cruise ship officers to avoid losses due to the kind of uncoordinated responses that occurred during the Angelina Lauro fire. These communication links should be rehearsed frequently and improved as technology improves.

Earthquake and tsunami predictions are still poorly developed arts. However, in the US Virgin Islands, even existing technology has not been applied to the fullest. For example, the few functioning sea level gauges in the

Figure 5. One of several shopping alleys in Charlotte Amalie, within 100 meters of the sea. Originally serving as pathways to shoreline piers, today they are filled with busy shoppers and merchants.

US Virgin Islands are not set up for tsunami detection and not liked into any aspect of an emergency warning system. No plans have been announced to emplace bottom-mounted pressure gauges offshore in order to detect and announce the approach of a tsunami with the ability to distinguish tsunamis from large storm waves.

The US Virgin Islands are served by the Virgin Islands Territorial Emergency Management Agency (VITEMA) which communicates efficiently with the National Weather Service in San Juan, Puerto Rico, which takes the lead on storm warnings, and with the Puerto Rico Seismic Network (PRSN). PRSN is further set up to accept advisement from the Pacific Tsunami Warning Network, which operates around the clock. However, the present channels among the regional emergency managers and with PRSN are only guaranteed during office hours on work days. Although PRSN might detect a tsunamigenic earthquake, procedures have yet to be developed for emergency service providers in the US Virgin Islands to translate this advisory into an emergency warning for the public.

Public awareness of earthquakes and of tsunamis, to a much lesser extent, has begun. There is, among emergency planners, the understanding that the best response to a sudden hazard is a well educated citizenry. VITEMA has used awareness days to share information about hurricanes and earthquakes by radio, television and with displays and information centers in public places. With regard to tsunamis, however, brochures have not been distributed and emergency routes for tsunami evacuation have not been publicized. Of concern also is that presently safeguarding against tsunamis, specifically, is not considered in choosing the locations of public buildings. Questions about tsunami safety do not impede the granting of building permits.

## 8. Conclusions

Although tsunamis are quite rare, one similar to the tsunami of 1867 is sure to occur at some time in the future and may occur at any time. Improved communication and articulation among the emergency managers of the region and building of tsunami detection technology can do much to improve safety from crippling tsunami impacts. However, for the local tsunami, such as the one of 1867 in the Virgin Islands, warning times will be very short. This suggests that well-developed emergency action plans must include identified and publicized escape routes, procedures for altering traffic and plans for minimizing devastation after the event. In addition, Virgin Islands decision makers may find it useful to study the experiences and practices followed in tsunami-plagued communities of Hawaii and in Japan in establishing new guidelines for the permitting of building in the tsunami-prone zones. Most effective will be action plans that have been developed for and invested in every citizen well in advance of any emergency situation. To a large extent the citizenry will have to take responsibility for its own safety when a sudden tsunami threatens.

## 9. References

Bryant, Edward, *Tsunami, The Underrated Hazard*. Cambridge University Press, 320 pp., 2001.

Bureau of Economic Research, United States Virgin Islands, On-line report on Cruise Ship Calls, Cruise Passenger Arrivals, Civilian Employment, Total Construction Permit Value, Air Visitor Arrivals, and Hotel Room Nights Occupied, http://www.usviber.org/CS94-895.pdf, 2003. COBEX, Hurricane Marilyn Boat List as amended 11/18/95. http://www. usvi.net/ usvi/mari/boats.html, 1995.

Esannason, Angela, Party time near as tug arrives. *Virgin Islands Daily News*, July 28, 1979, Vol 49, No. 12405, St. Thomas, p.1., 1979.

Dookhan, Isaac, *A History of the Virgin Islands of the United States*. Caribbean Universities Press, Essex, UK, 321 pp., 1974.

Island Resources Foundation, Mitigating the Impacts of Natural Hazards in the U.S. Virgin Islands, http://www.irf.org/irhazmit.htm, 1996.

Lander, James F. and P.A. Lockridge, *United States Tsunamis (Including United*

*States Possessions) 1690-1988*, Publication 41-2, National Geophysical Data Center, Boulder, 265 pp., 1989.
McCann, William R. On the Earthquake Hazard of Puerto Rico and the Virgin Islands. In: *A Workshop on "Earthquake Hazards in the Virgin Islands Region"*, Paula L. Gori and W.W. Hays (ed.). United States Department of Interior Geological Survey Open File Report 84-762, Reston, 23-42., 1984.
Oil and Gas Journal, Hess Oil Virgin Islands Corp., *Oil and Gas Journal,* September 25, 1995, **93**, No. 9, p. 46, 1995.
Sabino, Mark, Personal communication with the safety officer for the West Indian Company, which owns and manages the Havensight dock and shopping center, St. Thomas, U.S. Virgin Islands, 2004.
Singer, Gerald, St. John Beach Guide, http://www.stjohnbeachguide.com/timeline.htm, 2004.
Smith, Martin S. and J.B. Shepard, Explosive submarine eruptions of Kick `em Jenny volcano: preliminary investigations of the potential tsunami hazard in the Eastern Caribbean Region. In: Ambeh, W.B. (ed.) *Proceedings, Caribbean Conference on Natural Hazards: Volcanoes, Earthquakes, Windstorms, Floods*. University of the West Indies, Mona Campus, 249-260, 1994.
Tansill, Charles Callan, Letter of William H. Seward quoted in Pierce Papers (Harvard University), *The Purchase of the Danish West Indies*, Baltimore, The Johns Hopkins Press, p. 145, 1932.
ten Brink, Uri and E. Geist, Submarine slides north of Puerto Rico and their tsunami potential. Paper presented at the NSF Caribbean Tsunami Workshop, March 30-31, 2004, San Juan. See also this publication. 2004.
U.S. Coast Guard, *Marine Casualty Report M/V Angelina Lauro*, USGS No. 16732/01282, 42 pp., 1981.
U.S. Office of Insular Affairs, *1999 Report on the State of the Islands, US Virgin Islands*. United States Department of Interior, http://www.viaccess.net/politics/1999vi.html, 1999.
van Housel, Louis, An earthquake experience. *Scribner's Monthly*, **15**, 662-672, 1878.
V.I. Department of Finance, Exhibit I-1, General Fund Statement of Revenues. Government of the US Virgin Islands, http://www.usvi.org/oit/5yrplan, 1999.
Watlington, Roy A. and S.H. Lincoln, *Disaster and Disruption in 1867: Hurricane, Earthquake and Tsunami in the Danish West Indies*. Eastern Caribbean Center, St. Thomas, 134 pp., 1997.
Witt, James L., Remarks made at the National Press Club Newsmaker Luncheon, November 10, 1998. Federal Emergency Management Agency, 1998.
Zahibo, Narcisse and E. Pelinovsky, The 1867 Virgin Island tsunami:

Observations and Modeling. Paper presented at the NSF Caribbean Tsunami Workshop, March 30-31, 2004, San Juan. See also this publication. 2004.

# THE UNESCO/IOC INTERNATIONAL TSUNAMI INFORMATION CENTER AND ITS ROLE IN THE CARIBBEAN

LAURA S. L. KONG
*Director, UNESCO/IOC International Tsunami Information Center*

The Tsunami Warning System in the Pacific (TWSP) is a global international early warning system dedicated to the monitoring of tsunami hazards in the Pacific and dissemination of timely warnings on potentially destructive distant tsunamis. The International Tsunami Information Center (ITIC), established in 1965 by the UNESCO Intergovernmental Oceanographic Commission (IOC), oversees many of its functions. Among these are the monitoring and recommendation of improvements to the TWSP, the facilitation of technology transfer to countries establishing national warning systems, and the maintenance of a tsunami library and clearinghouse for the development and distribution of educational and awareness materials in different foreign languages. TWSP warning operations are headquartered at the Richard H. Hagemeyer Pacific Tsunami Warning Center (PTWC); the USA hosts both ITIC and PTWC. The 26-member International Coordination Group for the TWSP (ICG/ITSU), a subsidiary body of the UNESCO/IOC formed in 1968, serves as the coordinating body for the system. The Tsunami Programme and ICG/ITSU are unique among the IOC activities as they oversee a fully operational and successful early warning system solely dedicated to saving lives and property.

In recent years, the ITIC has been involved in a number of tsunami warning system initiatives efforts outside the Pacific. Although the IOC Tsunami Programme has been centered in the Pacific for many years, there is no special regional significance to this beyond the proneness of the Pacific region to large earthquakes that can generate tsunamis. Recently, the Member States of the IAS, the South Pacific and Indian Ocean, and the Mediterranean have been requesting more attention to tsunami warning in their respective areas, and the ITIC and ICG/ITSU have contributed their expertise in the development of warning system proposals by these groups.

In the Caribbean, the ITIC has been a member of the IOCARIBE Tsunami Steering Group of Experts and the ICG/ITSU Working Group that put together the IAS Tsunami Warning System proposal that was fully endorsed by the Seventh Intergovernmental Session of the IOC Sub-Commission for the Caribbean and Adjacent Regions (IOCARIBE) in February 2002. Recommendation SC-IOCARIBE-VII.2, subsequently adopted by the 35$^{th}$ Session of the IOC Executive Council, highlighted the importance of addressing the tsunami risk and called on Member States to continue to coordinate regionally towards the implementation of the system. Recently, the ITIC has encouraged the development of the Central American – Pacific Coast Regional Tsunami Warning System, and has been working with Nicaragua and El Salvador to improve their sea level monitoring capabilities and to provide in-country training on tsunamis and tsunami warning systems. Additionally, through the ISDNR Early Warning Conference consultations, the ITIC continues to work to elevate awareness of the Caribbean and Central American tsunami risk among regional disaster preparedness organizations such as CEPREDENAC and the OAS.

# THE RICHARD H. HAGEMEYER PACIFIC TSUNAMI WARNING CENTER AND ITS ROLE IN THE CARIBBEAN

CHARLES S. MCCREERY
*Director, Richard H. Hagemeyer Pacific Tsunami Warning Center*

The United States National Weather Service (NWS) operates two warning centers to help protect U.S. Pacific coasts from the adverse effects of the tsunami hazard -- the Richard H. Hagemeyer Pacific Tsunami Warning Center (PTWC) in Ewa Beach, Hawaii, and the West Coast / Alaska Tsunami Warning Center (WC/ATWC) in Palmer, Alaska. There is no warning center for U.S. Atlantic or Caribbean coasts primarily because the hazard there is much smaller and no recent impacts have occurred. Due to ever-increasing coastal vulnerability, however, some form of mitigation and warning program for the Caribbean and Atlantic tsunami threat is needed. The NWS has recently been taking steps to address this issue. At a January 2003 meeting hosted by the University of Puerto Rico at Mayagüez, NWS personnel joined with representatives of other federal agencies and with regional scientists and emergency management representatives to formulate a tsunami warning plan for Puerto Rico and the Virgin Islands. As a result of those discussions, the following preliminary protocol was devised to provide limited tsunami warning capabilities for Puerto Rico and the Virgin Islands. The Puerto Rico Seismic Network (PRSN) will produce a rapid evaluation of any large nearby earthquake and will subsequently advise the Puerto Rico Emergency Management Agency (PREMA) if there is a potential tsunami threat. If PREMA concurs it will, in turn, sound sirens and advise the San Juan Weather Forecast Office (WFO) to issue a public message and activate the Emergency Alert System (EAS). Since PRSN is only manned during normal business hours, however, if the earthquake occurs at night or on the weekend it may not be able to respond in a timely way. To support the protocol, PTWC, backed up by WC/ATWC, will issue a Tsunami Information Bulletin for any large earthquake in the Caribbean or Atlantic, advising if there is a tsunami threat. The bulletin will be received by the San Juan WFO that will advise PREMA and take appropriate action. Although this protocol will provide some tsunami warning protection, there are many remaining issues to address including 24x7 capabilities, false warnings, no sea level data or evaluation, cancellations, evacuation plans and public education. In addition, the protocol does not provide guidance to other coastal regions in the Caribbean that may be at risk. PTWC, based on its more than 50 years experience with tsunami warning in the Pacific, is willing to help with this effort in whatever way it can.

# NEES PROGRAM'S SUPPORT FOR TSUNAMI EXPERIMENTAL RESEARCH

CHERRI M. PANCAKE
*Northwest Alliance for Computational Science & Engineering*
*Oregon State University*
*The George E. Brown Jr. Network for Earthquake Engineering Simulation*

The National Earthquake Engineering Simulation is an NSF-sponsored program that seeks to revolutionize the way in which earthquake engineering research is carried out through shared facilities supporting collaborative research at unprecedented scales. A total of 15 state-of-the-art experimental facilities are being linked together with advanced computing and networking technology. Managed by NEES Consortium, Inc., the entire system will open for business on October 1, 2004. A key feature of the NEES program is the new tsunami basin located at Oregon State University. Like the other NEES facilities, it was designed specifically to support distributed, collaborative research. Colleagues and students can participate remotely in experiments from their own institutions, using a unique, web-based electronic notebook that actually enhance observation through real-time display of sensor data, instant replay, and access to information on experiment design. This is made possible by a comprehensive data archive that also allows other groups to study or reuse data from previous tests as well. The ability to integrate numerical and physical experiments will allow researchers not only to validate their numerical models through experimentation, but also to use numerical methods to steer the course of lab experiments. Finally, it will be possible to design experiments that span physical laboratories, making it possible to tie in factors such as soil liquefaction or landslide-based generation of tsunamis. This presentation will describe the remarkable new capabilities offered by the NEES infrastructure, focusing on examples from the NEES tsunami basin.

# INTEGRATED TSUNAMI SCENARIO SIMULATION

Harry Yeh
*Oregon State University*

Earthquake scientists and engineers are well aware of the importance of detailed field surveys. However, it is difficult to collect sufficient and accurate data because the data/information is ephemeral and difficult to obtain. An alternative is to perform repeatable and precisely controlled "scenario" simulations. A scenario simulation means a case study. Tsunami phenomena and effects are simulated for given geographical, seismological, geological, and societal conditions. Simulations must be comprehensive and integrate not only tsunami generation, propagation, runup motion and its effects, but also other types of simulations such as warning transmission to the public, evacuation, environmental impacts, and rescue tactics. The simulation exercises should include physical models, numerical models, informatics, human behavior, communication simulations, and other exercises that will integrate the tsunami source with its eventual effects on communities and the environment.

What we intend to do is to set up a scenario for a hypothetical situation. The simulation exercise based upon a hypothetical but realistic coastal situation would result in realistic engineering evaluation, but it would not cause potential social and political concerns of a direct study of a specific location. An important benefit of the use of a scenario is to investigate a common problem from different angles by researchers in a wide variety of disciplines. Another benefit is to provide a better means for technology transfer from basic research. Results from research in academia generally end up in journal publication, but not be extended beyond in spite of their potential usefulness. Exercise of scenario simulations would make a researcher to effectively disseminate his/her findings to the broader community. Involvement in the scenario simulation will lead each researcher in the specialized area to interact closely with researchers in other areas, hence the researchers will obtain a broader picture for research, and identify critical problems and linkages to be examined for improvement.

Two workshops have been held; the first in Seattle WA, 2002, and the second in Corvallis OR, 2003, and two more are planned. A series of the forthcoming workshops will provide a coherent integrated community collaborative arena through a common task of tsunami scenario simulations, and will induce synergistic advancement in a wide variety of the simulation models. The ultimate goal is to achieve a widely accessible platform for simulation of virtual tsunamis through a community effort. When fully implemented, the tsunami scenario simulations will be useful for hazards planning in coastal communities, and for the coordination and development of scientific research.

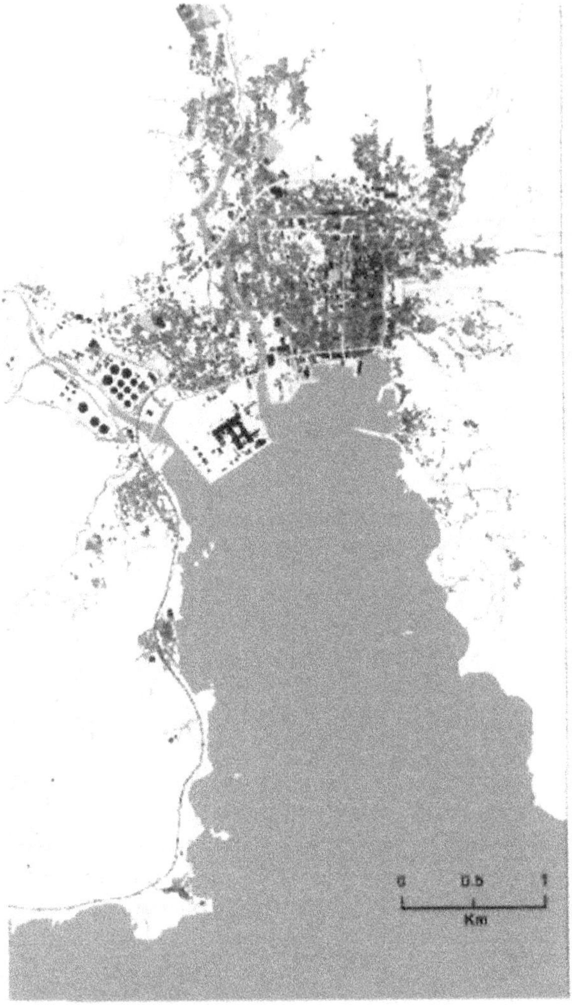


An example of a hypothetical but realistic coastal community in the GIS framework utilized for tsunami scenario simulations.

# TSUNAMI PHYSICS AND MODELING EFFORTS

# TSUNAMI FORCES IN THE RUNUP ZONE

HARRY YEH

*Department of Civil Engineering, Oregon State University*
*Corvallis, Oregon 97331, USA*

Structural damage by tsunami can be caused by direct water forces, impact forces by a water-born missile, fire spread by floating materials (including burning oil), scour and slope/foundation failure, and winds induced by the wave motion -- this last effect is considered to be very local and minor. As the first step, the first and fundamental cause, direct water force, is examined. The simplest case to study is runup on a uniformly sloping beach in a one-dimensional configuration. The existing analytic-numeric hybrid model of the fully nonlinear non-dispersive shallow-water wave equation is used to determine the temporal and spatial variations of water depth and flow velocity in the tsunami runup region, from which the linear-momentum flux is evaluated. The momentum flux can be interpreted as the drag force for a surface-piercing stationary object being placed vertically over the flow depth, but without the constant multiplier, i.e. the drag coefficient. The envelope of the extreme momentum-flux in the runup/rundown process is computed. The result will provide the tsunami force distribution in the runup zone. The actual force acting on a specific object must be computed with its size and shape, and the drag coefficient. Some example computations are given.

## 1. Introduction

Because it is impractical to give warning and evacuate people from the direct seismic effects of an earthquake, structures and infrastructures must be built to withstand the strong ground motion. On the other hand, there is usually a short lead-time for forecasting tsunami attack after receiving a seismic signal, which makes effective warning and evacuation possible. Combined with the fact that tsunami is a rare event, the primary mitigation measures have been to develop effective warning system and evacuation strategies. For this reason, research in the past has not focused on the development of methodologies for tsunami-resistant structures and infrastructures.

While this strategy – emphasis on warning and evacuation – makes sense, there are emerging needs to consider tsunami-resistant designs for man-made structures, particularly those whose failures can have severe effects. For example, destruction of a LNG or oil offshore berth terminal, jetty, and its storage tank can trigger giant fires, as well as severe environmental damages. In other cases, failure of critical coastal structures by tsunamis could cause enormous economic and military setbacks: examples are major coastal bridges and naval bases. Considering rapid utilization of the coastal areas in recent years, such disasters are waiting to happen.

There is another, and equally important need to evaluate tsunami-resistant structures. In some localities, evacuating people to high ground might be difficult. People living on a long spit or an island with no high ground nearby may not have a sufficient time to reach a safe ground. The only choice to minimize human casualty might be evacuation to upper floors of tsunami resistant buildings, i.e. tsunami shelters. Such buildings must be evaluated based on the design requirements to survive strong seismic ground shaking as well as subsequent tsunami effects.

There are some design guidelines for building standards available for the conditions under floods and wind-generated-wave loading: they are FEMA's "Coastal Construction Manual" (CCM) and ASCE's design standard "Minimum Design Loads for Buildings and Other Structures" (ASCE-7-02). Such design guidelines are inadequate for tsunami loading. There are several factors that make the development of tsunami construction guidelines formidable.

Unlike floods and storm waves, the characteristics and behavior of tsunamis are still uncertain, partly because of the lack of sufficient data due to the rare and short-lived nature of the event. Significant coastal tsunami inundation typically subsides within a few hours, and it happens at the same vicinity once every few hundred years or longer. It is suggested that the last significant tsunami triggered by faulting the Cascadia subduction (off the Washington-Oregon Coasts) was approximately 300 years ago (e.g. Atwater, 1987; Atwater et al., 1991; Satake et al., 1996). Standard coastal and oceanic instrumentation is often inadequate for tsunami measurements, due to tsunamis' distinct characteristics from storm waves.

In some events, tsunami runup heights deviate significantly within neighboring areas; such variations are not adequately predictable. The central portion of the town of El Transito was completely destroyed by the 1992 Nicaragua Tsunami but only minor damage resulted in the neighboring area: see Fig. 1. Field and modeling efforts reveal that local bathymetry often has a major influence on tsunami amplification (Yeh, 1998; SCOR 107, 2002). Another example is concentration of tsunami energy in a narrow region of Sissano Lagoon resulting from the 1998 Papua New Guinea Tsunami. It was caused primarily by tsunami refraction owing to the convex shaped nearshore bathymetry (Matsuyama et al., 1999). The exceptionally high tsunami runup (32 m) measured at a pocket beach in Monai, Okushiri Island, by the 1993 Hokkaido Nansei-Oki Tsunami resulted from interference of the reflected and incident tsunamis directed by the bathymetry and coastal topography

a) b)

Figure 1. Scenes of the tsunami damages at the town of El Transito, caused by the 1992 Nicaragua Tsunami. b) beach house only 300 m north of the completely destroyed scene shown in (a).

(Matsuyama and Tanaka, 2001). Tsunami effects in the three-dimensional coastal bathymetry and topography are far from being understood sufficiently.

It is difficult to establish a design guideline for tsunami loading for buildings and infrastructures in the coastal zone. The difficulties stem from the following factors:

- Because it is an extremely rare event, no probabilistic analysis on tsunami effects would be rational, and the design incident tsunami condition cannot be determined reliably.
- The fluid force is a strong function of the cross-shore distance: the force diminishes quickly away from the shore. The fluid force is also a function of the along-shore location due to the three-dimensional bathymetry and topography effects as discussed above.
- The fluid force exerted on a structure is also influenced by the surrounding buildings and infrastructure.
- The buoyant force due to inundation must be considered for the analysis: the rapidly rising water often causes a building to float.
- Tsunami runup and drawdown are a transient process: for a typical tsunami, the water surface fluctuates with runup amplitude of several meters during a period of a few to tens of minutes. This timescale is intermediate between the hours to days typical of river flood problems, and the tens of seconds or less associated with cyclic wind-wave loading on coastal structures.
- Building and infra-structure damage by tsunami can be caused not only by direct fluid forces, but also by impact forces by water-born missiles, as well as foundation failure due to scour.

In this paper, I address only one of the factors necessary for the tsunami-load predictions: the estimation of horizontal fluid force resulting from tsunami runup/drawdown processes onto a uniformly sloping beach. No other factor listed above, including along-shore variations, is considered. The fully nonlinear shallow-water-wave theory is used to predict the distribution of fluid force in the cross-shore distance. The theoretically based prediction should yield a solid basis for the development of a design guideline for tsunami loading.

## 2. Methodology

The solution algorithm of tsunami runup onto a uniformly sloping beach was developed by Carrier et al. (2003). The algorithm is based on the fully nonlinear shallow-water theory with an arbitrary initial condition. The nonlinear equations of mass conservation and linear momentum are first transformed to the single linear hyperbolic equation. Using the Fourier-Bessel transform and its inversion, the equation is further transformed to the Green function representation. The solutions in the physical time and space domains are then obtained by numerical integration. The following is a concise sketch of the derivation; for details, see Carrier et al. (2003).

We start with the fully nonlinear shallow-water-wave equations for propagation over a plane beach with a uniform slope α:

$$\begin{aligned} &\frac{\partial \eta'}{\partial t'} + \frac{\partial}{\partial x'}\left[u'(\alpha x' + \eta')\right] = 0, \\ &\frac{\partial u'}{\partial t'} + u'\frac{\partial u'}{\partial x'} + g\frac{\partial \eta'}{\partial x'} = 0 \end{aligned} \tag{1}$$

where $u'$ is the horizontal flow velocity that is assumed uniform over the depth, $\eta'$ is the vertical departure of the water surface from its quiescent position, $g$ is gravitational acceleration, $x'$ is the horizontal coordinate pointing offshore from the shoreline of the quiescent state, and $t'$ is time. The pair of shallow-water equations is nondimensionalized with the following scaling parameters:

$$u' = \sqrt{g\alpha L}\, u; \quad \eta' = \alpha L\, \eta; \quad x' = L\, x; \quad t' = \sqrt{L/\alpha g}\, t, \tag{2}$$

where $L$ is any convenient horizontal length scale, which may be the distance from the shoreline to the middle of the postulated initial wave condition. Then with the transformation by the distorted coordinates $\lambda = t - u$ and $\sigma = \sqrt{x + \eta}$, (1) can be combined to a single linear differential equation of a cylindrical wave:

$$4\sigma\frac{\partial^2\varphi}{\partial\lambda^2}-\frac{\partial}{\partial\sigma}\left(\sigma\frac{\partial\varphi}{\partial\sigma}\right)=0\cdot \tag{3}$$

Note that $u$ and $\eta$ can be obtained once (3) is solved for $\varphi$:

$$u=-\frac{1}{2\sigma}\frac{\partial\varphi}{\partial\sigma},\text{ and }\eta=\frac{\partial\varphi}{\partial\lambda}-\frac{u^2}{2}. \tag{4}$$

Equation (3) is a second-order partial differential equation, which requires two independent initial conditions at $\lambda = 0$ to solve for a unique solution: the general initial conditions can be expressed as

$$\varphi(\sigma,0)=P(\sigma)\ \text{ and }\ \frac{\partial\varphi(\sigma,0)}{\partial\lambda}=F(\sigma), \tag{5}$$

in which

$$P(\sigma)=-\int_0^\sigma 2\sigma' u(\sigma',0)\,d\sigma',\ \text{ and }\ \mathrm{F}(\sigma)=\eta(\sigma,0)+\frac{u^2(\sigma,0)}{2}. \tag{6}$$

Utilizing the Fourier-Bessel transform of (3) and (5), the above initial-value problem can be represented with the Green function:

$$\varphi(\sigma,\lambda)=2\left\{\int_0^\infty F(b)\,G(b,\sigma,\lambda)\,db+\int_0^\infty P(b)\,G_\lambda(b,\sigma,\lambda)\,db\right\}, \tag{7}$$

where

$$G(b,\sigma,\lambda)=\begin{cases}0 & \text{for } \frac{\lambda}{2}<|\sigma-b| \\ \frac{1}{\pi}\sqrt{\frac{b}{\sigma}}K\left(\frac{\lambda^2-4(\sigma-b)^2}{16\sigma b}\right) & \text{for } |\sigma-b|<\frac{\lambda}{2}<|\sigma+b| \\ \frac{4}{\pi}\frac{b}{\sqrt{\lambda^2-4(\sigma-b)^2}}K\left(\frac{16\sigma b}{\lambda^2-4(\sigma-b)^2}\right) & \text{for } \frac{\lambda}{2}>|\sigma+b|\end{cases} \tag{8}$$

is the Green function which would permit the determination of the $\varphi(\sigma, \lambda)$ that would ensue from the initial condition. In (8), $K(\bullet)$ is the Complete Elliptic Integral of the first kind. Equation (7) can be integrated numerically with standard integration software (e.g. the IMSL library) for any regular initial conditions in (5).

Unlike the ordinary time-stepping numerical methods with grids or elements, the present method can solve the fully nonlinear shallow-water-wave equations for arbitrarily selected times and locations (or selected temporal and spatial domain) without computing for other domains. This is advantageous especially for the computations for tsunami runup: only near $\sigma = 0$ and a certain range of $\lambda$ needs to be computed. Also the present method will cause no numerical error accumulation for large $t$, because it does not rely on time-stepping computation.

Using this algorithm, Carrier et al. (2003) analyzed the runup and rundown motions with four initial water displacements: a) the Gaussian shape, b) the negative Gaussian shape, that is the inverse profile of Case a, c) the leading depression N-wave shape, typically caused by a seismic fault dislocation by subduction earthquake, and d) the leading depression N-wave shape, typically caused by an offshore submarine landslide. The initial waveforms are shown in Fig. 2. The detailed temporal and spatial variations of water-surface and velocity associated with the runup and rundown were computed. They also computed the linear-momentum flux per unit breadth; that is, for the shallow-water theory, the product of fluid density, inundation depth, and square of the flow velocity. The momentum flux may be interpreted as the drag force per unit breadth for a surface-piercing stationary object being placed vertically over the flow depth. Hence it provides a measure of the net force exerted on such an object (e.g. a building).

Carrier et al. (2003) found that for all four cases, the maximum momentum flux occurs near the extreme drawdown location, and immediately prior to or after the flow reversal. For predominantly positive initial waves (Cases a and c), the maximum momentum flux results during the drawdown phase and points offshore. On the other hand, for the cases of predominantly negative initial waveform (Cases b and d), the maximum momentum flux occurs during the runup immediately after the extreme drawdown, acting in the inshore direction.

## 3. *Tsunami Force Distribution*

There are many reports on discoveries of boulders and rocks that are considered being transported by tsunamis. Some cases are speculative, but some field observations made by the recent reconnaissance tsunami surveys show clear evidence of such a phenomenon: see Fig. 3. Utilizing Case d of Carrier et al. (2003), i.e. the model of submarine-landslide induced tsunami, I demonstrate that the realistic tsunami action is capable of transporting large boulders onto the shore. The linear momentum flux of Case d was computed and presented in the

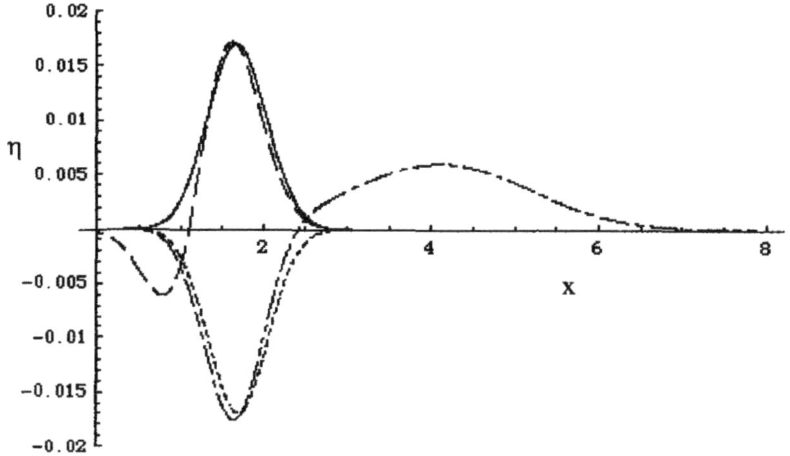


Figure 2. Initial waveforms used by Carrier et al. (2003): ——, Gaussian shape; - - - -, negative Gaussian shape; — — —, leading depression N-wave by fault dislocation; — - — -, leading depression N-wave caused by submarine landslide.

dimensionless form in the space-time plane in Fig. 4. As mentioned earlier, the maximum force occurs at the initial runup phase immediately after the drawdown, and it is in the inshore direction. From the results in Fig. 4, the maximum force distribution in space was found as shown in Fig. 5 (the maximum force shown in the figure is in the inshore direction everywhere).

Taking the scaling parameter $L = 5$ km, Case d represents the submarine landslide generated on the beach slope of 1/10 with the maximum water-surface depression of approximately 9 m at 8.5 km offshore, which reflects a realistic but very large event; the computed maximum runup height is 16.4 m. With these parameters, the maximum force per unit breadth is estimated by

$$F = \frac{1}{2} C_d \rho h u^2 \tag{9}$$

where $C_d$ is the drag coefficient (set unity), $h$ is the total water depth, and $\rho$ is the fluid density. Figure 6 shows the force distribution in the cross-shore location. Similar distribution curves were presented by Dean (2000) for wind- wave attenuation in coastal flood zones. Note that the drag coefficient is a function of Reynolds number and the shape of the object, and is for steady flows. Actual tsunami action is not steady and the shape of boulders is not uniform. None the less, the value of $C_d$ for a 3-D object like a boulder should not deviate significantly from O(1). For a 2-D object, Arnason (2005) found in his

Figure 3. A beach rock (boulder) transported by the 1993 Okushiri Tsunami. The rock is located approximately 30 m inshore at 5 m height from the shoreline.

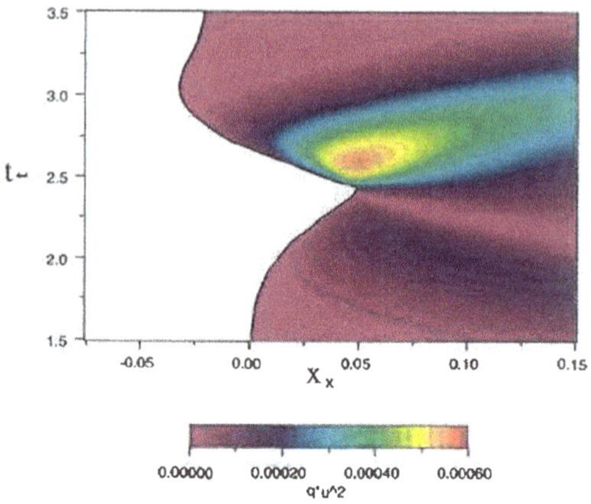


Figure 4. Temporal and spatial variation of the linear momentum flux for Case d (the wave form caused by the submarine landslide).

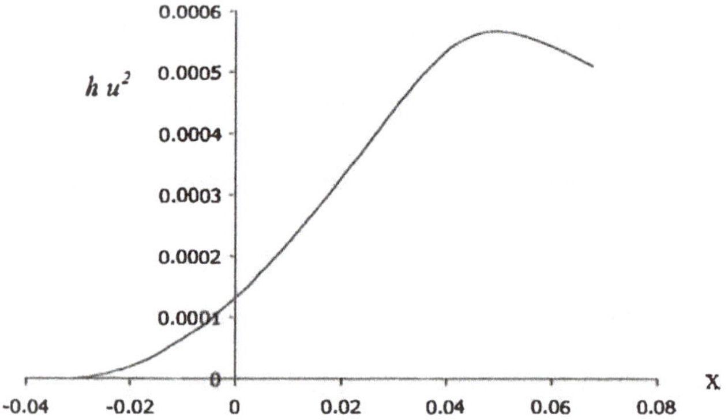


Figure 5. The spatial distribution of the maximum momentum flux taken from Fig. 4.

laboratory experiments that $C_d \approx 2$ for a surface piercing vertical square column under the flow conditions simulating the tsunami action on the horizontal bed; the value is approximately the same for steady 2-D flows. Arnason (2004) also demonstrated that the initial tsunami impact force (transient) is slightly higher than the subsequent drag-dominated force (quasi-steady).

Consider a boulder of 10 m$^3$ ($\approx$150 kN of its weight force in the water) initially located at 100 m offshore (x = 100m; depth = 10 m); the boulder is larger than the one shown in Fig. 3. According to Fig. 6, the maximum fluid forceat this location is approximately 400 kN/m. If the boulder has the frontal

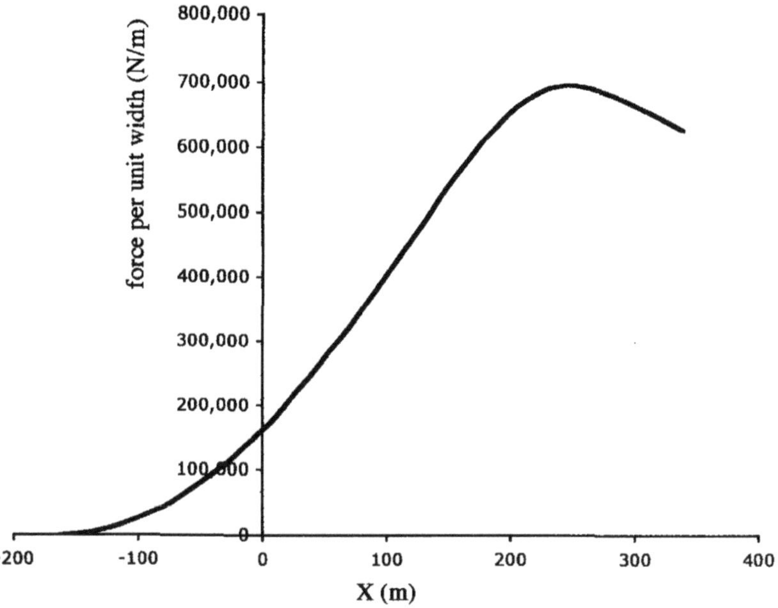


Figure 6. The spatial distribution of the maximum momentum flux (fluid force per unit breadth) for $L = 5$ km and $\alpha = 1/10$ based on Case d.

area of $10^{2/3} \approx 4.6\,m^2$ then, the force exerting on the boulder would be 185 kN. The ratio of the exerted force to the weight ($\approx 1.23$) clearly exceeds the reasonable value of the critical kinetic friction coefficient: the ratio represents the condition equivalent to the static condition that the boulder is placed on the slope of more than 50°! This clearly demonstrates that a significant tsunami is capable of transporting a large boulder.

The maximum force distribution in the runup zone is also computed for Cases a, b and c. Note that as seen in Fig. 2, Carrier et al. (2003) selected the examples so that the initial waveform for each case has approximately the same maximum amplitude as well as the breadth of the primary displacement. To include a wider set of the initial tsunami conditions, two additional cases of the Gaussian shape are computed: one with a smaller and narrower displacement than Case a, the other with a displacement farther offshore. Figure 7 shows the maximum fluid force distributions for all six cases by translating the origin (x =0) at the maximum runup location: the short vertical bars along the abscissa indicate the initial shoreline locations. The maximum fluid force occurs during the runup phase for Cases b and c, and during the drawdown phase in the rest of the cases. In spite of a variety of the initial waveforms, amplitudes, and the source locations, the distribution curves in the runup zone (i.e. the initial

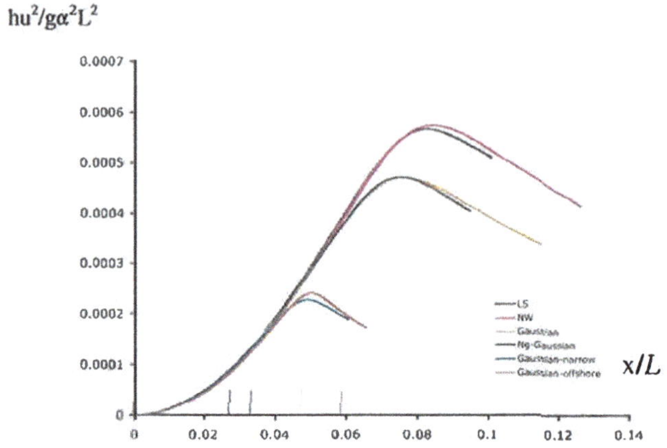


Figure 7. The spatial distribution of the maximum momentum flux for various initial tsunami conditions. The short vertical lines along the abscissa show the locations of the initial shoreline for the first four cases.

shoreline to the maximum runup) appear to collapse to a single curve in the present choice of the non-dimensionalization. I also computed for additional cases with much weaker initial conditions, i.e. smaller displacement amplitude and broader displacement breadth. Those fluid force curves do not coincide with the others, but fall under, hence the rate of increase in the force may represent the envelope of the maximum tsunami force in the runup zone.

To make the results in Fig. 7 more useful, the results are rescaled with the parameter $L$. Note that the length scale $L$ in (2) can be any horizontal length scale. It is more convenient to represent the tsunami force distribution with the runup distance as shown in Fig. 8, where $\ell$ is the distance between the initial shoreline and the maximum runup locations, i.e. the inundation distance.

As an example, suppose the predicted inundation distance to be 250 m on the beach with the slope 1/50 – this is equivalent to the maximum runup height, 5 m. Assuming $C_d \approx 2.0$ for a building of 10 m wide, the predicted water force exerted on the building can be presented in Fig. 9. According to Fig. 9, the tsunami force exerted on the building located at 50 m inshore from the shoreline would be 200 kN.

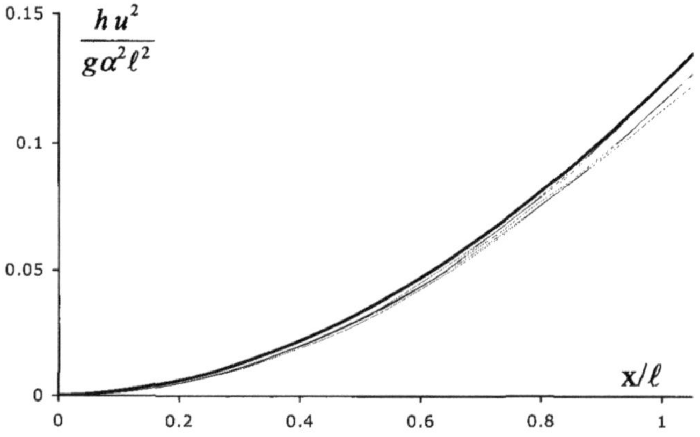


Figure 8. The spatial distribution of the maximum momentum flux for various initial tsunami conditions, scaled with the inundation distance *l*. The envelope is drawn in a thick line.

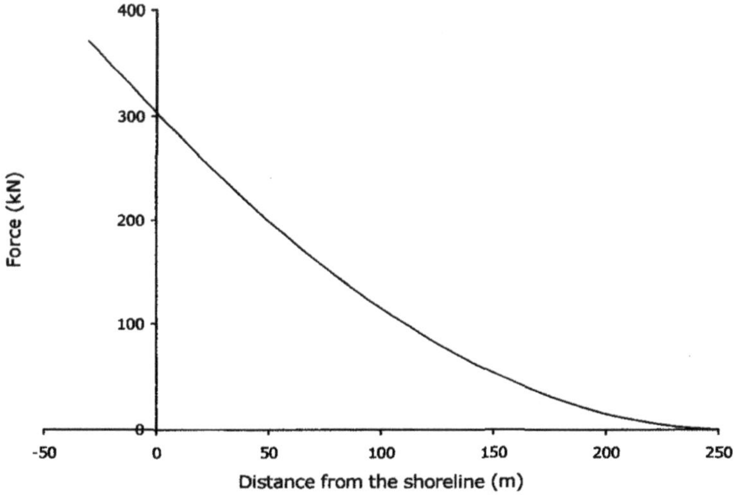


Figure 9. Tsunami forces exerted on a 10-m wide building when the inundation distance is 250 m on a beach slope 1/50.

## 4. Discussion

Tsunami forces in the runup zone were evaluated using the algorithm recently developed by Carrier et al. (2003). While the model is theoretically sound – the exact solution of the fully nonlinear shallow-water-wave theory – the assumptions involved in the algorithm make the force evaluation valid only for ideal and simplified situations: i.e. 1-D spatial variation on a uniformly sloping beach. While the simplified conditions do not accurately represent a real coastal environment, the present results – in particular, Fig. 8 – should provide a convenient estimate of tsunami-force attenuation from the initial shoreline to the maximum runup location. Since the estimation is theoretically based with unambiguous assumptions, the results in Fig. 8 can be a basis for a design guideline for tsunami resistant buildings and infrastructures located in the runup zone. To estimate the fluid force exerted on a specific object, one must determine the value of drag-coefficient $C_d$ in (9). A recent laboratory study by Arnason (2004) suggests $C_d \approx 1 \sim 2$ for partially submerged columns, depending on the column shape and the Reynolds number.

## 5. Acknowledgments

This work was supported by the National Tsunami Hazard Mitigation Program through Washington State Military Department. The theoretical development of the prediction model was supported by the US National Science Foundation (CMS-9978399 and CMS-0245206).

## 6. References

Arnason, H. 2005. Interactions between an incident bore and a free-standing coastal structure. Ph.D. thesis, University of Washington, 172pp.

Atwater, B. F. (1987). "Evidence for Great Holocane Earthquakes Along the Outer Coast of Washington State." Science 236, 942-944.

Atwater, B. F., M. Stuiver, and D. K. Yamaguchi. (1991). "Radiocarbon test of earthquake magnitude at the Cascadia subduction zone." Nature 353, 156-158.

Carrier, G.F., Wu, T.T. and Yeh, H. (2003). Tsunami Runup and Drawdown on a Plane Beach, Journal of Fluid Mechanics, 475, 79-99.

Dean, R.G. (2000). "Storm damage reduction potential via beach nourishment " Coastal Engineering 2000, Conference Proceedings, ASCE, 3305-3318.

Coastal Construction Manual, FEMA 55, 3rd Ed. Federal Emergency Management Agency.

Matsuyama, M., and Tanaka, H. (2001). An experimental study of the highest run-up height in 1993 Hokkaido Nansei-oki earthquake tsunami. International Tsunami Symposium: Seattle, August 7-10.

Matsuyama, M., Walsh, J.P., and Yeh, H. (1999). The effect of bathymetry on tsunami characteristics at Sissano Lagoon, Papua New Guinea. Geophysical Research Letters, 26, 3513-3516.

MinimumDesign Loads for Buildings and Other Structures, ASCE-7-02.

Satake, Kenji, K. Shimazaki, Y. Tsuji, and K. Ueda. (1996). "Time and size of a giant earthquake in Cascadia inferred from Japanese tsunami records of January 1700." Nature 379, 247-248.

SCOR 107, (2002). Improved global bathymetry. IOC Technical Series No 63, IOC, UNESCO, Paris, 122pp.

Yeh, H. (1998). Tsunami Researchers Outline Steps for Better Data. EOS, Trans. Amer. Geophys. Union, 79, 480 & 484.

# WAVE GENERATION, RUNUP AND RUNDOWN FROM THREE-DIMENSIONAL SLIDING MASSES

Philip L.-F. Liu
*Cornell University*

Subaerial and submarine landslides may generate sizeable offshore and onshore propagating waves leading to significant danger of coastal inundation with little warning time. To better understand the waves and runup, several small and large-scale scale experiments have been conducted. On the other hand, due to the limited computational capability and the difficulties of treating complex turbulent surface waves and the moving boundary, not many simulation models based on the 3D Navier-Stokes equations have been reported. In this study, we develop a Large Eddy Simulation (LES) model by adopting the traditional Smagorinsky Sub-Grid Scale (SGS) model. The LES is coupled with a moving boundary algorithm for simulating the slide movement. The Volume of Fluid (VOF) method similar to used to track the surface location. 45 simulations have been conducted with different initial elevations and specific weight of solid slides. The results are compared with the experimental data. Many physical phenomena observed from the numerical solutions will be addressed and discussed. The simulation model can also be used to calculate wave forces on structures.

# THREE-DIMENSIONAL RUNUP DUE TO SUBMERGED AND SUBAERIAL LANDSLIDES*

P. LYNETT

*Department of Civil and Ocean Engineering*
*Texas A&M University*
*College Station, TX 77843, USA*
*E-mail: plynett@civil.tamu.edu*

P. L.-F. LIU

*School of Civil and Environmental Engineering*
*Cornell University*
*Ithaca, NY 14853, USA*
*E-mail: pll3@cornell.edu*

A large set of numerical experiments are designed to examine the maximum runup generated by 3D submerged and subaerial, solid body landslides. A depth-integrated numerical model is utilized, allowing for the efficient simulation of landslides in shallow and intermediate water. Six dimensionless parameters are introduced: the slide thickness, the slide wavenumber, a slide shape parameter, the horizontal aspect ratio of the slide, the specific gravity of the slide mass, and the slope of the beach. Six sets of simulations are first presented, wherein one of the six dimensionless parameters are singularly varied. This allows for the identification of parameter dependence on maximum runup. After combining the dependencies, a number of relationships appear. Most notably, a very clear division between the near and far field is observed. For submerged slides, a nondimensional estimation of the maximum runup just landward of the slide results, as well as the location and magnitude of the secondary runup peak. This secondary peak is due to the propagation of edge waves, and is in some cases larger than the peak immediately landward of the slide. The results presented in this paper may be useful for preliminary hazard assessment, where a simple and quick estimation of the maximum runup height and locations are required. Additionally, the formulas developed will be particularly beneficial to those developing 3D landslide experiments.

*The research reported here is partially supported by grants from National Science Foundation (cms-9528013, cts-9808542, and cms 9908392)

## 1. Introduction

To date, very few studies of landslide generated tsunamis include two horizontal dimension (2HD) effects. Numerically, although there are numerous 2HD landslide model presentations (e.g. Jiang & LeBlond, 1994 and Grilli *et al.*, 2002) there is little in the way of model application for tsunami characterization. From an experimental perspective, 2HD landslide tsunami studies are particularly daunting, because of difficultly in minimizing reflection off three tank walls, such that uncontaminated runup measurements can be obtained. Despite this hurdle, recent experiments (Synolakis & Raichlen, 2002) have been performed with a tank width of 4 slide widths. This type of experiment allows for detailed physical study of the leading tsunami waves, but wall reflections quickly contaminate secondary behavior. Without examination of the secondary behavior, it becomes possible to miss one of the important phenomenon controlling landslide tsunami runup - edge waves.

Edge waves have been studied by coastal engineers and scientists for many decades, largely motivated by Ursell's (1952) analytical solution for edge waves on a constant slope. Edge waves, while rarely observed directly in typical coastal environments (Schaffer & Jonsson, 1992), can be easily discriminated for transient disturbances, such as tsunamis (Gonzalez *et al.*, 1995). It seems reasonable to expect that submerged landslide tsunamis, and especially subaerial landslide tsunamis, with their fair amount of oblique wave energy, would generate shore-trapped, edge wave modes. In fact, this expectation has already been observed in experimental testing (e.g. Liu *et al.*, 1998; Chang, 1995) and numerical simulation of landslide tsunamis (e.g. Tinti *et al.* 1999).

This paper is organized as follows. First, a description of the theoretical model to be used is presented, followed by the method of numerical simulation. Next, the landslide geometry and motion profiles are given. It is in this section that the non-dimensional parameters governing the physical slide characteristics are developed. After the setup description, results from the numerical simulations are examined. For each of the six non-dimensional parameters, a subset of simulations is run, varying that parameter while holding the other five constant. Finally, all of the scaling laws extracted from the six subsets of simulations are combined, yielding global scalings.

## 2. Model Equations

The model to be applied here is the 2HD, depth-integrated, multi-layer model. The multi-layer concept represents a different approach to developing a depth-integrated model with high-order dispersive properties. The multi-layer derivation consists of a piecewise integration of the primitive equations of motion through $N$ constant-density layers of arbitrary thickness. Within each layer, an independent velocity profile is determined. With $N$ separate velocity profiles, matched at the interfaces of the layers, the resulting set of equations have $N+1$ free parameters, allowing for an optimization with known analytical properties of water waves. The optimized two-layer model equations, which are used in the study presented here, show good linear wave characteristics up to $kh \approx 8$, while the second-order nonlinear behavior is well captured to $kh \approx 6$. Details of the multi-layer approach can be found in Lynett (2002) and Lynett & Liu (2004a, 2004b), while applications can be found in Basterretxea *et al.* (2004) and Ryu *et al.* (2003).

The simplest of the multi-layer models is the one-layer model, which is equivalent to the model of Liu (1994) and the fully nonlinear extension of Nwogu's (1993) model presented by Wei & Kirby (1995). An analysis of the one-layer model's ability to predict the waves created by submarine landslides was done in Lynett & Liu (2002). All of the accuracy restrictions given in that paper, in the form of minimum slide length scale to depth of submergence ratios, were closely followed in this study.

## 3. Numerical Model

The finite difference algorithm presented in Lynett & Liu (2002, 2004a) is used to solve the model equations. A high-order predictor-corrector scheme is utilized, employing a third order in time explicit Adams-Bashforth predictor step, and a fourth order in time Adams-Moulton implicit corrector step (Press *et al.*, 1989). Spatial derivatives are differenced to fourth order accuracy, yielding a model which is numerically accurate to $(\Delta x)^4, (\Delta y)^4$ in space and $(\Delta t)^4$ in time.

Runup and rundown of the waves generated by the landslide disturbance will be examined. The moving boundary scheme employed here is the technique developed by Lynett *et al.* (2002). Founded around the restrictions of the high-order numerical wave propagation model, the moving boundary scheme utilizes linear extrapolation of free surface and velocity through the shoreline, into the dry region. This approach allows for the five-point finite difference formulas to be applied at all points, even those neighboring dry points, and thus eliminates the need of conditional statements.

## 4. Landslide Description

### 4.1. *Slide Motion*

All of the simulations undertaken for this study use a solid slide mass traveling down an infinite slope. When fully submerged, the slide motion is described by the formulation given by Watts (1997). This formulation requires the specification of a number of coefficients, and those used in this paper are identical to the set employed by Grilli & Watts (1999).

Many of the simulations performed, however, are for initially subaerial slides. The expression for slide velocity must be altered to include the aerial acceleration. The approach taken here is to formulate the slide velocity as a weighted average of the aerial and submerged velocity, where the weighting is based on the fraction of the landslide submerged. Thus the slope-parallel velocity of the slide is given by

$$f_s u_s + f_a g t sin\theta \tag{1}$$

where $u_s$ is the time-dependant velocity of a submerged landslide, and can be found in Grilli *et al.* (2002), $g$ is gravity, $t$ is the time, and $\theta$ is the slope angle. The coefficients $f_s$ and $f_a$ represent the volume fractions of the landslide submerged and aerial, respectively, and of course must fall between 0 and 1, summing to 1.

### 4.2. *Landslide Geometry*

The graphical definitions of the variables described in this section are shown in Figure 1. The spatial profile of the time-variable seafloor profile can be given in the general form:

$$h(x, y, t) = h_o(x) - h_S(x, t)G(y) \tag{2}$$

where $h_o$ is the planar "baseline" profile with slope angle $\theta$, $h_S$ is the shore-normal profile of the sliding mass, and $G$ is a specified transverse (shore-parallel) profile. The shore-normal profile is similar to Lynett & Liu (2002):

$$h_S(x, t) = \Delta h \frac{\left[1 + tanh\left(\frac{x - x_l(t)}{S_l}\right)\right]\left[1 - tanh\left(\frac{x - x_r(t)}{S_r}\right)\right]}{\left[1 + tanh\left(\frac{b cos\theta}{2S_l}\right)\right]\left[1 - tanh\left(-\frac{b cos\theta}{2S_r}\right)\right]} \tag{3}$$

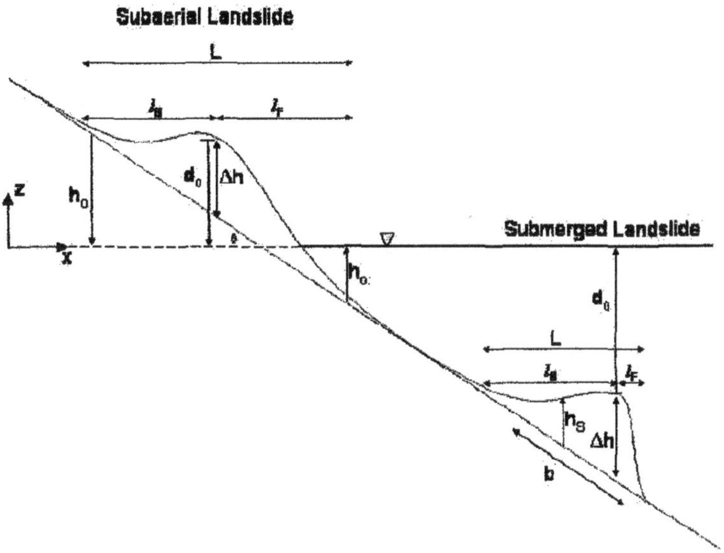


Figure 1. Definitions for a subaerial (left) and submerged (right) sliding mass. Note that the subaerial slide is symmetric in the horizontal plane ($\kappa = l_F/l_B = 1$) while the submerged slide is not ($\kappa \approx 0.15$).

where $\Delta h$ is the maximum vertical height of the slide, $x_l$ is the location of the $tanh$ inflection point of the left side of the slide, $x_r$ is the location of the inflection point on the right side, $b$ is the slope-parallel distance between the inflection points, and $S_l, S_r$ are shape factors controlling the steepness of the left and right slide sides. The right and left boundaries, and steepness factors are given by

$$x_l(t) = x_c(t) - \frac{b}{2}cos\theta, \quad x_r(t) = x_c(t) + \frac{b}{2}cos\theta$$

where $x_o$ is the horizontal location of the center point of the slide, determined using (1). Shown in Fig. 1 are typical slide shapes created by the above formulation. It is remarked that the slightly awkward form of (3) is employed so that the effects of slide asymmetry, which can be altered through $S_l$ and $S_r$, can be examined.

For the transverse profile, a Gaussian is employed:

$$G(y) = exp\left[-\left(\frac{y - y_o}{0.23W}\right)^2\right] \tag{4}$$

where $y_o$ is the centerline coordinate of the slide, $W$ is the width of the slide, defined as the distance between locations where the vertical slide thickness is 1% of the maximum thickness. For the simulations to be presented in the following sections, there are eight parameters to be varied:

$\Delta h$: the maximum vertical thickness of the slide.

$d_o$: the vertical distance from the centerpoint of the slide to $z$=0; positive for the submerged slides and negative for the subaerial slides presented here.

$l_F, l_B$: the horizontal length of the front slide side (forward-facing) and the back slide side, respectively, in the shore-normal direction. Defined as the distance from the location of maximum vertical slide thickness, $\Delta h$, to the location where vertical slide thickness is $0.01\Delta h$. These lengths are varied through changing $S_l$ and $S_r$.

$L$: $= l_F + l_B$ the total horizontal length of the slide in the shore-normal direction. Additionally, we define a wavenumber for the slide, $k_S$, such that $k_S = 2\pi/L$.

$W$: the horizontal width of the slide in the shore-parallel direction. Defined as the distance between locations where the vertical slide thickness is $0.01\Delta h$.

$\rho_S$: the density of the landslide material.

$\theta$: the slope angle of the "baseline" depth profile, $h_o$.

Within this dimensional set, six characteristic dimensionless parameters are defined:

$\epsilon = \Delta h/d_o$: the dimensionless slide thickness. This parameter is related to the dimensionless amplitude of the generated wave, and thus is an indicator of the importance of shallow water nonlinearity.

$\kappa = l_F/l_B$: the ratio of the front and back slide lengths. Within this value is a measure of the shape, or symmetry of the slide mass, where a value of 1 represents a symmetric slide in the horizontal plane.

$\mu = k_S\Delta h$: a measure of the steepness of the slide. Constructed as such to carry a relevance to the steepness of a water wave, $ka$.

$A = L/W$: the aspect ratio of the slide.

$\gamma = \rho_S/\rho_{water}$: the specific gravity of the landslide material.

$S = tan\theta$: slope of planar beach.

Each of these dimensionless parameters will first be varied independently, and the individual effects of each on the 2HD runup profile will be discussed. The accuracy restrictions given in Lynett & Liu (2002) can be expressed in terms of the above parameters, and requires:

$$\mu\left(max\left[\kappa,\frac{1}{\kappa}\right]+1\right)\left(\frac{1}{\epsilon}+1\right)\leq\frac{4\pi}{7} \tag{5}$$

Additionally, slides with large $\epsilon$ (relatively small $d_o$) are not simulated. These are submerged slides that are nearly initially penetrating the still water surface, and numerical simulations indicated that turbulent effects were likely very important. Note that for this analysis, all of the slides denoted "subaerial" are characterized by $d_o < 0$, while all "submerged" slides have $d_o > 0$. The large $\epsilon$ slides represent a transitional class of slides between the submerged and subaerial slides simulated in this paper.

## 5. 2HD Runup Created by Submerged and Subaerial Landslides

Before analyzing the individual effects of the dimensionless parameters through groups of simulations, a single simulation is examined with

$$\epsilon = 0.4, \quad \kappa = 1.0, \quad \mu = 0.01, \quad A = 1.0, \quad \gamma = 2.0, \quad S = 1/10$$

The above $\epsilon$ indicates that nonlinear effects are likely important, $\mu$ tells that this slide is thin (or long), and the slope is relatively steep, although still in the validity range of the depth-integrated model used here. This simulation, as well as all to be discussed, employed a spatial domain that was large enough such that reflections off the side walls do not affect any of the results presented. Eight spatial snapshots of the free surface elevation, $\zeta$, are plotted in Figure 2. As the slide motion initiates, at $t\sqrt{gd_o}/L$=$t'$=0.3, an elevation wave is generated and traveling offshore, while the depression wave created by the landward slide face is already interacting with the shoreline. As the slide continues into deeper water, the offshore directed waves develop into a oscillatory train, trailed by short waves resembling a "V". By $t'$=1.0, edge waves begin to dominate the wave field near the shoreline. These trapped waves exhibit their characteristic dispersive behavior, as the longer waves travel away from $y$=0 quickly, leaving a dispersive tail of shorter waves behind. As the edge wave energy leaks offshore, the nonlinear occurrence of Mach stem is evident (at $|y/W| = y' \approx 7$, $t'$=2.4), adding further complexity to this complicated phenomenon. It is clear from this simulation that 2HD effects can be extremely important to the accurate prediction of landslide generated shoreline movement.

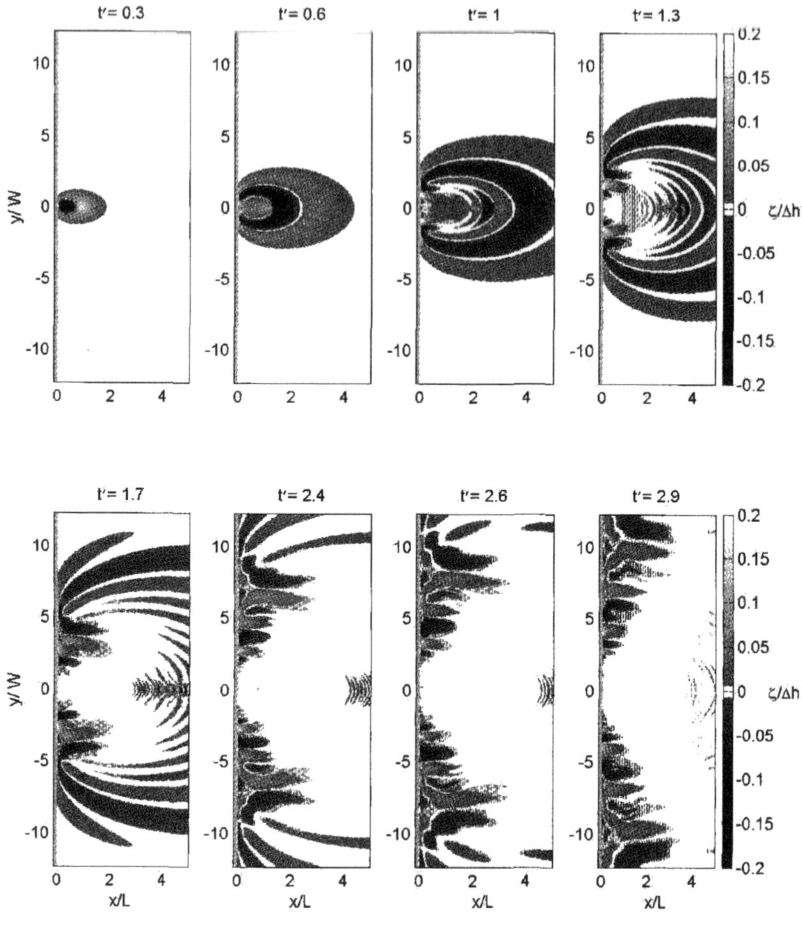


Figure 2. Spatial snapshots of the free surface.

The complete picture of shoreline motion for this simulation is shown in Figure 3. This figure gives the shoreline elevation as a function of time for all $y'$ shoreline locations. At the earliest times, $t' < 0.2$, the only section of shoreline experiencing water motion is that directly landward of the slide due to the backface-generated depression wave. The frontface elevation wave quickly wraps around the depression such that immediately outside of the slide centerline (i.e. $y' > 0.5$), shoreline motion is characterized by

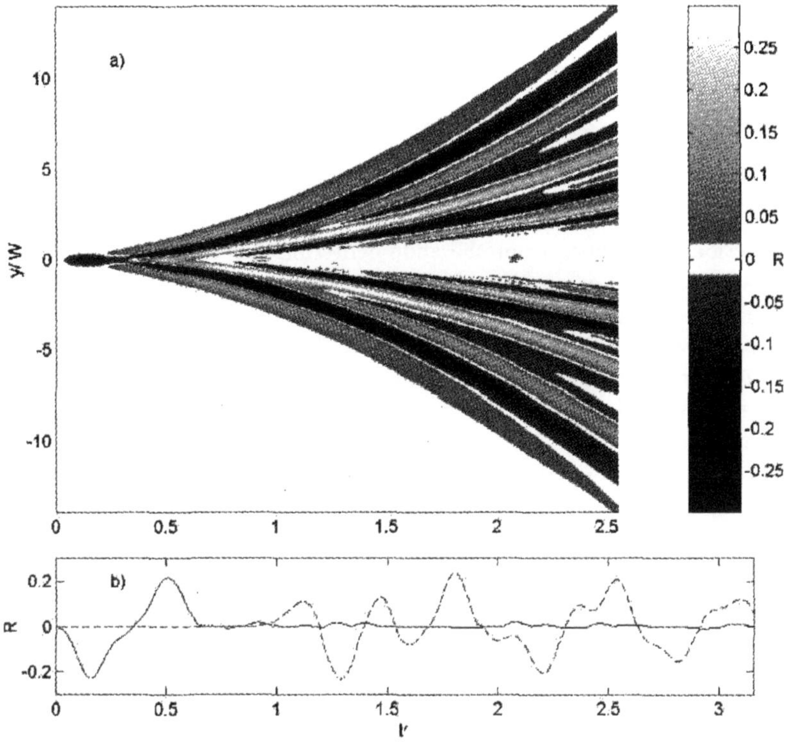


Figure 3. In the upper plot a), shoreline elevation as a function of time and transverse coordinate. In b), runup time series at $y/W = 0$ (solid line) and $y/W = 3$ (dashed line).

a leading elevation wave. Progressing further in time, the edge waves lead to numerous significant oscillations of the shoreline, for example there are a series of large runup maxima near $y'$=3, with the third being the largest (runup time series are also shown in Fig. 3). This is in interesting contrast to the location where one might expect the motion to be most extreme; at $y'$=0 there is only a single elevation peak, with a magnitude less than that for $1 < y' < 5$. Again, 2HD effects, in particular edge wave formation, drive maximum runup patterns.

In the following subsections, the 2HD dependance on six dimensionless parameters is examined.

### 5.1. *Determination of Parameter Dependance*

First, $\epsilon$ is varied through the range $0.125< |\epsilon| <1.0$, for both positive and negative $\epsilon$ values. The parameters which are constant for all of these simulations are

$$\kappa = 1.0, \quad \mu = 0.1, \quad A = 1.0, \quad \gamma = 2.0, \quad S = 1/10$$

The dimensionless maximum shoreline elevations or runups, $R_{max}$, as a function of distance from the slide centerline for both submerged and subaerial landslides are given in Figures 4a) and b), respectively. The dimensional runup has been scaled by $\Delta h$ to yield $R_{max}$. For the submerged slides, there is a clear trend of decreasing $R_{max}$ with decreasing $\epsilon$. This is not the case for the subaerial slides, as from roughly $y'$=0.5 to 4.0, the curves fall very close together. The secondary maximum in runup, which is due to edge wave formation, occurs between $y'$=1.5 to 2.5 with a magnitude $0.55< R_{max} <0.65$ for the subaerial slide range tested here.

Now, trends within the runup profiles are sought. As there is only one variable parameter, $\epsilon$, for all the simulations in Fig. 4, it must be used in any attempt to condense the data into single trends. First, the submerged runup profiles in Fig. 4a) were multiplied by the factor $1/\epsilon$; this modified plot is given as Figure 5a). While no unique trend could be achieved across the range $0.1< y' < 4.5$, it is evident that the maximum runup directly landward of the slide ($y' \approx 0$) falls in the range $0.18< R_{max}/\epsilon <0.21$. Perhaps more interesting is that all of the runup profiles condense into a single trend for $y' >4.5$. This trend is very slowly decreasing, with a mean value of $R_{max}/\epsilon \approx 0.11$. The subaerial slide runup can also be condensed for $y' >4.5$. Shown in Fig 5b) is $R_{max}$ multiplied by $\epsilon^{1/4}$. The result is a very high correlation trend for $y' >4.5$ such that $R_{max}\epsilon^{1/4}$=0.2. These two data manipulations strongly indicate that there is a transition near $y'$ =4.5, where for lesser values an accurate description of the source is extremely important, but for values larger, not important at all. Thus $y' \approx 4.5$, or 4.5 slide widths from the centerline, likely divides the near and far fields.

For a final analysis of these simulations, the submerged slide profiles are once more examined. By multiplying $y'$ by $\epsilon$, all of the secondary runup peaks, a result of the edge waves, occur at the same location of $\epsilon y' \approx 0.7$, with a magnitude $0.22< R_{max}/\epsilon^{4/3} < 0.26$. This data is shown in 5c). This result indicates that even in the very near field area, where nonlinear and dispersive effects are important, certain aspects of the slide behavior will follow simple scaling laws.

Following the same procedure, the other five dimensionless parameters

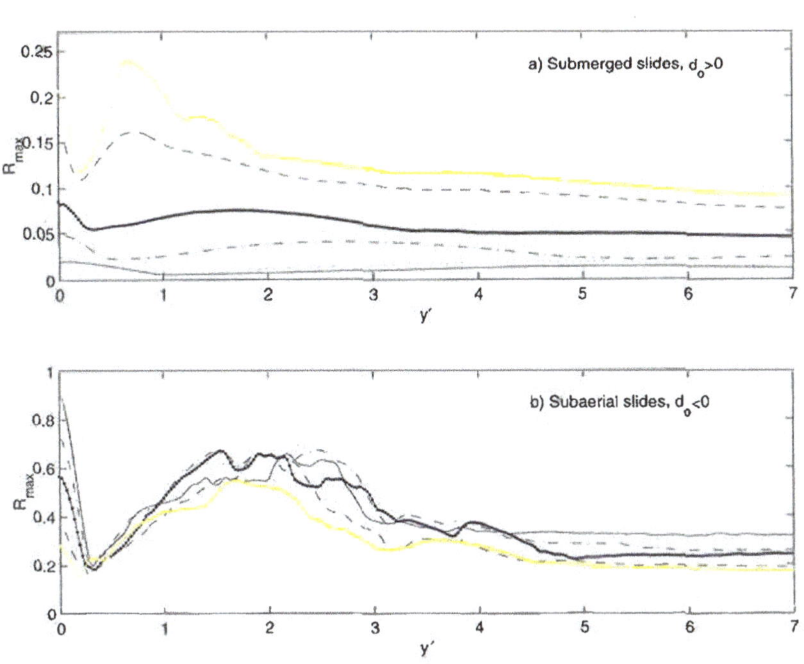


Figure 4. Maximum shoreline elevation at a function of distance from slide centerline for slides with variable $\epsilon$ values. The different line types correspond to different $\epsilon$ values in subplots [a,b]: solid line [0.125,-0.125], dotted line [0.15,-0.15], dashed-dotted line [0.25,-0.25], dark thick line [0.4,-0.4], dashed line [0.75,-0,75], light thick line [1.0, -1.0].

are varied independently. The combined trends are discussed in the next section.

## 5.2. *Combined Trends*

### 5.2.1. *Far Field*

With the parameter dependencies for each of the six sets of numerical experiments described above, we can combine these dependencies to determine global trends. In addition, 18 more simulations were run with dimensionless parameters lying within the ranges examined above, and are included in the plots discussed now. Looking first to the far field behavior of submerged landslide runup, each of the six independant scalings are combined, and the resulting plot is given in Figure 6a). For far-field, submerged landslide

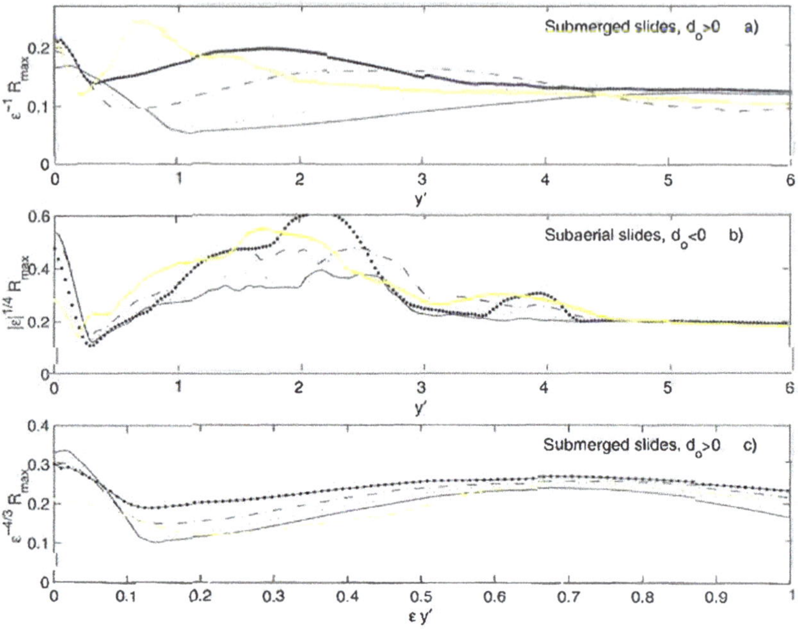


Figure 5. Various trends of the maximum runup for slides with variable $\epsilon$ values. Figure setup same as in Fig. 4.

runup, the important parameter is:

$$C_1 = \epsilon \mu^{-6/5} A^{5/6} S^{4/3} \kappa^{-1/4} \gamma^{4/3} \tag{6}$$

Examination of Figure 6a) shows that $C_1$ influences the far field aspects of submerged landslide runup. Looking to the far field, which is reasonably defined as $y' > 5A^{-4/5}$, we see the convergence of runup curves. This convergence starts with a maximum runup of $R_{max} \approx 0.07C_1$ and decreases slowly, exhibiting decreasing scatter within the trend, reaching $R_{max} \approx 0.06C_1$ by $y' = 7A^{-4/5}$.

For far-field, subaerial landslide runup, the parameter of interest is:

$$C_2 = \epsilon^{-1/4} \gamma \tag{7}$$

An extremely clear confluence of subaerial runup curves for $y' > 5$ is shown in Figure 6b). This is an interesting result, indicating that for subaerial slides, the far field runup can be very closely estimated with only knowledge of the thickness, initial location, slide width, and density. The maximum

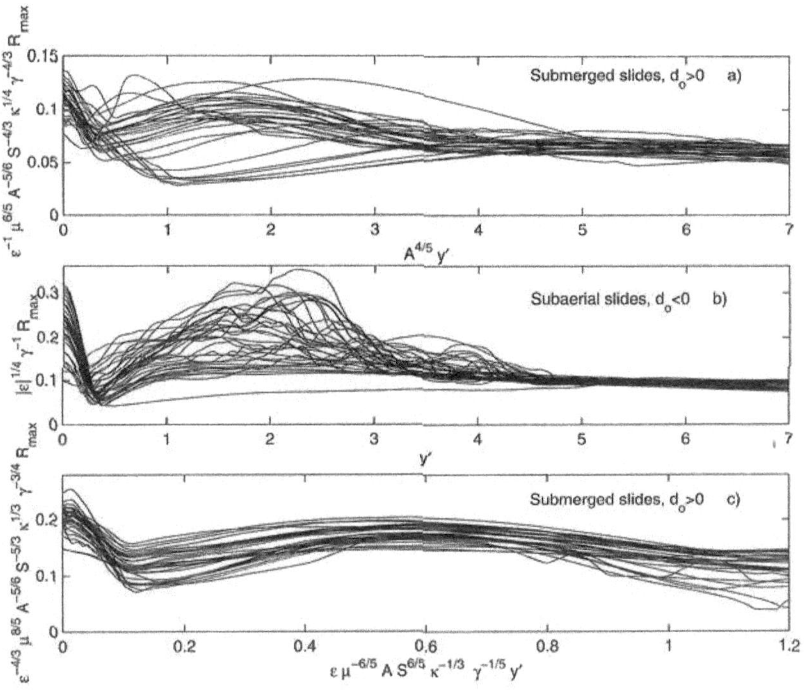


Figure 6. Scaled maximum shoreline elevation as a function of distance from slide centerline for all slides.

runup is $R_{max} \approx 0.10C_2$ at $y' = 5$, slowly decreasing to $R_{max} \approx 0.09C_2$ at $y' = 7$.

Finally, the secondary runup peak, due to edge wave propagation, is checked for submerged slides. Here, the important parameters impacting the properties of secondary runup are:

$$C_3 = \epsilon^{4/3}\mu^{-8/5}A^{5/6}S^{5/3}\kappa^{-1/3}\gamma^{3/4} \tag{8}$$

$$C_4 = \epsilon^{-1}\mu^{6/5}A^{-1}S^{-6/5}\kappa^{1/3}\gamma^{-1/5} \tag{9}$$

*With these parameters, the secondary runup peak can be described approximately as being located at* $y' \approx 0.60C_4$, with a magnitude of $R_{max} \approx 0.17C_3$. The error for the predictions presented in this section is roughly $\mp 15\%$

While the entire focus of this paper up to now has been on runup, rundown can be an equally important, and destructive, aspect of the waves

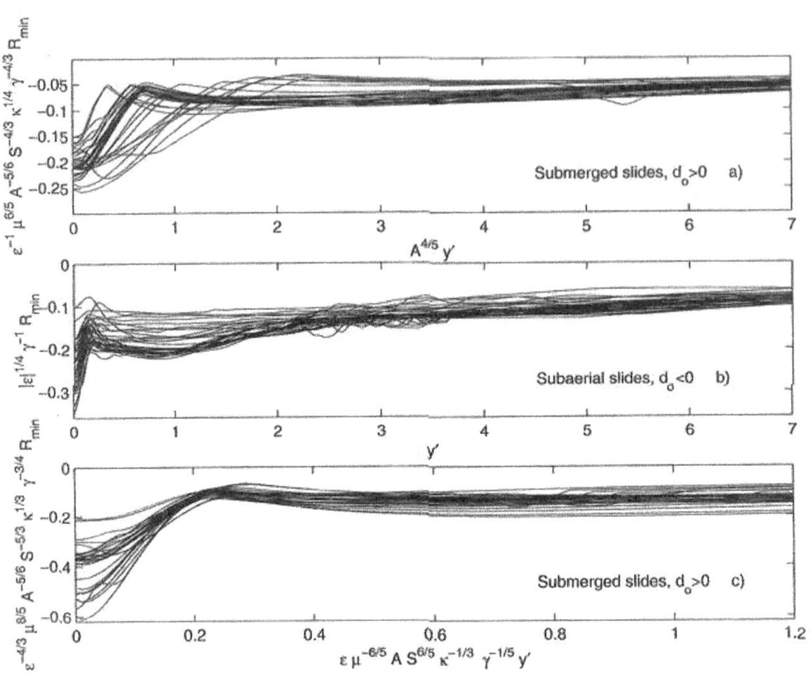


**Figure 7.** Scaled minimum shoreline elevation as a function of distance from slide centerline for all slides.

generated by landslides. Applying the scalings found above for runup, the rundown is given in Figure 7, presented in the same form as Fig. 6. Just as with the runup, we see the far field collapse of the submerged and subaerial rundown, shown in Fig.7a) and b), at roughly the same shoreline locations. This is an indication that the scalings are in fact a good basic representation of the physical aspects of the landslide problem; they were determined for runup, but work equally well for rundown.

### 5.2.2. *Near Field*

All of the previous analysis examined far field scalings, and the search for asymptotic runup trends. The entire analysis can easily be repeated in an attempt to uncover the proper runup scalings for a specific location in the near field. In particular, dimensionless expressions are sought for runup and rundown at $y' \approx 0$, which for many slide setups will be the location of

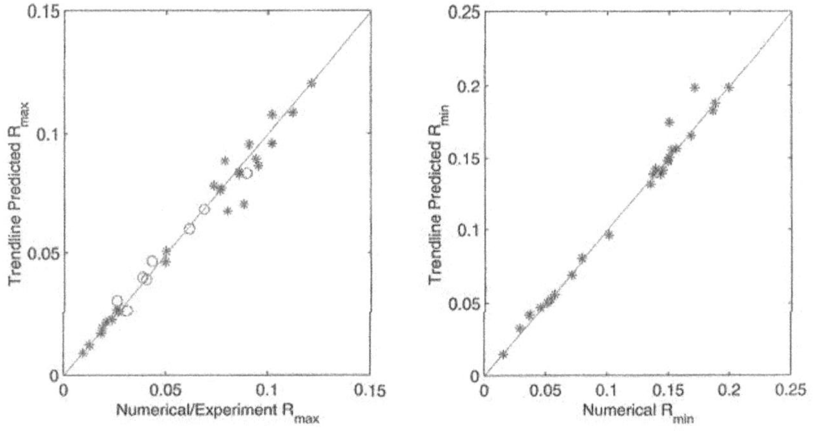


Figure 8. Accuracy of predicted submerged slide runup/rundown versus results obtained by numerical simulation (squares) and experiment (circles).

the largest shoreline movement. Additionally, these $y' \approx 0$ runup/rundown scalings will allow for comparisons with existing experimental data, and validation of the trend-finding approach utilized here.

Repeating the analysis of sections 5.1-5.6, the runup at $y' \approx 0$ for submerged slides can be given by $R_{max} \approx 0.30C_5$, where

$$C_5 = \epsilon^{1.16}\mu^{-1.40}A^{0.83}S^{1.71}\kappa^{-0.015}\gamma^{0.80} \tag{10}$$

Rundown for submerged slides is described by $R_{min} \approx -0.23C_6$, where

$$C_6 = \epsilon\mu^{-1.19}A^{0.87}S^{1.14}\kappa^{-0.23}\gamma^{0.62}. \tag{11}$$

It is reiterated here that the dimensional runup and rundown has been scaled by $\Delta h$ to yield $R_{max}$ and $R_{min}$. Figure 8 shows the excellent fit for the runup and rundown predictions. Also included in the plots are experimental data, discussed in the next section.

## 6. Comparison with Experimental Data

Due to the experimental constraints on measuring the 3D waves created by landslides, such as accurate measurement of the slide time history and the extremely long and wide tanks required to eliminate reflection effects, very little data exists. Only recently has some data been presented, for example the sliding body experiments of Synokalis & Raichlen (2002). In this paper, some of the data from the Raichlen & Synokalis experiments, where shoreline location was measured, will be employed. For these cases, a spherical hemisphere slid down a 1:2 slope. The hemisphere had a radius of 1.5 $ft$; the initial depth of submergence and the density of the mass were varied. To be consistent with the work done in this paper, when calculating the length scales of the slide, we look at the horizontal projection of the slide when on the slope. When examining the projection, $L \approx 2.6ft$, $\delta h \approx 1.5ft$, $l_B \approx 1.8ft$, and $l_F \approx 0.8ft$.

| Run # | $\epsilon$ | $\gamma$ | Pred. $R_{max}$ (cm) | Meas. $R_{max}$ (cm) | % Error |
|---|---|---|---|---|---|
| 9 | 2.3 | 2.3 | 3.1 | 3.1 | 0% |
| 10 | 1.5 | 2.3 | 2.0 | 2.1 | -5% |
| 11 | 1.1 | 2.3 | 1.4 | 1.2 | 15% |
| 23 | 2.3 | 3.2 | 4.1 | 3.8 | 10% |
| 25 | 0.8 | 3.2 | 1.2 | 1.4 | -15% |
| 26 | 1.1 | 3.2 | 1.8 | 1.8 | 0% |
| 34 | 2.3 | 2.0 | 2.8 | 2.7 | 5% |
| 35 | 1.5 | 2.0 | 1.8 | 1.8 | 0% |

Table 1 summarizes both the predicted runup and the measured experimental runup. Experiment runs 8,22, and 33 were not included in the comparison, as these cases exhibited significant breaking and turbulence in the laboratory. The predicted runup is calculated from the scalings developed in this paper: $R_{max} \approx 0.30C_5$. The accuracy of the predictions very good. For the $\gamma = 2.0$ & 2.3 runs, the predictions are excellent. For the $\gamma = 3.2$ runs, the predictions are not quite as good, are are indicative of the error range of the runup prediction formula. Possible explanations for the decrease in accuracy for larger $\gamma$ is increased slide-bottom friction, or increased level of turbulence due to faster slide motion. These physical aspects are not considered in the numerical model. It should also be noted that the experimental $\epsilon$, $\mu$, $\kappa$, and $S$ values are all outside of the range examined numerically in this paper, yet good agreement is still achieved.

In addition to the hemisphere slide experiments compared above, Raichlen *et al.* (2004) also present wave data for tsunamis generated by triangular block shaped slides. Due to the sharp corners and very large slopes of these shapes, the Boussinesq-type numerical model is incapable of simulating the triangular block slide. Additionally, comparison of the experimental triangular block slide runup at $y = 0$ with (10) shows very poor agreement with errors on the order of 100%. As with the hemisphere experiments, $\epsilon, \mu, \kappa$, and $S$ values for the triangular slide are all outside of the range examined numerically in this paper. It must therefore be stated that the trends developed in this paper may only be applicable for smooth shaped slides.

## 7. Conclusions

A comprehensive set of numerical experiments, including over 75 numerical simulations requiring more than 100 days of desktop CPU time, has been undertaken. The submerged and subaerial slides examined here created non-breaking waves in shallow to intermediate water. Through analysis of the runup from these slides, a number of dimensionless relationships have been determined. For submerged slides, the maximum runup immediately landward of the slide, as well as the magnitude and location of the secondary runup peak, can now be reasonably estimated. For both subaerial and submerged slides, the division between near and far field has been shown, along with maximum runup at these locations. In dimensional form, the key relationships are summarized as follows:

(1) Submerged slides

Maximum runup immediately landward of slide centerline

$$R/\Delta h \approx 0.30\epsilon^{1.16}\mu^{-1.40}A^{0.83}S^{1.71}\kappa^{-0.015}\gamma^{0.80} \tag{12}$$

Maximum rundown immediately landward of slide centerline

$$R/\Delta h \approx -0.23\epsilon\mu^{-1.19}A^{0.87}S^{1.14}\kappa^{-0.23}\gamma^{0.62} \tag{13}$$

Near-far field division

$$y/W \approx 5A^{-4/5} \tag{14}$$

Runup at near-far field division

$$R/\Delta h \approx 0.07\epsilon\mu^{-6/5}A^{5/6}S^{4/3}\kappa^{-1/4}\gamma^{4/3} \tag{15}$$

Location of secondary runup peak due to edge waves

$$y/W \approx 0.6\epsilon^{-1}\mu^{6/5}A^{-1}S^{-6/5}\kappa^{1/3}\gamma^{-1/5} \tag{16}$$

Secondary peak runup

$$R/\Delta h \approx 0.17\epsilon^{4/3}\mu^{-8/5}A^{5/6}S^{5/3}\kappa^{-1/3}\gamma^{3/4} \tag{17}$$

(2) Subaerial Slides

Near-far field division

$$y/W \approx 5 \tag{18}$$

Runup at near-far field division

$$R/\Delta h \approx 0.1\epsilon^{-1/4}\gamma \tag{19}$$

Comparisons with experiment show that the scaling relationships yield reasonable predictions. These relationships may be useful for preliminary hazard assessment, where a simple and quick estimation of the maximum runup height and locations are required. Additionally, the formulas developed in this paper will be particulary beneficial to those developing 3D landslide experiments. Finally, it should be noted that the simulations performed for this paper all make use of a simple planar slope, and do not take into account irregular bathymetry effects. It is known from PNG, as well as from many tsunami events, that bathymetric focusing can greatly impact the shoreline runup profiles (e.g. Matsuyama *et al.*,1999), and thus one must always be mindful of site specific seafloor contours when estimating runup.

## References


1. Basterretxea, G., Orfila, A., Jordi, A., Casas, B., Lynett, P., Liu, P. L.-F., Duarte, C. M., and Tintor, J. 2004. "Evolution of an Embayed Beach with Posidonia Oceanica Seabeds (Mallorca, Balearic Islands)," in press for Journal of Coastal Research.
2. Chang, K.-T. 1995. "Evolution of landslide-generated edge wave packet, " Ph.D. thesis, University of Washington.
3. Gonzalez, F., Satake, K., Boss, E., Mofjeld, H.. 1995. "Edge Wave And Non-Trapped Modes Of The 25 April 1992 Cape-Mendocino Tsunami," Pure Appl Geophys 144 (3-4): 409-426.
4. Grilli, S.T, and Watts, P. 1999. "Modeling of Waves Generated by a Moving Submerged Body. Applications to Underwater Landslides," Engng. Analysis with Boundary Elements , 23, 645-656.
5. Grilli, S.T., Vogelmann, S. and Watts, P. 2002. "Development of a 3D Numerical Wave Tank for modeling tsunami generation by underwater landslides," Engng. Analysis Boundary Elemt. 26(4), 301-313
6. Jiang, L., and P. H. LeBlond. 1994. "Three-dimensional modeling of tsunami generation due to a submarine landslide," J. Phys. Oceanog., 24, 559- 572.
7. Liu, P. L.-F. 1994. "Model equations for wave propagation from deep to shallow water," In *Advances in Coastal Engineering* (ed. P. L.-F. Liu), vol. 1, pp.125-157. World Scientific.
8. Liu, P. L.-F., Yeh, H., Lin, P., Chang. K., and Cho, Y.-S. 1994. "Generation and evolution of edge-waves packets," Physics of Fluids, v.10(7), p. 1635-1657.
9. Lynett, P. 2002. "A multi-layer approach to modeling nonlinear, dispersive waves from deep water to the shore," PhD thesis, Cornell University. Available online at http://ceprofs.tamu.edu/plynett/ publications/publications.htm

10. Lynett, P. and Liu, P. L.-F. 2002, "A numerical study of submarine landslide generated waves and runup," Royal Society of London A (485(2028)), 2885-2910.
11. Lynett, P., Wu, T.-R., and Liu, P. L.-F. 2002. "Modeling Wave Runup with Depth-Integrated Equations," Coastal Engineering, v. 46(2), p. 89-107.
12. Lynett, P., Borrero, J., Liu, P. L.-F., and Synolakis, C.E. 2003. "Field Survey and Numerical Simulations: A Review of the 1998 Papua New Guinea Tsunami," Pure and Applied Geophysics, v.160, p. 2119-2146.
13. Lynett, P. and Liu, P. L.-F. 2004a. "A Two-Layer Approach to Water Wave Modeling," in press for Proc. Royal Society of London A
14. Lynett, P. and Liu, P. L.-F. 2004b. "Linear Analysis of the Multi-Layer Model," in press for Coastal Engineering
15. Matsuyama, M., Walsh, J.P., and Yeh, H. 1999. "The effect of bathymetry on tsunami characteristics at Sissano Lagoon, Papua New Guinea," Geophysical Research Letters, 26, 3513-3516.
16. Nwogu, O. 1993 "Alternative form of Boussinesq equations for nearshore wave propagation," Journal of Waterway, Port, Coastal and Ocean Engng. 119(6), 618-638.
17. Raichlen, F., Synolakis, C.E., Wu, T.R, and Liu, P. L.-F. 2004. "Runup and rundown from three-dimensional sliding masses" in preparation.
18. Ryu, S., Kim, M.H., and Lynett, P. 2003. "Fully Nonlinear Wave-Current Interactions and Kinematics by a BEM-based Numerical Wave Tank," Journal of Computational Mechanics, 32, 336-346.
19. Schaffer, H. and Jonsson, I. 1992. "Edge waves revisited," Coastal Engineering, Volume 16, Issue 4, Pages 349-368
20. Synolakis, C. E., and Raichlen, F. 2002. "Wave and run-up generated by a three-dimensional sliding mass," 28th Intl Conf. Coastal Engng, Abstracts, Paper 299.
21. Tinti S, Bortolucci E, Armigliato. 1999. "A Numerical simulation of the landslide-induced tsunami of 1988 on Vulcano Island, Italy" Bulletin of Volcanology 61 (1-2): 121-137.
22. Ursell, F. 1952. "Edge waves on a sloping beach," Proceedings of the Royal Society London A 214, 79-97.
23. Watts, P. 1997. "Water Waves Generated by Underwater Landslides," Ph. D. Thesis, California Institute of Technology, 1997
24. Wei, G. and Kirby, J. 1995. "A time-dependent numerical code for extended Boussinesq equations," Journal of Waterway, Port, Coastal and Ocean Engng. 120, 251-261.

# PHYSICAL MODELING OF LANDSLIDE GENERATED TSUNAMI

HERMANN M. FRITZ

*School of Civil and Environmental Engineering, Georgia Institute of Technology*
*210 Technology Circle*
*Savannah, Georgia 31407, USA*

Subaerial landslide impact generated tsunami waves were investigated in a two-dimensional physical laboratory model based on the generalized Froude similarity. Digital particle image velocimetry (PIV) was applied to the landslide impact and wave generation. At high impact velocities flow separation occurred on the slide shoulder resulting in a hydrodynamic impact crater. The recorded wave profiles were extremely unsteady and non-linear. Four wave types were determined: weakly non-linear oscillatory wave, non-linear transition wave, solitary-like wave and dissipative transient bore. Most of the generated impulse waves were located in the intermediate water depth wave regime. Nevertheless the propagation velocity of the leading wave crest closely followed the theoretical approximations for a solitary wave. The main wave characteristics were related to the landslide parameters driving the entire wave generation process. The slide Froude number was identified as the dominant parameter.

## 1. Introduction

Examples of historical tsunamis in the Caribbean generated by submarine landslides include the disastrous 1692 Port Royal tsunami in Jamaica, caused by a slump and grain flow into Kingston harbor from beneath Port Royal itself (Elliot 1892; Pawson and Buisseret, 1975), and a smaller tsunami associated with the 1907 Kingston earthquake (Fuller, 1907). A lahar generated a tsunami in the early stages of the 1902 eruption of Mont Pelee (Chretien & Brousse 1989) causing damage along the waterfront at St. Pierre a few days prior to the destructive pyroclastic flow of May 8$^{th}$. Larger prehistoric examples of landslide events, that may have generated tsunamis, include subaerial and submarine strato-volcano collapses (the former documented onshore and offshore from Montserrat, Martinique, Dominica and St. Lucia (Wright et al. 1984; Smith & Roobol 1990; Deplus et al. 2001; Le Friant et al. 2002, 2003, 2004), ocean trench wall landslides exemplified by giant slump scars north of Puerto Rico (Scanlon & Masson 1996), and carbonate platform margin collapses from the Bahamas carbonate platform (Mullins et al., 1992). Historic examples in similar geologic settings encompass the Ritter Island collapse in 1888, Papua New Guinea (Ward and Day, 2003). Recent examples of mass flows impacting into open and confined water bodies are shown Figure 1.

The present study focused on the wave generation process and the near field wave characteristics in a two dimensional physical model. Some aspects of characteristics were controlled by means of a novel pneumatic landslide the wave propagation and run-up in a three dimensional physical model were investigated by Huber (1980) and Müller (1995), respectively. The slide impact

Figure 1. (a) Pyroclastic flow impacting into the ocean on Montserrat (copyright: Montserrat Volcano Observatory/Government of Montserrat and British Geological Survey; photo used by permission of the Director, MVO); (b) Landslide scar and deposit in the Guaitara valley, Colombia (copyright: Fritz, 2003).

generator (Fritz and Moser, 2003). Instantaneous flow fields of the impact area were determined with particle image velocimetry (Fritz et al., 2003a,b). The physical model results were compared to the giant rockslide generated impulse wave which struck the shores of the Lituya Bay, Alaska, in 1958 (Miller, 1960). A cross-section of Gilbert Inlet was rebuilt and the measured wave run-up matched the trimline of forest destruction (Fritz et al., 2001). The near field wave characteristics were determined (Fritz et al., 2004). A synthesis of this experimental work conducted at VAW (ETH Zurich) is presented.

## 2. Experimental Set-up

The granular rockslide impact experiments were conducted in a rectangular prismatic water wave channel (L×W×H=11m×0.5m×1m) with varying stillwater depths $h$ = 0.30, 0.45 and 0.675 m. At the front end of the channel a 3 m long hill slope ramp was built into the channel as shown in Figure 2. The hill slope angle $\alpha$ was variable from 30° to 90°, but only the angle $\alpha = 45°$ and the slide masses $m_s = 27$, 54 and 108 kg were considered. A total of 137 experimental runs were conducted (Fritz, 2002b).

Three different measurement techniques were built into the physical model: laser distance sensors (LDS), particle image velocimetry (PIV) and capacitance wave gauges (CWG). The PIV-system consisted of a twin cavity Nd.YAG-laser and a high-resolution CCD-camera (Fritz, 2002a). The details of the PIV-data analysis and accuracies of velocity components and derivatives are given in Fritz et al. (2003a). PIV provided instantaneous velocity vector fields and gave insight into the kinematics of wave generation (Fritz et al., 2003b).

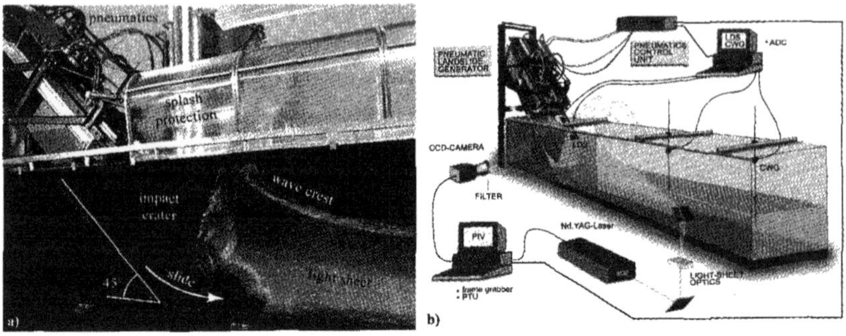


Figure 2. (a) Landslide impact experiment at $\mathsf{F} = 3.3$, $m_s = 108$ kg, $h = 0.45$ m and $\alpha = 45°$; (b) measurement setup with pneumatic installation and the three measurement systems: laser distance sensors, capacitance wave gages and laser based digital particle image velocimetry (PIV).

The slide profiles $\xi\ (x_{ramp},t)$ before impact were scanned with two laser distance sensors (LDS) as shown in Figure 3a). The wave profiles $\eta\ (x,t)$ in the propagation area were determined with seven capacitance wave gauges (CWG) positioned along the channel axis as shown in Figure 3b). The wave channel was relatively short with its length of 11m, but adequate to study the near-field wave characteristics (Walder et al., 2003). Kamphuis and Bowering (1970) as well as Huber (1980) investigated the wave propagation up to $x/h \approx 100$.

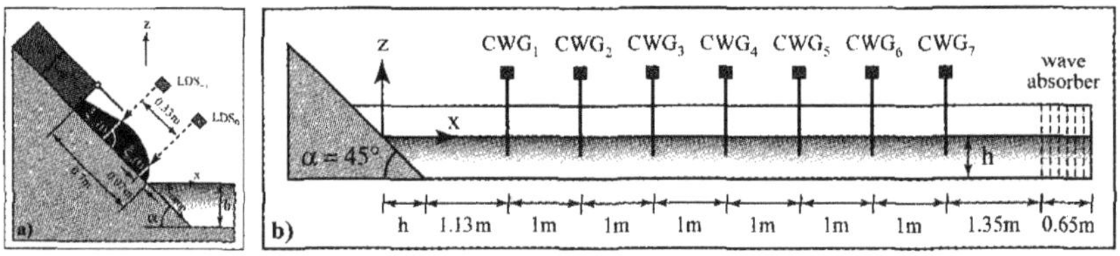


Figure 3. (a) Positioning of the two laser distance sensors (LDS) and (b)the seven capacitance wave gauges (CWG) along the channel axis.

The landslides were modeled with a granular material (PP-$BaSO_4$) shown in Figure 4a). The granulate properties were: grain density $\rho_g = 2.64$ t/m$^3$, grain diameter $d_g =$ 4 mm, bulk slide density $\rho_s = 1.62$ t/m$^3$, bulk slide porosity $n_{por} = 39$ %, effective internal friction angle $\phi' = 43°$, and dynamic bed friction angle $\delta = 24°$. Typical superimposed slide profiles $\xi\ (x_{ramp},t)$ of 5 slides with corresponding slide velocities, mass and release shape are shown in Figure 4b).

The pneumatic landslide generator controlled the dynamic slide impact characteristics (Fritz and Moser, 2003). The following four relevant parameters governing the wave generation were varied: granular slide mass $m_s$, slide

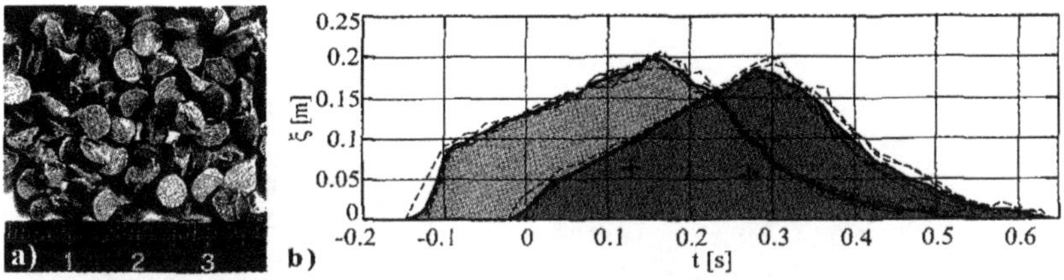


Figure 4 (a) landslide material: PP-BaSO$_4$ granulate with grain diameter $d_g = 4$ mm, grain density $\rho_g = 2.64$ t/m$^3$, porosity $n_{por} = 39$ % and slide density $\rho_s = 1.62$ t/m$^3$; (b) typical slide profiles $\xi(t)$ at locations $x_{ramp}$ = (▒) –0.4m and (■) –0.07m, (——) mean and (— —) individual profiles.

impact velocity $v_s$, stillwater depth $h$ and slide thickness $s$. The parameters definitions and the wave crest and trough annotations are shown in Figure 5a) and b), respectively. The main slide parameters are: thickness $s$, length $l_s$, centroid velocity $v_s$ at impact and density $\rho_s$. The water body topography is characterized by the still water depth $h$ and the hill slope angle $\alpha$. The wave characteristics are described by the wave profile $\eta$, wave height $H$ or amplitude $a$, the wave propagation velocity $c$ and the wavelength $L$ or period $T$.

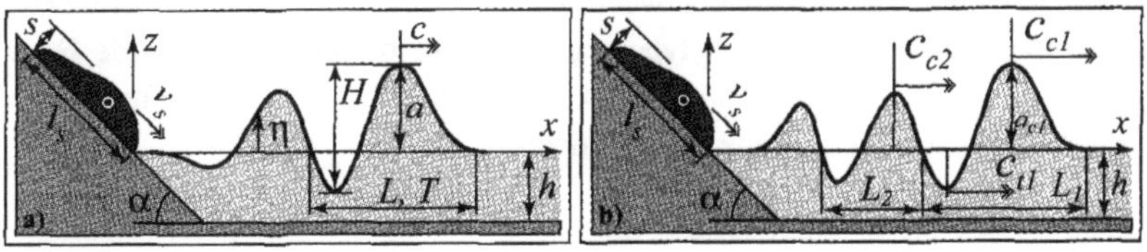


Figure 5. Definitions of (a) main slide, water body and tsunami parameters as well as (b) individual wave crest and trough annotations.

## 3. Tsunami generation by landslide impact

### 3.1. *Main flow regimes*

The predominant visual flow classification criterion was flow separation on the back of the landslide forming a hydrodynamic impact crater. Characteristic examples of an unseparated and a separated flow around penetrating landslides are shown in Figure 6a) and b), respectively. The flow separation has a dominant influence on the displaced water volume, which can exceed the landslide volume significantly. The separation was governed by the slide Froude number $\mathsf{F} = v_s/(gh)^{1/2}$. Flow separation was determined by

$$\mathsf{F} > [(5/3) + (1/2)S] \tag{1}$$

with the relative slide thickness $S = s/h$ (Fritz, 2002b).

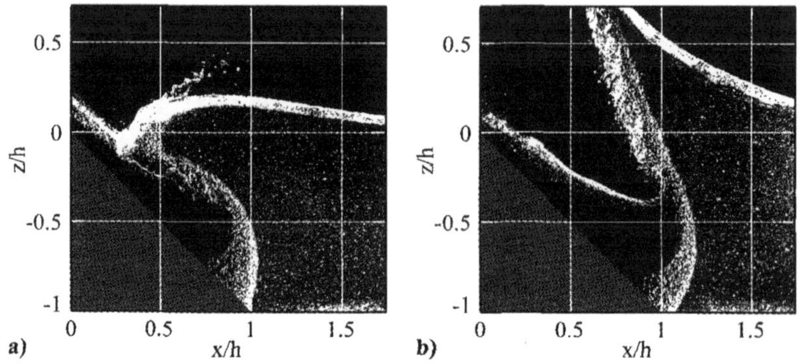


Figure 6. Flow separation on the slide shoulder at $h = 0.45$m: (a) unseparated flow at F = 1.4, $V = 0.35$, $S = 0.23$ and $t(g/h)^{1/2} = 1.13$; (b) separated flow at F = 2.6, $V = 0.35$, $S = 0.21$ and $t(g/h)^{1/2} = 0.79$.

Two hydrodynamic impact crater types were observed in the separated flow regime defined by (1). The subsequent crater collapse can lead to a closure resulting in an instantaneous cavity (Birkhoff and Zarantonello, 1957). The closure allowed to distinguish between backward and outward collapsing craters shown in Figure 7. The crater type regimes were defined by the slide Froude number $\mathsf{F} = v_s/(gh)^{1/2}$ and the relative slide thickness $S = s/h$. The outward collapsing water crater regime was defined as

$$\mathrm{F} > [4 - (5/2)S]. \tag{2}$$

Most relevant events of landslide generated impulse waves were observed within the Froude number range $0.5 < \mathsf{F} < 4$ (Slingerland and Voight, 1979). Hence all discussed flow types are expected to occur in real events.

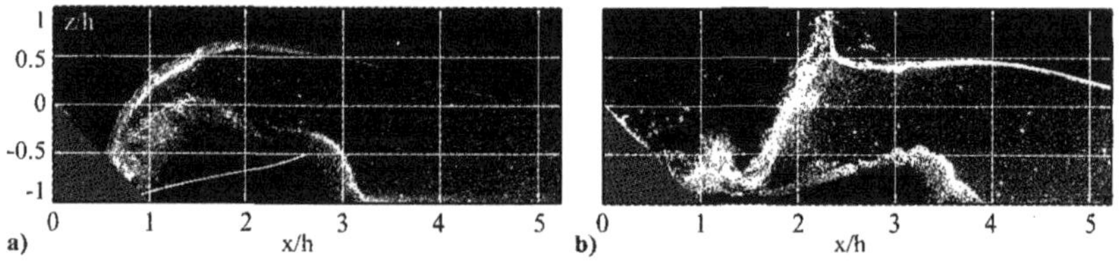


Figure 7. (a) Backward collapsing impact crater at F = 2.8, $V = 0.79$, $S = 0.34$, $h = 0.3$m; (b) outward collapsing impact crater at F = 3.2, $V = 0.79$, $S = 0.31$, $h = 0.3$m.

### 3.2. *Unseparated flow type*

A characteristic example of an unseparated flow around a penetrating landslide at F = 1.7 is shown in Figure 8. The set of figures includes the original PIV-images and the velocity components. The slide thickness increased during the slide penetration (Figure 8Aa). A massive phase mixing occurred in the wake of the slide (Figure 8Ac). The detrainment of the air included in the pore volume of the granular slide affected the first trough (Figure 8Ad). The second

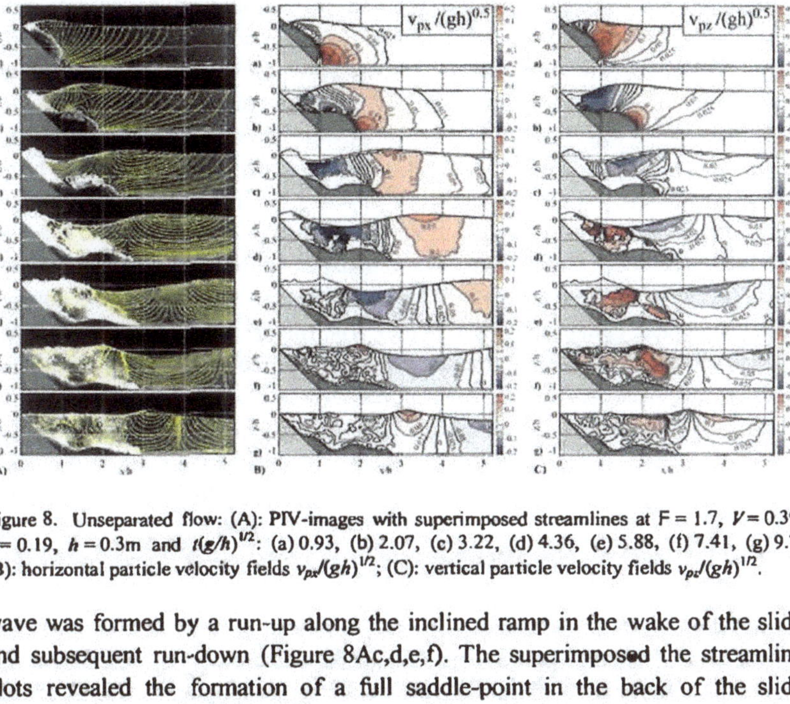


Figure 8. Unseparated flow: (A): PIV-images with superimposed streamlines at F = 1.7, $V$ = 0.39, $S$ = 0.19, $h$ = 0.3m and $t(g/h)^{1/2}$: (a) 0.93, (b) 2.07, (c) 3.22, (d) 4.36, (e) 5.88, (f) 7.41, (g) 9.7; (B): horizontal particle velocity fields $v_{px}/(gh)^{1/2}$; (C): vertical particle velocity fields $v_{pz}/(gh)^{1/2}$.

wave was formed by a run-up along the inclined ramp in the wake of the slide and subsequent run-down (Figure 8Ac,d,e,f). The superimposed the streamline plots revealed the formation of a full saddle-point in the back of the slide (Figure 8Aa,b,c). The half saddles propagated outward along the channel bottom with the wave pattern.

The largest absolute water particle velocities were measured locally around the slide front during slide penetration. The horizontal particle velocities were zero along vertical lines through the half-saddles, whereas the largest values were below the wave crest (Figure 8B). The water particle velocity was only fractions of the shallow water wave velocity $(gh)^{1/2}$. The vertical particle velocities were zero along vertical lines through the wave crests (Figure 8C). The largest values were encountered at the location of the largest water surface slope. The computed vorticity in the pure water flow area was roughly zero

(Fritz, 2002b). The PIV-data confirmed the assumption of water wave *irrotationality*. Vorticity was observed on the slide surface due to the shear flow and in the mixing zone above the landslide deposit due to the dissipative, three dimensional turbulence. The wave profile continuously stretched apart over the short sequence. The leading wave is an intermediate water depth wave, whereas the trailing wave exhibits the classic Stokesian wave profile.

### 3.3. *Outward collapsing impact crater flow type*

A characteristic example at the lower end of the outward collapsing crater regime is shown in Figure 9. The water flow separated on the slide shoulder (Figure 9Aa,b,c). The water crater exposed the ramp and the back of the landslide to the atmosphere. The displaced water volume obviously exceeded the landslide volume significantly. As the process of crater growth had terminated outward bulk motion of water was still present as a residuum. In contrast to the backward collapse no closure occurred in this case (Fritz et al., 2003b). The outward collapsing crater resulted in a main positive leading wave and negative base surge. The backward motion was initiated at the bottom of the crater wall. The leading wave crest was issued by the crater rim and propagated outward during the crater collapse (Figure 9Ad,e). The run-up of the base surge on the inclined ramp and the subsequent run-down formed the secondary wave system. The streamline plots revealed the formation of a half saddle in the water uplift separating the outward from the inward flow (Figure 9Ac,d,e).

The largest velocities were measured locally around the slide front during slide penetration and in the splash. The water particle velocity below the wave crests was only fractions of the shallow water wave velocity $(gh)^{1/2}$, but increased compared to the previous examples due to the increase in wave height. The largest negative values of the horizontal velocity components were observed in the inward rush and run-up along the inclined ramp during the collapse of the impact crater (Figure 9Bd,e). Negative vertical velocities were observed at the beginning of the crater collapse along the crater walls (Figure 9Cc,d). The water displaced by the landslide at no instant reached a state near a static uplift. The kinetic energy of the landslide imparted on the water body was only partially converted into the potential energy of the uplift, whereas a significant part prevailed as kinetic energy in the form of the velocity field imposed onto the water body. The computed in-plane divergence was roughly zero in the pure water flow area confirming the two-dimensional model assumption (Fritz et al., 2003b). The pure water flow was irrotational. Significant vorticity was observed on the slide surface due to the shear flow and in the mixing zone above the landslide deposit due to the dissipative turbulence.

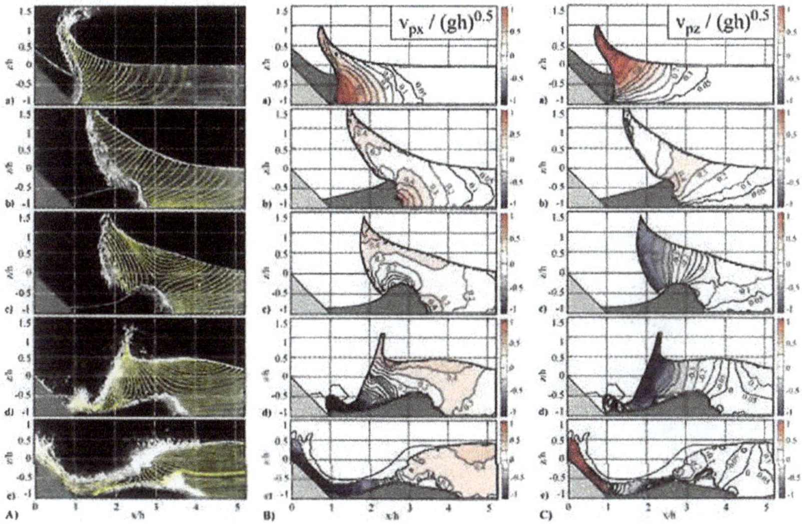


Figure 9. Outward collapsing impact crater: (A): PIV-images with superimposed streamlines at F = 3.2, $V$ = 0.79, $S$ = 0.31, $h$ = 0.3m and $t(g/h)^{1/2}$: (a) 0.58, (b) 1.73, (c) 2.49, (d) 3.25, (e) 4.01; (B): horizontal particle velocity fields $v_{px}/(gh)^{1/2}$; (C): vertical particle velocity fields $v_{pz}/(gh)^{1/2}$.

### 3.4. *Trailing bore formation regime*

A characteristic example of a transient bore is shown in Figure 10. The presented sequence begins with the first wave trough after the leading wave crest had left the area of view to the right and the backward run-up along the inclined ramp had reached its highest point (Figure 10Aa). The inward motion along the channel bottom collides with the run-down forming a surge (Figure 10Ab,c). The surge propagating outward on top of the backward flow is analogous to a transient bore (Figure 10Ad,e,f,g). A key feature is the dissipation process. A considerable portion of the energy imparted to the fluid is lost as a result of turbulent mixing at the impact site from the rebound of the inward flow. The energy dissipated hydro-dynamically remained constant at roughly 40% (Fritz, 2002b). The bore propagated outward roughly at $(gh)^{1/2}$. When the underlying incoming flow reaches sub-critical conditions or the bore height becomes smaller than 60% of the incoming flow depth, the dissipative bore no longer exists (Favre, 1935). It is then transformed into a translatory non-dissipative undular bore.

The streamline plots and the velocity components reveal the internal flow structure (Figure 9A and B). The half saddle along the ramp marked the collision between the run-down and the inward rush resulting in a spike a the free surface (Figure 9Ba,b). The largest water particle velocities were measured locally along the ramp and in the breaking bore. The horizontal water particle velocity in the bore exceeded even the shallow water wave velocity $(gh)^{1/2}$ computed with the still water depth (Figure 9Bd,e,f). The computed out-of-plane vorticity fields are shown in (Figure 9C). In contrast to the previous mainly irrotational waves the bore formation involved large positive vorticity values similar to the wave breaking process (Figure 9c,d,e,f,g).

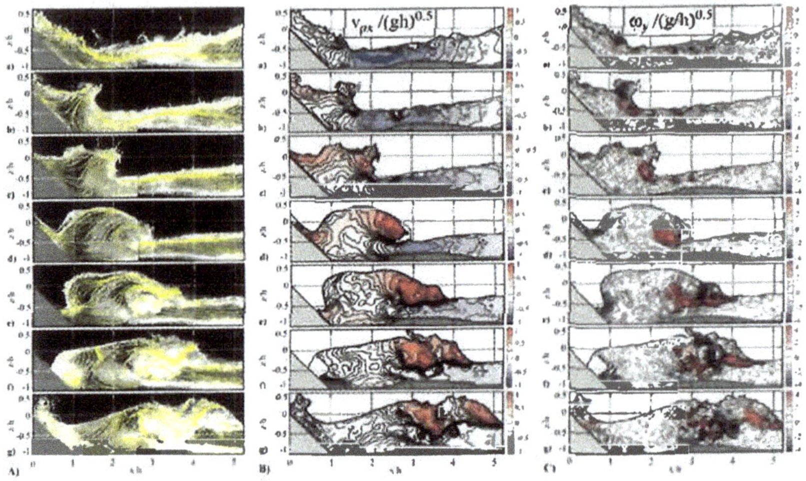

Figure 10. Secondary bore formation: (A): PIV-images with superimposed streamlines at F = 4.7, $V$ = 0.39, $S$ = 0.17, $h$ = 0.3m and $t(g/h)^{1/2}$: (a) 6.97, (b) 7.73, (c) 8.49, (d) 9.25, (e) 10.01, (f) 10.78, (g) 11.51; (B): horizontal particle velocity fields $v_{px}/(gh)^{1/2}$; (C): out-of-plane vorticity fields $\omega_y/(g/h)^{1/2}$.

## 3.5. *Water displacement*

The water displacement volume $V_d$ was extracted from the PIV-recordings of cases with flow separation (Fritz, 2002b). Scaling the time after impact $t$ with the time of the maximum displacement volume $t_D$ and the water displacement volume $V_d$ by the maximum water displacement volume $V_D$ collapsed all displacement volume curves. The evolutions of the water displacement volume $V_d/V_D$ is shown in Figure 11a). The water displacement estimates broke down during the collapse of the impact craters because of the massive phase mixing. The water displacement volume as a function of time may be approximated by

$$V_d/V_D = \sin[(\pi/2)(\ t/t_D)^{(3\pi/8)}] \quad (3)$$

The maximum water displacement volume $V_D$ and the time $t_D$ were governed by the slide Froude number $\mathsf{F} = v_s/(gh)^{1/2}$ and the dimensionless slide volume $V = V_s/(bh^2)$. The multiple regressions yield

$$V_D/(bh^2) = (0.8)\,\mathsf{F}^1\,V^{1/2} \tag{4}$$

$$t_D\,(g/h)^{1/2} = (1.7)\,\mathsf{F}^{0.3}\,V^{0.2} \tag{5}$$

with a correlation coefficients $r^2 = 0.93$ and $r^2 = 0.75$, respectively. The determined maximum water displacement volumes $V_D$ were within $1.8 \leq V_D/V_s \leq 8.1$, significantly exceeding the landslide volume $V_s$ per unit width. The time of maximum water displacement volume represented by $t_D$ was always reached well before the subaqueous slide motion was terminated (Fritz, 2002b).

The volume of the leading wave crest was determined by integration from the wave gauge recordings (Fritz, 2002b). The multiple regression for the leading wave crest volume $V_{c1}(x/h = 5)$ yields

$$V_{c1}/(bh^2) = (0.55)\,\mathsf{F}^{1.4}\,V^{0.4}\,S^{0.4} \tag{6}$$

with a correlation coefficient $r^2 = 0.97$. The slide Froude number $\mathsf{F} = v_s/(gh)^{1/2}$ is the dominant parameter but also the dimensionless slide volume $V = V_s/(bh^2)$ and thickness $S = s/h$ need to be included. The leading wave crest volume $V_{c1}$ was always smaller or equal to the maximum water displacement volume $V_D$ fulfilling the necessary condition $V_{c1} \leq V_D$ as shown in Figure 11b). The leading wave crest volume reached the maximum water displacement volume for the case of soliton-like waves.

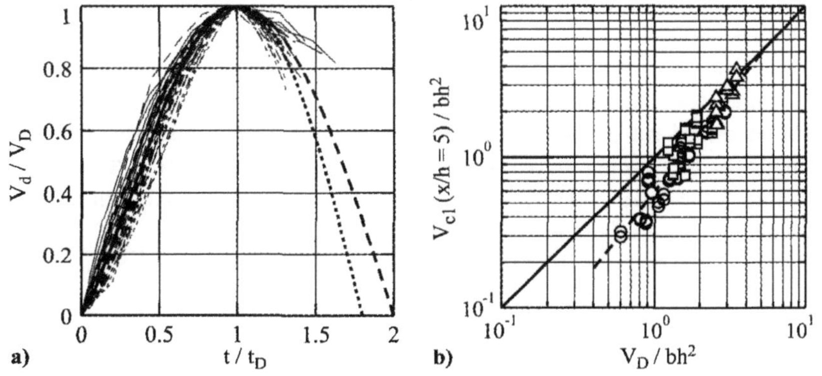


Figure 11. Impact crater formation: (a) water displacement $V_d/V_D$ with (—) Eq. (3) at $t/t_D \leq 1$; (b) leading wave crest volume $V_{c1}/(bh^2)$ versus maximum water displacement volume $V_D/(bh^2)$ with (– –) $V_{c1}/(bh^2) = (0.6)\,[V_D/(bh^2)]^{1.3}$ with $r^2 = 0.90$, (—) $V_{c1} = V_D$.

## 4. Tsunami near-field wave characteristics

### 4.1. *Wave profile recordings*

The wave gauge positions were shown in Figure 3b). The propagation of weakly and strongly non-linear waves are shown in Figure 12a) and b), respectively. All observed waves were non-linear with $H/h > 0.03$ (Fritz, 2002b). Weakly non-linear oscillatory waves were characterized by frequency

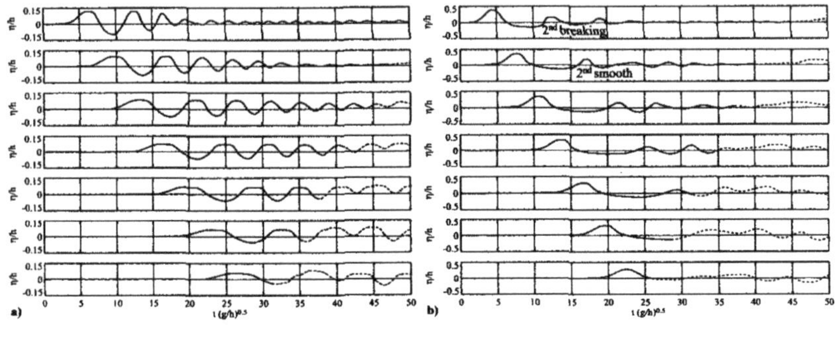


Figure 12. Wave gauge recordings: (a) non-linear oscillatory waves at F = 1.7, V = 0.39, S = 0.19, h = 0.3m with (___) valid wave recording and (....) wave recording affected by wave reflection; (b) non-linear transition waves at F = 2.8, V = 0.79, S = 0.34, h = 0.3m, symbols as in (a).

dispersion (Fritz and Liu, 2002). The nonlinear transition wave exhibits a leading wave crest and a long shallow trough followed by a dispersive wave train. The leading wave crest travels faster than the trailing waves. The wave profiles are non-steady in a moving reference frame in contrast to cnoidal waves (Wiegel, 1960).

In all cases the wavelengths decreased towards the back of the wave train. The measured wavelengths increased with the propagation distance. The wave lengths of the leading waves were within $5 < L_1/h < 25$, whereas those of the trailing wave were within $2 < L_2/h < 8$. The wavelengths of the trailing wave were roughly a third of the leading wave.

### 4.2. *Wave type classification*

The observed wave profiles can be grouped into four wave types: weakly non-linear oscillatory waves, non-linear transition waves, solitary-like waves and dissipative transient bores. The wave profiles are of weakly non-linear oscillatory, strongly non-linear transition, solitary-like and dissipative transient bore, as shown in Figure 13a,b,c,d), respectively. This comparison shows the broad spectrum of observed wave types and different amplitudes. The transient dissipative bore shown in Figure 13d) was recorded prior to breaking. The wave type classification was based on the slide Froude number $\mathsf{F} = v_s/(gh)^{1/2}$ and the relative slide thickness $S = s/h$ (Fritz, 2002b). A non-linear oscillatory wave pattern was observed if $\mathsf{F} < (4 - 8S)$.(3). The non-linear transition wave region may be determined by $(4 - 8S) < \mathsf{F} < (6.6 - 8S)$. The solitary-like wave region

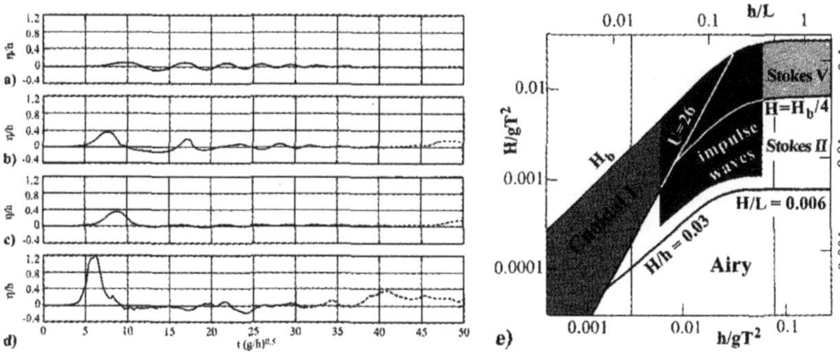


Figure 13. Wave types: (a) non-linear oscillatory waves at F = 1.7, V = 0.39, S = 0.19, h = 0.3m; (b) non-linear transition waves at F = 2.8, V = 0.79, S = 0.34, h = 0.3m; (c) solitary-like wave at F = 1.8, V = 1.57, S = 0.61, h = 0.3m; (d) dissipative transient bore at F = 4.1, V = 1.57, S = 0.56, h = 0.3m; (e) comparison of (■) measured waves with wave theories after LeMéhauté (1976).

may be given by $(6.6 - 8S) < \mathsf{F} < (8.2 - 8S)$. A dissipative transient bore as the leading wave was observed if $(6.6 - 8S) < \mathsf{F} < (8.2 - 8S)$. A graphical representation of the wave type classification is given by Fritz (2002b).

All measured impulse waves were non-linear and classified as intermediate water depth waves. Contrary to the linear wave theory a nonlinear wave theory which is accurate over the whole range from shallow to deep water waves does not exist. Of practical relevance are mainly the Stokes theory for finite amplitude deep water waves and the cnoidal and solitary wave theories for finite amplitude shallow water waves (LeMéhauté, 1976). The transition occurs in the intermediate water depth regime at $\mathsf{U} = aL^2/h^3 = 26$ as shown in Figure 13b). The cnoidal wave theory may be applicable to most leading wave crests, whereas the trailing waves fall into the Stokesian regime (Fritz, 2002).

### 4.3. *Maximum wave crest amplitude*

In most cases the leading wave crest was the highest wave crest. The maximum positive wave amplitude $a_C$ is independent the position within the wave group. The maximum crest amplitudes exceeded the maximum trough amplitudes $a_C \geq a_T$ (Fritz, 2002b). The partition of the total wave height $H$ between the crest and trough amplitudes strongly depended on the wave type. The multiple regression for the maximum wave crest amplitude $a_C/h$ yields

$$a_C/h = (1/4)\, \mathsf{F}^{1.4}\, S^{0.8} \tag{7}$$

with a correlation coefficient $r^2 = 0.93$. The slide Froude number $\mathsf{F} = v_s/(gh)^{1/2}$ is the dominant parameter, while neglecting the dimensionless slide volume $V = V_s/(bh^2)$ had almost no effect. The measured crest amplitudes were within $0.05 < a_C/h < 1.25$. The observed breaker heights matched the breaking criterion $a_C/h = 0.78$ (McCowan, 1894).

### 4.4. *Wave celerity*

The wave propagation velocity was determined from gauge to gauge. The wave celerity definitions are shown in Figure 5b) with the crest velocity $c_{c1}$ and the trough velocity $c_{t1}$ of the leading wave. The wave propagation velocity decreased from the front to the back of the wave trains due the decreasing wavelength of the trailing waves (Fritz 2002b). The first wave crest celerity $c_{c1}/(gh)^{1/2}$ and second wave crest celerity $c_{c2}/(gh)^{1/2}$ are shown in Figure 14a).

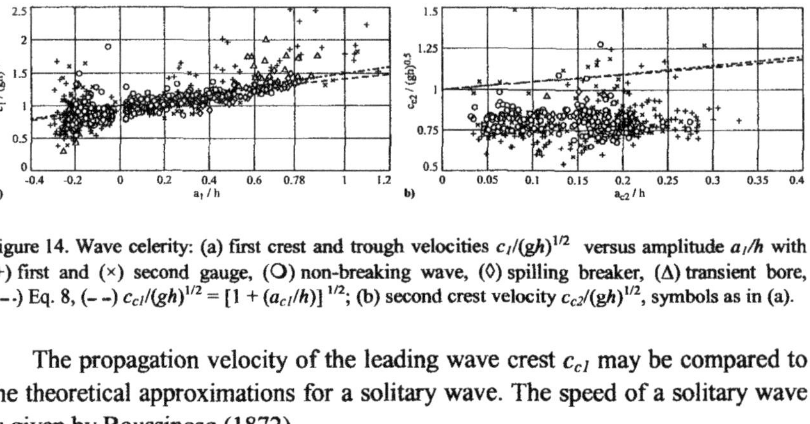


Figure 14. Wave celerity: (a) first crest and trough velocities $c_1/(gh)^{1/2}$ versus amplitude $a_1/h$ with (+) first and (×) second gauge, (O) non-breaking wave, (◊) spilling breaker, (Δ) transient bore, (-.-) Eq. 8, (- -) $c_{c1}/(gh)^{1/2} = [1 + (a_{c1}/h)]^{1/2}$; (b) second crest velocity $c_{c2}/(gh)^{1/2}$, symbols as in (a).

The propagation velocity of the leading wave crest $c_{c1}$ may be compared to the theoretical approximations for a solitary wave. The speed of a solitary wave is given by Boussinesq (1872)

$$c_{c1}/(gh)^{1/2} = 1 + (a_{c1}/2h). \tag{8}$$

This approximation is nearly equal to $c_{c1}/(gh)^{1/2} = [1+(a_{c1}/h)]^{1/2}$ as determined by Russell (1844). The propagation velocity of the leading wave crest $c_{c1}$ follows closely the theoretical approximations for a solitary wave given by Eq. (8). The wave propagation velocity was strongly affected by the wave non-linearity. The amplitude dispersion characterized the propagation velocity of the leading wave crest $c_{c1}$. LeMéhauté and Khangoankar (1992) showed that the propagation velocity of the leading wave crest of an explosion generated impulse wave train matched the solitary wave propagation velocity. The second wave crest propagation velocity concentrated around $(0.8)c_{c2}/(gh)^{1/2}$ lagging the linear shallow water wave propagation velocity by 20% on average. The reduction in celerity from the leading to the trailing wave is attributed to a decrease in wavelength resulting in frequency dispersion (Fritz, 2002b).

### 4.5. *Impact energy conversion*

The impact energy is primarily determined by the kinetic slide energy $E_s = (1/2)m_s v_s^2$, with the centroid velocity $v_s$ upon impact onto the water body. A significant fraction of the impact energy is dissipated during the energy conversion and wave generation. The total wave energy may be estimated to $E_{tot} \approx 2E_{pot}$ assuming equipartition between the kinetic and the potential wave energy (Williams, 1985). The total wave energies were determined from the wave gauge records for two integration ranges resulting in the leading energy of the leading wave crest $E_{c1}$ and the total wave train energy $E_{wt}$ comprising all waves. The energy conversion for the leading wave crest energy and the entire wave train was estimated to

$$E_{c1}/E_s = 0.10\mathrm{F}^{0.8}V^{-0.3}S \qquad (9)$$

$$E_{wt}/E_s = (1/8)\mathrm{F}^{0.5}(V/S)^{-1} \qquad (10)$$

with correlations coefficients of $r^2 = 0.81$ and $r^2 = 0.73$, respectively. The slide thickness $S$ dominates the energy transmission onto the leading wave crest energy. Between 2 and 30% of the kinematic slide impact energy were transferred to the leading wave crest. Between 4 and 50% of the kinetic slide impact energy was transferred onto the entire wave train. The overall wave generation efficiency decreases with increasing dimensionless slide length $V/S$.

## 5. Conclusions

Landslide impact generated tsunamis were investigated in physical laboratory model based on the generalized Froude similarity. The results matched the largest landslide generated tsunami observed in Lituya Bay. The pneumatic landslide generator controlled the granular slide impact characteristics. The main flow features in the slide impact and wave generation zone were determined with PIV. Flow separation and the subsequent impact crater resulted in instantaneous water displacement volumes exceeding the landslide volume by up to an order of magnitude. Four different wave types were observed and the near-field wave characteristics determined. The slide Froude number was the dominant parameter. The measured landslide generated tsunami waves were classified as intermediate water depth waves. The results are mainly limited to two dimensions. The data may be used to validate advanced numerical models.

**Acknowledgments**

This research work was supported by the Swiss National Science Foundation, grant number 2100-050586.97 and the Swiss Federal Institute of Technology (ETH). The author expresses his gratitude to Prof. Dr. H.-E. Minor, director of VAW, and Prof. Dr. W.H. Hager for their continued advice and support.

**References**

Birkhoff, T., Zarantonello, F.H. (1957). *Jets, wakes and cavities*. Academic Press, New York.

Boussinesq, J. (1872). Théorie des ondes et des remous que se propagent le long d'un canal rectangulaire horizontal, en communiquant au liquide contenu dans ce canal des vitesses sensiblement pareilles de la surface au fond. *J. Math. Pures Appl.* **17**:55-108.

Chretien, S.; Brousse, R. (1989). Events preceding the great eruption of 8 May, 1902 at Mount Pelee, Martinique. In: Mount Pelee, eds. Boudon, G., Gourgaud, A. *J. Volcanology and Geothermal Research*, 38(1-2):67-75.

Deplus, C., Le Friant, A., Boudon, G., Komorowski, J.-C., Villemant, B., Harford, C.; Segoufin, J., Cheminee, J.-L. (2001). Submarine evidence for large-scale debris avalanches in the Lesser Antilles Arc. *Earth and Planetary Science Letters*, 192(2):145-157.

Elliott, A.B. (1892). The great earthquake of Port Royal (Jamaica). *Popular Science* (New York): 774-784.

Favre, H. (1935). *Étude théorique et expérimentale des ondes de translation dans les canaux découverts*. Dunod, Paris.

Fritz, H.M., Hager, W.H., Minor, H.-E. (2001). Lituya Bay case: rockslide impact and wave run-up. *Science of Tsunami Hazards* **19**(1):3-22.

Fritz, H.M. (2002a). PIV applied to landslide generated impulse waves. In: Adrian, R.J. et al. (eds) *Laser techniques for fluid mechanics*, 305-320. Springer, New York, Berlin, Heidelberg.

Fritz, H.M. (2002b). Initial phase of landslide generated impulse waves. In: Minor, H.-E. (ed) *VAW Mitteilung 178*. Versuchsanstalt für Wasserbau, Hydrologie und Glaziologie, ETH Zürich.

Fritz, H.M., Liu, P.C. (2002). An application of wavelet transform analysis to landslide generated impulse waves. *Proc. 4th International Symposium on Ocean Wave Measurement and Analysis*, San Francisco 2001, Eds. Edge, B.L. et al., ASCE, **2**:1477-1486.

Fritz, H.M., Moser, P. (2003). Pneumatic landslide generator. *Int. J. Fluid Power* **4**(1):49-57.

Fritz, H.M., Hager, W.H., Minor, H.-E. (2003a). Landslide generated impulse waves, part 1: instantaneous flow fields. *Exp. Fluids* **35**:505-519.

Fritz, H.M., Hager, W.H., Minor, H.-E. (2003b). Landslide generated impulse waves, part 2: hydrodynamic impact craters. *Exp. Fluids* **35**:520-532.

Fritz, H.M., Hager, W.H., Minor, H.-E. (2004). Near-field characteristics of landslide generated impulse waves. *J. Waterway, Port, Coastal, and Ocean Engrg.*, ASCE (in press).

Fuller, M.L. (1907). Notes on the Jamaica earthquake. *J. Geology*: 696-721.

Huber, A. (1980). Schwallwellen in Seen als Folge von Bergstürzen. *VAW-Mitteilung* 47, ed. Vischer, D., Versuchsanstalt für Wasserbau, Hydrologie und Glaziologie, ETH Zürich (in German).

Kamphuis, J.W., Bowering, R.J. (1970). Impulse waves generated by landslides. *Proc. 12th* Coastal Engineering Conf. ASCE **1**:575-588.

Le Friant, A., Harford, C.L., Deplus, C., Boudon, G., Sparks, R.S.J., Herd, R.A., Komorowski, J.C. (2004). Geomorphological evolution of Montserrat (West Indies); importance of flank collapse and erosional processes. *J. the Geological Society of London*, 161, Part 1:147-160.

Le Friant, A., Heinrich, P., Deplus, C. Boudon, G. (2003). Numerical simulation of the last flank-collapse event of Montagne Pelee, Martinique, Lesser Antilles. *Geophysical Research Letters*, 30(2):4pp.

Le Friant, A., Boudon, G., Deplus, C., Villemant, B. (2003). Large-scale flank collapse events during the activity of Montagne Pelee, Martinique, Lesser Antilles. *J. Geophysical Research*, 108(B1):15 pp.

Le Friant, A., Boudon, G., Komorowski, J.-C., Deplus, C. (2002). L'ile de la Dominique, a l'origine des avalanches de debris les plus volumineuses de l'arc des Petites Antilles. *Comptes Rendus - Academie des sciences. Geoscience*, 334(4):235-243.

LeMéhauté, B. (1976). *An introduction to hydrodynamics and water waves*. Springer, New York.

LeMéhauté, B., Khangoankar, T. (1992). Generation and propagation of explosion generated waves in shallow water. *Technical Report* DNA-TR-92-40, Defense Nuclear Agency, Washington D.C.

McCowan, J. (1894). On the highest wave of permanent type. *Phil. Mag. J. Sci.* Series 5 **38**:351-358.

Miche, R. (1944). Mouvements ondulatoires de la mer en profondeur constante ou décroissante. *Ann. des Ponts et Chaussées* 114(2): 25-78 ; 114(7): 131-164 ; 114(13): 279-292 ; 114(19): 369-406.

Miller, D.J. (1960). Giant waves in Lituya Bay, Alaska. *Geological Survey Professional Paper* 354-C. U.S. Government Printing Office, Washington D.C.

Müller, D. (1995). Auflaufen und Überschwappen von Impulswellen an Talsperren. *VAW-Mitteilung* 137, ed. Vischer, D., Versuchsanstalt für Wasserbau, Hydrologie und Glaziologie, ETH Zürich (in German).

Mullins, H.T., Breen, N., Dolan, J., Wellner, R.W., Petruccione, J.L., Gaylord, M., Andersen, B., Melillo, A.J., Jurgens, A.D., Orange, D. (1992). Carbonate platforms along the southeast Bahamas - Hispaniola collision zone. *Marine Geology* 105:109-209.

Pawson M., Buisseret, D. (1975). *Port Royal, Jamaica*. Clarendon Press, Oxford (GBR).

Russell, J.S. (1844). Report on waves. *Report of the 14th Meeting of the British Association for the Advancement of Science*, York, **14**:311-390.

Savage, S.B. (1979). Gravity flow of cohesionless granular materials in chutes and channels. *J. Fluid Mech.* **92**:53-96.

Scanlon, K.M., Masson, D.G., (1996). Sedimentary processes in a tectonically active region; Puerto Rico north insular slope. *Geology of the United States' seafloor; the view from GLORIA*, 123pp., eds.: Gardner, J.V. et al., Cambridge University Press, Cambridge (GBR).

Slingerland, R.L., Voight, B. (1979). Occurrences, properties and predictive models of landslide-generated impulse waves. Rockslides and avalanches **2**:317-397, Ed. Voight, B. *Developments in geotechnical engineering* **14B**. Elsevier, Amsterdam.

Smith, A.L.; Roobol, M.J. (1990). Mt. Pelee, Martinique; a study of an active island-arc volcano. *Memoir - Geological Society of America*, 175:105pp.

Walder, J.S., Watts, P., Sorensen, O.E., Janssen, K. (2003). Water waves generated by subaerial mass flows. *J. Geophys. Res.* **108**(B5):2236-2254.

Ward, S.N., Day, S. (2003). Ritter Island Volcano – lateral collapse and the tsunami of 1888. *Geophys. J. Int.* **154**:891-902.

Wiegel, R.L. (1960). A presentation of cnoidal wave theory for practical application. *J. Fluid Mech.* **7**:273-286.

Wiegel, R.L. (1964). *Oceanographical engineering*. Prentice-Hall, Englewood Cliffs, N.J.

Williams, J.M. (1985). *Tables of progressive gravity waves*. Pitman, Boston.

Wright, J.V., Roobol, M.J., Smith, A.L., Sparks, R.S.J., Brazier, S.A., Rose, W.I.Jr., Sigurdsson, H. (1984). Late Quaternary explosive silicic volcanism on St. Lucia, West Indies. *Geological Magazine*, 121(1):1-15.

# GENERATION AND PROPAGATION OF TSUNAMIS TRIGGERED BY EARTHQUAKES AND LANDSLIDES: A THEORETICAL AND A SIMULATION VIEWPOINT

EMILE A. OKAL
*Department. of Geological Sciences*
*Northwestern University, Evanston, IL 60208, USA*

COSTAS E. SYNOLAKIS
*Department of Civil Engineering*
*University of Southern California, Los Angeles, CA 90089*

We apply both theoretical and simulation techniques to identify critical properties of tsunamis generated by both earthquake dislocations and landslides, in order to define robust discriminant of their respective sources. In the far field, we derive asymptotic expressions for the energy, spectral amplitude and directivity of tsunami waves generated by double-couples obeying universal scaling laws, and by landslides modeled as single forces. We discuss the influence of Dahlen's [1993] quadrupolar corrections for the latter, which we find crucially important in the case of tsunami waves. We show that the spectrum of landslide tsunami waves is significantly offset to high frequencies (10 mHz), where dispersion becomes important thus diminishing far-field amplitudes. We reproduce the results of Kajiura [1981] on the growth of the energy of earthquake tsunamis as the power 4/3 of the seismic moment. We show that the total energy generated into a tsunami wave by a landslide of acceptable size remains dwarfed by that generated by a mega-thrust dislocation. We discuss the far-field directivity pattern due to source finiteness for both dislocation and landslide models, and conclude that strong directivity, i.e., a narrow azimuthal lobe of maximal amplitudes, requires generation by a dislocation source in the case of tsunami waves, and can thus be used a source discriminant. In the near field, we use numerical simulations to explore the influence of earthquake source parameters on observable properties of tsunami run-up amplitudes. These include earthquake geometry (source depth, strike, dip, slip angles), size (seismic moment), location (distance from shore and centroid depth), as well as beach geometry (water depth and beach slope). We focus both on maximum run-up and on the distribution of computed run-up along the beach, which leads to two dimensionless parameters, namely (i) the ratio of maximum run-up to the amount of slip on the fault; and (ii) the aspect ratio of a bell curve fitted to the lateral distribution of run-up. We find that the first parameter remains of order 1, while the second one remains smaller that the maximum strain released during the earthquake, itself an invariant. By contrast, in the case of landslide sources, the aspect ratio of the lateral distribution of run-up along the beach can be 1 to 2 orders of magnitude greater, suggesting its use as a potential discriminant in the near field. We apply these discriminants to the case of the 1946 Aleutian tsunami, and suggest that its source must have involved both a large slow earthquake (to explain the far field tsunami) and a coeval landslide (responsible for the near field tsunami).

# MODELING OF TSUNAMIS IN THE EASTERN MEDITERRANEAN AND COMPARISON WITH CARIBBEAN

**AHMET CEVDET YALCINER**
*Assoc. Prof. Dr., Middle East Technical University, Department of Civil Engineering, Ocean Engineering Research Center, 06531 Ankara Turkey*
yalciner@metu.edu.tr

**HULYA KARAKUS**
*Res. Assist., Middle East Technical University, Department of Civil Engineering, Ocean Engineering Research Center, 06531 Ankara Turkey*
khulya@metu.edu.tr

**UGUR KURAN**

[1]*Ministry of Construction and Resettlement, General Directorate of Disaster Affairs, Earthquake Research Center, Ankara*
ugurkuran@ttnet.net.tr

Numerous tsunamis have been observed in the Caribbean and Mediterranean Seas in history. The oceanographical, geographical, geological and geophysical characteristics of the Caribbean and Mediterranean Seas are presented and discussed briefly. For comparison the general characteristics of historical tsunamis occured in both seas documented in literature are tabulated. The estimated zones of tsunami sources in relation to the recent earthquake activities are presented. As illustrative examples showing the characteristics of tsunami propagation and coastal amplification in Eastern Mediterranean, the simulations of different tsunami scenarios are also presented and discussed for Eastern Mediterranean.

## 1. INTRODUCTION

The Caribbean Sea is one of the biggest marginal seas in the planet and located in between lattitudes 7°N and 23°N, and longitudes -88°E and -60°E. It is bordered on the North and East by the West Indies archipelago, on the South by the South America mainland, and on the West by the Central American isthmus. The Caribbean is linked to the Gulf of Mexico by the Yucatán Channel; to the Atlantic by many straits, of which the Windward Channel and Mona Passage are the most important; and to the Pacific Ocean by the Panama Canal. The Caribbean sea is well known and center of attraction with its tropical climate, storms, rain, clear waters, sun and reefs.

The Mediterranean Sea is one of the biggest marginal seas in the planet and located between latitudes 30°N and 47°N, and longitudes -5°E and 43°W. It is bordered on the North and West by Europe, on the south by Africa, on the East by Asia. Siciliy divides the sea in to eastern and western basins. There is a single connection with the ocean: the Strait of Gibraltar. The other connection with the Red Sea is Suez Canal. *The Meditemaean Sea is well known and is the center of* attraction with its breeze, clear waters, sun and remnants of ancient civilizations.

Numerous earthquakes and associated tsunamis in history in the Caribbean and Mediterranean seas seem as the precursor of future similar events. In this study oceanographical, geographical, geological and geophysical characteristics of the Caribbean and Mediterranean Seas are presented and discussed briefly. For comparison, the general characteristics of historical tsunamis occurring in both seas, and documented in the literature, are tabulated. The estimated zones of tsunami sources in relation to the recent earthquake activities are presented. Simulations of different tsunami scenarios are also presented and discussed for illustration of the propagation and coastal amplification of tsunamis in Eastern Mediterranean.

## 2. GENERAL CHARACTERISTICS OF THE CARIBBEAN SEA

The Caribbean region, located in the tropics, extends in a broad arc of over 4000 km from the Bahamas in the north to Guyana and Suriname in the south. The region also includes, based on historical factors, Belize which is in Central America bordering Mexico and Guatemala. The countries are widely different in physical and population size as well as landforms and geology. The area of the Caribbean sea is 2,512,950 $km^2$ . The maximum depth is 7535 m in the Ocean (north of Puerto Rico) and 2500m in the Caribbean Sea (west of Jamaica). The Caribbean's waters are clear, warm, and less salty than the Atlantic; the basin has a very low tidal range about 0.3 m.

The Caribbean Sea has a counterclockwise current; water enters through the Lesser Antilles, is warmed, and exits via the Yucatán Channel, where it forms the Gulf Stream. Volcanic activity and earthquakes are common in the Caribbean, as are destructive hurricanes that originate over the sea or in the Atlantic. The general view of topography and bathymetry is shown in Figure 1.

There are three main geographical features of the Caribbean Sea, which are i) Outer ring of mainly coral islands with white sand beaches from the Bahamas, to the Turks and Caicos, the Cayman Islands, ii) Inner ring of more volcanic islands, with black sands, including the islands of the Greater Antilles - Cuba, Haiti, Dominican Republic and Puerto Rico. iii) The alluvial coast of Guyana, Suriname and French Guiana (Cayenne) on the South American mainland between the vast estuaries of the Orinoco and the Amazon.

'Caribbeanness' began to form when the first human groups set foot on the islands. According to the latest archaeological data, that historical legacy has a tradition of almost 10,000 years in Cuba, 7,000 years in Trinidad and 5,000 in Puerto Rico.

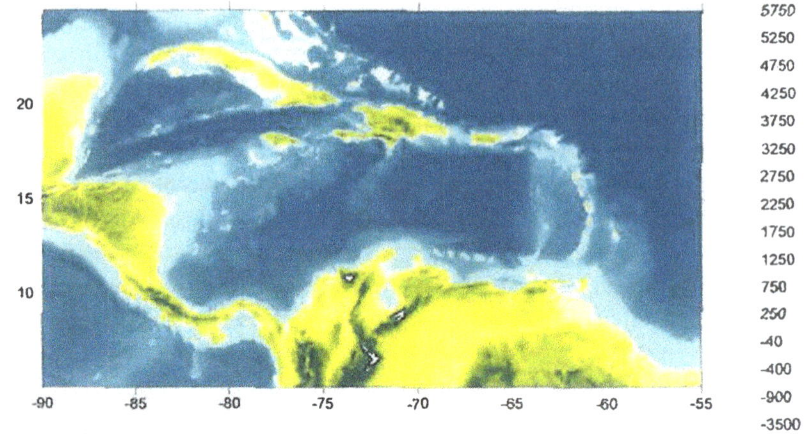


Figure 1. General view of topography and bathymetry of the Caribbean Sea

Many of the Caribbean countries are subject to earthquakes, volcanic and small-scale geothermal activity. Earthquakes are more frequent in the Northern Caribbean islands. Some earthquakes have been quite strong. In Jamaica, Port Royal was completely destroyed by an earthquake in 1692, which killed many persons. The most recent volcanic activity has occurred in Montserrat during the 1990's, with consequences for social and economic life of the island including the relocation of many on the island's population. The most widely known volcanic eruption in the Caribbean preceding Monsterrat was Mt. Pelee in Martinique in 1902 when volcanic ash and lava killed over 30,000 persons.

In terms of tsunami activities, the detailed information is given in Lander *et al.* 2002. Table 1 shows the tsunami events which have occurred in the Caribbean Sea, with their probable epicenters and efect areas. More information about each event is available in Lander et. al. (2002)

Table 1: The tsunami events occured in the Caribbean sea with their probable epicenters and efect areas (Lander et al 2002)

| No | Verified | Date (Yr mo da hr) | Lat. Long. | Effect Area |
|---|---|---|---|---|
| 1 | No | 1498 08 02 or 3 | 9,9N 62,3W | Venezuela |
| 2 | Yes | 1530 09 01 14:30 UT | 10,7N 64,1 W | Venezuela |
| 3 | No | 1539 11 24 (23:00 LT) | 15,0 N 86,5 W | Northern Honduras |
| 4 | No | 1541 12 25 | 10,8 N 64,2 W | Venezuela |
| 5 | No | 1543 | 10,7 N 64,1 W | Venezuela |
| 6 | No | 1688 03 01 | 17,6 N 76,5 W | Jamaica |
| 7 | Yes | 1690 04 16 | 17,5N 61,5 W | Leeward Is. |
| 8 | Yes | 1692 06 07( 11:43 LT ) | 17,8 N 76,7 W | Jamaica |
| 9 | No | 1726 | 10,6 N 64,2 W | Venezuela |
| 10 | No | 1750 | 10,7 N 64,1 W | Venezuela |
| 11 | No | 1751 09 15 (19:00 UT) | 18,5 N 70,7 W | Hispaniola |

| | | | | |
|---|---|---|---|---|
| 12 | Yes | 1751 10 18 | 18,5 N 70,7 W | Hispaniola |
| 13 | No | 1751 11 21 | 18,3 N 72,3 W | Haiti |
| 14 | Yes | 1755 11 01( 9:50 LT ) | 36,0 N 11,0 W | Lisbon, Portugal |
| 15 | Yes | 1755 11 18( 9:50 LT ) | 42,7 N 70,3 W | Cape Ann, Massachusetts |
| 16 | Yes | 1761 03 31( 12:05 LT ) | 37,0 N 10,0 W | Lisbon, Portugal |
| 17 | No | 1766 06 12(4:45 UT ) | 20,0 N 75,5 W | Santiago de Cuba and Bayamo, Cuba |
| 18 | No | 1766 10 21( 9:00 UT ) | 7,4 N 62,5 W | Venezuela |
| 19 | Yes | 1767 04 24(6:00 UT) | 14,4 N 61,0 w | Martinique and Barbados |
| 20 | No | 1769 | 18,5 N 72,3 W | Haiti |
| 21 | Yes | 1770 06 03( 19:15 LT ) | 18,3 N 72,2 W | Haiti |
| 22 | No | 1775 02 11 | 19,0 N 72,4 W | Hispaniola and Cuba |
| 23 | No | 1775 03 | 19,0 N 72,3 W or 20,0 N 15,8 W | Hispaniola |
| 24 | No | 1775 12 18 | 9, N 70,3 w | Hispaniola and Cuba |
| 25 | No | 1780 10 03( 22:00 LT ) | 18,1 N 78,1 W | Jamaica |
| 26 | No | 1781 08 01 | 18,2 N 78,1 W | Jamaica |
| 27 | No | 1787 10 27( 14:20 LT ) | 18,4 N 77,9 W | Jamaica |
| 28 | No | 1798 02 22 | 10,2 N 82,9 W | Costa Rica |
| 29 | No | 1802 03 19 | 10,4 N 68,0 W (5-20) 10,3 N 64,0 W (8-15 ) | Leeward Island |
| 30 | Yes | 1802 05 05 | 9,2 N 61,5 W | Venezuela |
| 31 | No | 1812 03 26 | 8,6 N 71,1 W | Venezuela |
| 32 | No | 1812 11 11 | 18,0 N 76,5 W | Jamaica |
| 33 | No | 1822 05 07 | 10,0 N 84,0 w | Costa Rica |
| 34 | Yes | 1823 11 30( 3:10 LT) | 14,4 N 61,0 W | Martinique |
| 35 | No | 1824 09 13 | 16,7 N 62,2 W | Guadeloupe |
| 36 | No | 1824 11 30 | 14,4 N 61,0 W | Martinique |
| 37 | No | 1825 02 | 11,2 N 74,2 W | Honduras |
| 38 | No | 1825 09 20( 1:45 UT ) | 10,4 N 61,3 W | British Guiana |
| 39 | No | 1831 12 03( 23:40 UT) | 12,4 N 61,5 W | Trinidad and St.Chritopher |
| 40 | No | 1837 07 26 | 18,2 N 64,5 W | Martinique |
| 41 | Yes | 1842 05 07( 17:30 LT ) | 19,7 N 72,8 W | Haiti |
| 42 | Yes | 1843 02 08( 14:50 UT ) | 16,5 N 62,2 W | Guadeloupe |
| 43 | No | 1843 02 17 | 17,0 N 61,8 W | Antigua |
| 44 | No | 1852 07 17 | 20,0 N 75,8 W | Cuba |
| 45 | Yes | 1853 07 15 | 12,1 N 63,6 W | Venezuela |
| 46 | No | 1855 09 25( 10:45 LT ) | 15,9 N 86,0 W | Honduras |
| 47 | Yes | 1856 08 09 | 16,0 N 88,0 W | Honduras |
| 48 | Yes | 1860 03 08 | 19,0 N 72,0 W | Hispaniola |
| 49 | No | 1867 ( or 09-10) | 18,0 N 65,5 W | Venezuela |
| 50 | Yes | 1867 11 18(18:45 UT ) | 18,0 N 65,5 W | St.Croix and St.Thomas, U.S. Virgin Islands |
| 51 | No | 1868 | 10,6 N 66,9 W | Venezuela |
| 52 | No | 1868 08 13 | 18,5 N 70,3 W | Venezuela |
| 53 | No | 1873 10 13( 18:05 LT ) | 30,1 N 85,6 W | Panama |
| 54 | No | 1874 03 11( 4:30 LT ) | 18,3 N 64,9 W | Lesser Antilles |
| 55 | No | 1881 08 12 | 19,9 N 76,8 W | Jamaica |
| 56 | Yes | 1882 09 07(7:50 UT ) | 7,3 N 77,8 W | Panama |
| 57 | Yes | 1883 08 27 | 5,8 S 106,3 E | Krakatoa, Indonesia |

| | | | | |
|---|---|---|---|---|
| 58 | Yes | 1887 09 23( 12:00 UT ) | 19,7 N 74,4 W | Haiti |
| 59 | Yes | 1900 10 29 | 10,9 N 66,8 W | Venezuela |
| 60 | No | 1902 05 08 | 14,4 N 61,0 W | Martinique |
| 61 | Yes | 1902 08 30( 21:25 LT ) | 14,4 N 61,0 W | Martinique |
| 62 | No | 1906 | 10,6 N 66,9 W | Venezuela |
| 63 | Yes | 1906 01 31(21:36 UT ) | 8,1 N 76,7 W | Jamaica |
| 64 | Yes | 1907 01 14(21:36 UT ) | 8,1 N 76,7 W | Jamaica |
| 65 | Yes | 1911 11 03 | 10,5 N 61,2 W | Trinidad |
| 66 | Yes | 1916 04 24( 8:02 UT ) | 11,0 N 85,0 W | Panama |
| 67 | No | 1916 11 12 | 10,1 N 66,8 W | |
| 68 | Yes | 1918 10 11( 4:14 UT ) | 18,5 N 67,5 W | Puerto Rico |
| 69 | Yes | 1918 10 24( 3:43 UT ) | 18,5 N 67,5 W | Puerto Rico |
| 70 | No | 1922 05 02( 20:24 UT ) | 18,2 N 66,6 W | Puerto Rico |
| 71 | No | 1928 09 13 | 106, N 63,2 W | Venezuela |
| 72 | Yes | 1929 01 17( 11:52 UT ) | 10,6 N 65,6 W | Venezuela |
| 73 | No | 1931 10 01 | 21,5 N 80,0 W | Cuba |
| 74 | No | 1932 02 03( 06:16 UT ) | 19,5 N 75,6 W | Cuba |
| 75 | No | 1932 11 04 | 10,7N 64,1 W | Venezuela |
| 76 | Yes | 1939 08 15( 3:52 UT ) | 22,5 N 79,2 W | Cuba |
| 77 | Yes | 1946 08 04(17:51 UT ) | 19,3 N 68,9 W | Dominician Republic, Haiti and Puerto Rico |
| 78 | Yes | 1946 08 08( 13:28 UT ) | 19,5 N 69,5 W | Puerto Rico |
| 79 | No | 1950 08 03 | 9,8 N 69,7 W | Venezuela |
| 80 | No | 1953 05 31 | 19,7 N 70,7 W | Dominican Republic |
| 81 | No | 1955 01 18 | 11,3 N 69,4 W | Venezuela |
| 82 | No | 1961 06 16 | 9,7 N 71,5 W | Venezuela |
| 83 | No | 1968 09 20 ( 6:09 UT ) | 10,7 N 62,6 W | Venezuela |
| 84 | Yes | 1969 12 25( 21,32 UT ) | 15,8 N 59,7 W | Leeward Is |
| 85 | No | 1979 09 13(2:12 UT ) | 5,9 N 82,5 W | Venezuela |
| 86 | Yes | 1985 03 16 | 17,0 N 62,4 W | Leeward Is |
| 87 | Yes | 1989 11 01( 10:25 UT ) | 19,0 N 68,8 W | Puerto Rico |
| 88 | Yes | 1991 04 22(21:56 UT ) | 9,7 N 83,1 W | Costa Rica |
| 89 | Yes | 1997 07 09(19:24 UT ) | 10,6 N 63,5 W | Venezuela |
| 90 | Yes | 1997 12 26(3:00 LT ) | 16,7 N 62,2 W | Montserrat |

Table 1 shows that over 90 tsunamis have occurred in the Carebbean Sea during the period between 1498-1997. Among these, 38 events have been verified and 52 events has not been verified so far (Lander et. al. 2002). These observations reveal that the tsunamis are more often observed in Cuba, Puerto Rico, Venezuela, Jamaica and Haiti.

## 3. GENERAL CHARACTERISTICS OF THE MEDITERRANEAN SEA

The Mediterranean Sea is 3900 km long and its maximum width is 1600 km and greatest depth is 4400 m (Figure 2). The Mediterranean Sea extends to the Aegean Sea and connects with the Black Sea through the Dardanelles; the Bosporus, and the Sea of Marmara; with the Atlantic Ocean through the Strait of Gibraltar; and with the Red Sea through the Suez Canal. Adriatic, Ionian,

Tyrrhenian seas are its other main divisions. The Mediterranean Sea has a warm and dry climate and frequent earthquakes and volcanic activities are also observed in the region.

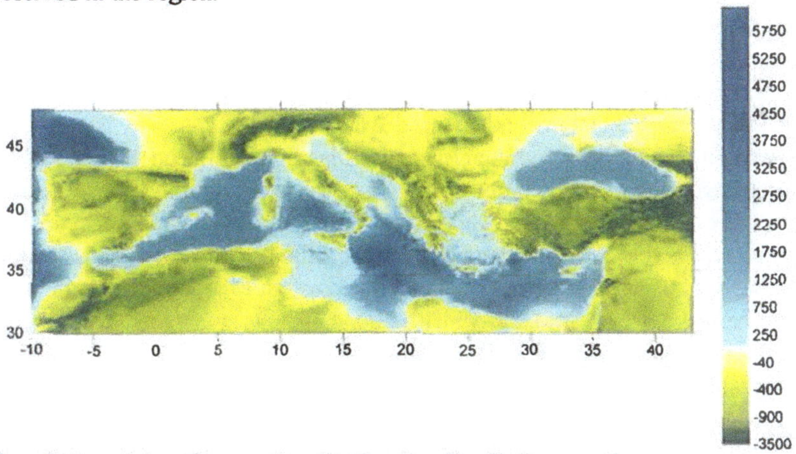


Figure 2. General view of topography and bathymetry of the Mediterranean Sea

Assyrians, Chaldeans, Hebrews, Hittites, Akkadians, Etruscans, Minoans, Mycenaeans, Greeks, Romans, Egyptians, Mesopotamians, Babylonians, Sumerians, Nubians, Persians. Or Alexander, Plato, Socrates, Hammurabi, Aristotle, the Pharaohs, Emperors, Kings, Sultans, Caesar, Cleopatra, Homer. Or dinosaurs, hunters, slaves, women, rulers, soldiers. Or the Iliad, the Aeneid, the Odyssey, the Olympics. Or culture, archaeology, art, music, theater, language, philosophy, politics, peace and war, life and living. All were found in the ruins, temples, forums, pyramids or in the remnants of the ordinary life of the Mediterranean.

The Mediterranean Civilization is one of the most ancient civilizations which flourished around the world and grew out of the Mediterranean breezes, sea and sun. The breeze, clear waters and sun in the Mediterranean connected peoples, ideas, patterns, developments, organizations, wars, religions, art, architecture, food and drink.

The sea was called *Mediterranean* (middle sea) by the surrounding *civilizations. It was used for commerce by merchants trading from Phoenicia. It* was called *Mare Nostrum* (our sea) after becoming a Roman lake.

The Mediterranean Basin consists of the Mediterranean Sea, Aegean Sea, Black Sea and the Sea of Marmara. With the opening of the Suez Canal (1869) the Mediterranean resumed its importance as a link on the route to the East. The Panama and Suez Canals are two similar human made passages in the Caribbean and Mediterranean, which caused very important biological, commercial, social, and cultural changes in both seas.

The sea is located in the Mediterranean part of Alpine-Himalayan orogenic belt. There are three main active fault zones nearby, and in the Mediterranen Basin, at the east, are the North Anatolian Fault Zone (NAF), East Anatolian Fault Zone (EAF) and the Hellenic Arc. The costal areas at the Eastern Mediterranean Basin have experienced tsunamis many times in history. The generation mechanisms and their characteristics have not been well described. According to the historical information, or distribution of fault zones, volcanos, and other probable tsunamigenic sea bottom formations, there are numerous source areas which may be considered responsible for those tsunamis. One of the important source areas of tsunamis in the eastern Mediterranean is the fault zone named the Hellenic Arc, which is a subduction zone, about 1000 km in length, that starts south west of the Greek mainland and follows a curve south of Crete and south east of Rhodes island, and trends towards Anatolia along a North East direction near Dalaman town. The deepest region of the Mediterranean, with a depth of 4000 m and more, is in between Rhodes and the Dalaman region near the Hellenic Arc. This region can be called the Rhodes - Dalaman trench and might be one of the most important tsunami prone areas in the whole Mediterranean Basin.

The Aegean Sea is one of the most active seismic regions in Europe, also containing a volcanic arc with active volcanoes. The Sea of Marmara is another inlad sea located northwest of Anatolia and near south of Istanbul. The North Anatolian Fault Zone is divided into two branches east of the Mamara Sea. One of the branches passes the sea along an east-west direction; another follows the line paralel to the south coast of the sea. Both branches enter the northern part of Aegean sea. The Black Sea is another sea in the Mediterranean basin and numerous tsunamis have also been observed there. The epicenters of earthquakes (with Magnitude M > 5) which have occurred since 1900 in the Eastern Mediterranean are shown in Figure 3.

In order to estimate possible zones of tsunami source areas, the instrumentally located epicenters of earthquakes since 1900 are shown in Figure 4 (for the Aegean Sea) and Figure 5 (for the Eastern Mediterranean) where the estimated zones of tsunami sources are also shown.

## 4. HISTORICAL TSUNAMIS IN EASTERN MEDITERRAEAN

The information about historical tsunamis obtained by examining the documents in the archives is given in Altinok and Ersoy, (2000). Table 2 shows the list of tsunamis occuring in the vicinity of Anatolia since 1630 B.C. (Doumas, 1980a,b, Hardy and Renfrew 1990a,b,c, Altinok et. al., 2001, Papadopulos, 2001) . It is

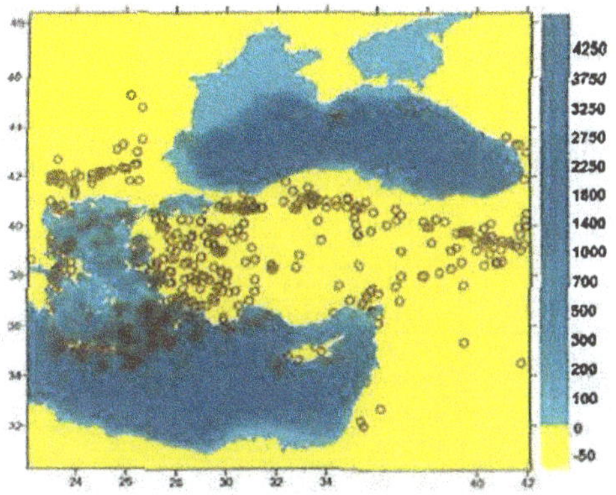


Figure 3. Epicenters of earthquakes (with Magnitude M > 5) occuring since 1900 in the Eastern Mediterranean

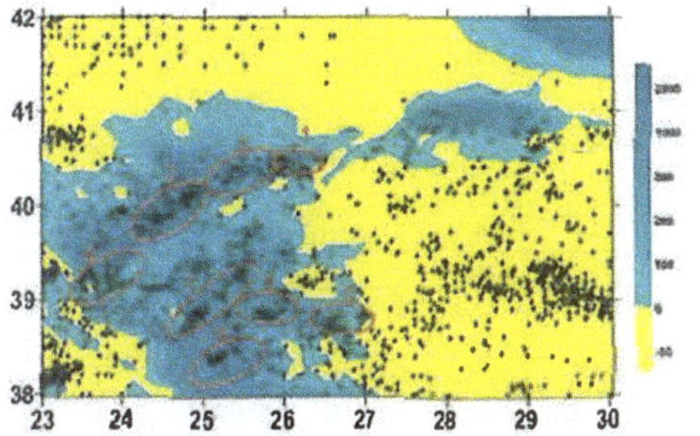


Figure 4. Estimated possible regions where tsunamis can be effective, directly or indirectly, in the Aegean Sea by using the epicenters (Magnitude M > 4) occurring in the last century. (KOERI, 2004)

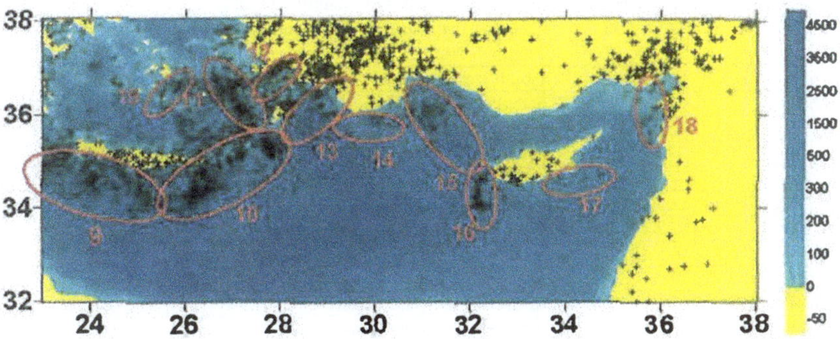


Figure 5. Estimated possible regions where tsunamis can be effective, directly or indirectly, in the Mediterenean coasts in Turkey by using the earthquake epicenters occurring in the last century (KOERI 2004)

Table 2: The list of tsunamis occurring in the vicinity of Anatolia since 1630 B.C. (Doumas, 1980a,b, Hardy and Renfrew 1990a,b,c, Altınok et. al., 2001, Papadopulos,2001)

| No | Verified | Date | Coastal Region |
|---|---|---|---|
| 1 | YES | 1631 B.C. | Santorini South Aegean |
| 2 | YES | 1300 | Dardanelles Coasts |
| 3 | YES | 330 | North East of Lemnos Island |
| 4 | YES | 22 | Rhodes, Cyprus, Corinth |
| 5 | YES | 140 | Acre, Tyre-Syria |
| 6 | YES | 26 | Paphos, Cyprus |
| 7 | YES | 46 A.D. | North east of Crete, Santorini Island |
| 8 | YES | 53/62/66 | Cnossos-Crete, Leben |
| 9 | YES | 68 | Demre, Patara-Lycia |
| 10 | YES | 76-78 | Larnaca, Paphos, Salamis-Cyprus |
| 11 | YES | 120/128 | Kapıdağ Peninsula, Iznik, Izmit |
| 12 | YES | 142 | Fethiye Gulf, Rhodes, Kos, Seriphos, Syme Islands |
| 13 | YES | 261-262 | South coasts of Anatolia |
| 14 | YES | 293-306 | Salamis-Cyprus |
| 15 | YES | 325 | Izmit Gulf |
| 16 | YES | 342 | Paphos, Famagusta- Cyprus |
| 17 | YES | 344 | Çanakkale Region, Thracian Coasts |
| 18 | YES | 358.08.24 | Izmit Gulf, Iznik, Istanbul |
| 19 | YES | 365.07.21 | East Mediterranean, Crete, Greece, Libyan, Sicilian, Alexandria, West Anatolia |
| 20 | YES | 368.10.11 | Iznik and its surrounding |
| 21 | YES | 407.04.01 | Istanbul |
| 22 | YES | 447.11.08 | Marmara Sea, Istanbul, Izmit Gulf, Marmara Isls.,Marmara and Çanakkale coasts |
| 23 | YES | 450.01.26 | Marmara Sea, Istanbul |
| 24 | YES | 477/480. 09.24/25/26 | Gelibolu, Istanbul, Çanakkale, Izmit, Bozcaada |
| 25 | YES | 488.09.26 | Izmit Gulf |
| 26 | YES | 524/525 | South coasts of Anatolia, Anazarba-Adana |
| 27 | YES | 529 winter | Thracian coasts of Marmara |
| 28 | YES | 542 winter | West coast of Thracia, Bandırma Gulf |
| 29 | YES | 543.09.06 | Kapıdağ Peninsula, Erdek, Bandırma |
| 30 | YES | 553.08.15 | Istanbul, Izmit Gulf |
| 31 | YES | 554.08.15 | South west coasts of Anatolia, Kos Isl., Mandalya Gulf |
| 32 | YES | 555.08.15/16 | Istanbul, Izmit Gulf |
| 33 | YES | 557.12.14 | Istanbul, Izmit Gulf |
| 34 | YES | 715 | Istanbul, Izmit Gulf |
| 35 | YES | 740.10.26 | Marmara Sea, Istanbul, Izmit, Iznik Lake |
| 36 | YES | 803.12.19 | İskenderun Gulf |
| 37 | YES | 859.11 | Syrian coasts and Near Samandag |
| 38 | YES | 975.10.26 | Istanbul, Thracian coasts of Marmara |
| 39 | YES | 989 | Istanbul, Marmara Coasts |
| 40 | YES | 990 | Istanbul, Marmara Coasts |
| 41 | YES | 1039.02.06 | Istanbul, Marmara Coasts |
| 42 | YES | 1064.09.23 | Iznik, Bandırma, Mürefte, Istanbul |
| 43 | YES | 1114.08.10 | Ceyhan, Antakya, Maraş |
| 44 | YES | 1157.07.15 | Hama-Homs, Chaizar Region |
| 45 | YES | 1202.05.22 | Cyprus, Syrian coasts, Egypt |

| | | | |
|---|---|---|---|
| 46 | YES | 1222.05,11 | Paphos, Limasol-Cyprus, Egypt |
| 47 | YES | 1304.08.08 | EastMediterranean, Rhodes, Crete, Peloponnesus |
| 49 | YES | 1344.10.14 | Marmara Sea, Istanbul, Gelibolu, Thracian Coasta |
| 48 | YES | 1332.02.02 | Marmara Sea, Istanbul |
| 49 | YES | 1344.10.14 | Marmara Sea, Istanbul, Thracian coasts, Gelibolu |
| 50 | YES | 1389.03.20 | İzmir , Chios ve Lesvos Isls. |
| 51 | YES | 1403.11.16 | South coasts of Anatolia, Syrian Coasts |
| 52 | YES | 1481.05.03 | Rhodes, South west coast of Anatolia, Crete |
| 53 | YES | 1489 | South coasts of Anatolia,Antalya |
| 55 | YES | 1494.07.01 | Harekleion-Crete |
| 55 | YES | 1509.09.10 | Istanbul , Marmara Coasts |
| 56 | YES | 1577.07.11 | Istanbul |
| 57 | YES | 1598 | Amasya, Corum |
| 58 | YES | 1609.04 | Rhodes , East Mediterranean |
| 59 | YES | 1612.12.08 | North of Crete |
| 60 | YES | 1646.04.05 | İstanbul |
| 61 | YES | 1659.09.29 | Santorini Patmos, SikinosIsls., Northern Crete |
| 62 | YES | 1667.11.30 | Izmir Gulf |
| 63 | YES | 1672.02.14 | Bozcaada, Kos Isl. |
| 64 | YES | 1688.07.10 | Izmir Gulf |
| 65 | YES | 1741.01.31 | Rhodes |
| 66 | YES | 1751.08.15 | Istanbul |
| 67 | YES | 1752.07.21 | Syrian coasts |
| 68 | YES | 1754.09.02 | Izmit Gulf, Istanbul |
| 69 | YES | 1766.05.22 | Istanbul, Marmara Sea |
| 70 | YES | 1822.08.13 | Antakya, Iskenderun,Kilis |
| 71 | YES | 1829.05.23 | Istanbul, Gelibolu |
| 72 | YES | 1851.02.28 | Fethiye, Kaya-Muğla , Rhodes |
| 73 | YES | 1851.04.03 | Fethiye Gulf |
| 74 | YES | 1851.05.23 | Rhodes, Dodecanese |
| 75 | YES | 1852.05.12 | Izmir |
| 76 | YES | 1852.09.08 | Izmir |
| 77 | YES | 1855.02.13 | Fethiye Gulf |
| 78 | YES | 1856.11.13 | Chios Island |
| 79 | YES | 1866.0131 | Santorini Island |
| 80 | YES | 1866.02.02 | Chios Isl. |
| 81 | YES | 1878.04.19 | Izmit, Istanbul, Marmara Sea |
| 82 | YES | 1878.05.10 | Izmit, Istanbul,Bursa |
| 83 | YES | 1886.08.27 | North Peloponnessus, Pylos, Izmir |
| 84 | YES | 1893.02.09 | Northern Aegean Sea, Samothrace Isl., Thracian coasts, Alexandroupolis |
| 85 | YES | 1894.07.10 | Istanbul |
| 86 | YES | 1926.06.26 | Rhodes, South west of Turkey, Archangelo, Fethiye, Karpathos, Harakleion |
| 87 | YES | 1928.03.31 | Izmir |
| 88 | YES | 1939.12.26-27 | Fatsa-Black Sea |
| 89 | YES | 1948.02.09 | Karpathos-Dodocanese |
| 90 | YES | 1949.07.23 | Eastern Aegean Sea, North Chios Isl. |
| 91 | YES | 1953.09.10 | South coasts of Turkey |
| 92 | YES | 1956.07.09 | Southern Aegean, Amorgos, Astypalaea Isls. |
| 93 | YES | 1963.09.18 | Eastern Marmara, Yalova, Karamürsel, Armutlu, Mudanya Gemlik |
| 94 | YES | 1968.02.19 | North Eagean Sea |
| 95 | YES | 1968.09.03 | Amasra-Black Sea |
| 96 | YES | 1999.08.17 | Izmit Bay |

seen from Table 2 that, during the last 36 centuries at least 96 tsunamis were documented in the Eastern Mediterranean. The coastal areas were not so densely populated and utilized in the past when these tsunamis occured. Therefore the effects of these tsunamis are limited to only the populations near the coasts. It must be clearly expected that any future similar tsunami will be much more destructive on the coasts. Besides examining the archive information, it is also helpful to use simulation technique to understand the generation, propagation and coastal amplification of tsunamis.

## 5. MODELING OF MEDITERRANEAN TSUNAMIS

Two different tsunami scenarios for the Mediterranean Sea have been simulated by using the numerical model TUNAMI N2, written in Tohoku University, Disaster Control Research Center by Prof. Imamura and distributed by the TIME project. The model was developed at the University of Southern California and registered by Imamura, Yalciner and Synolakis. The model simulates the propagation and coastal amplification of tsunami waves generated as the result of a combined mechanism of an earthquake, or sea floor subsidence, or collapse of the water surface in the computational domain.

It computes the propagation of the wave by calculating the water surface elevations and water particle velocities throughout the domain, at every time step during the simulation, according to given initial condition. The staggered leap-frog scheme (Shuto, Goto, Imamura, (1990)) is used to solve the shallow water equations in computation. The arrival time of the wave front, distribution of maximum positive amplitudes (runup) along the shores, the distribution of arrival times along shores at specified time steps, maximum negative amplitudes, time at which maximum positive amplitudes occurred in the domain, water surface and velocities at specified times in the domain, time histories of water surface at specified coastal locations, are other output parameters of the model.

### 5.1 Simulation of Minoan Tsunami

The eruption of the Thera volcano in late Minoan time (1600–1300 B.C.) is considered to have been the most significant Aegean explosive volcanism during the late Holocene. The last eruptive phase of Thera resulted in an enormous submarine caldera, which is believed to have produced tsunamis on a large scale. The collapse of the dome complex of the caldera is assumed as 8 × 9 km wide and ~700 m deep. Seawater rushed into the caldera, submerging it to an average depth of 380m (Minoura et. al., 2000). Sudden collapse of the volcano probably generated large waves, which formed the Minoan tsunami. The simulation of the Minoan Tsunami has been made by TUNAMI N2 with the grid size of 700 m and time step of 2.4 sec. The sea state at time 20 minutes after collapse, and the arrival of the wave at Crete Island, is shown in Figure 6. According to simulation results, the arrival time of waves at the western coast of Anatolia is one hour.

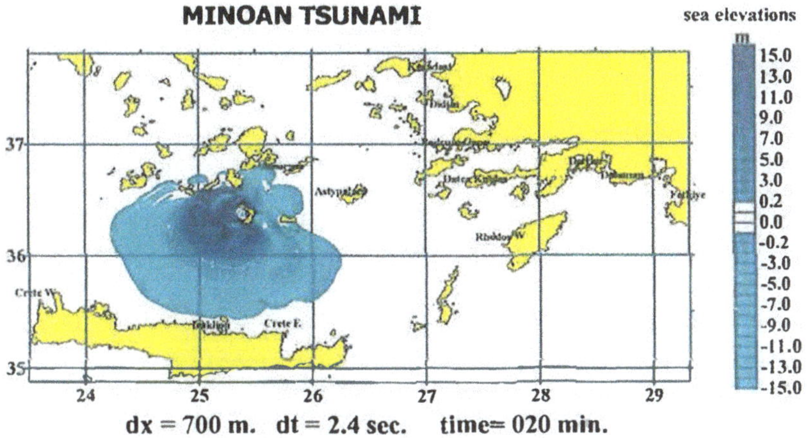


Figure 6. The sea state at time 20 minutes after collapse during the Minoan tsunami

### 5.2 Simulation of May 11, 1222 Tsunami

A destructive earthquake on May 11, 1222, in Cyprus is associated with a seismic sea wave that was felt in Egypt. Paphos and Limassol were completely destroyed, with loss of life. By using the available information about the epicenter coordinates, the fault line (axis of the initial form of the tsunami wave) is assumed in between the coordinates 32.45°E, 34.80°N and 31.88°E, 35.03°N with a fault length of 53 km and an angle of 290° from North. Since there is no sufficient information available about the source parameters of this earthquake, the width of the source area (minor axis of the ellipse) is assumed as 15km. The dip and slip angles of the fault are selected as 20° and -90°, respectively, the displacement and focal depth have been selected as 12m and 15 km. The crest and trough amplitudes are 2.6 m and -0.9 m. respectively [Okada, 1985]

For modelling this event, we have used the propagation model and computed the sea state at different times, and time histories of water surface oscillations and the maximum positive amplitudes at every grid point have been stored. Snapshots (at t = 0.5, 5, 10, 15 min) of the tsunami wave propagation are shown in Figure 7. According to simulation results, the tsunami arrives within 5 minutes to the nearest coast.

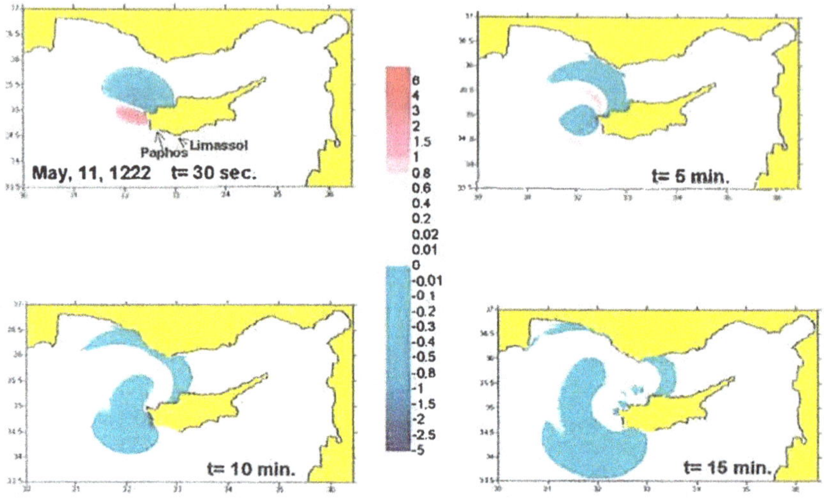


Figure 7: The sea state at different time steps (30 sec., 5 min., 10min., 15min) of the May 11, 1222 tsunami

## 6. Acknowledgement

Prof. Fumihiko Imamura, Prof. Costas Synolakis and Prof. Efim Pelinovsky are acknowledged because of their long term, valuable, cooperation, discussion and collaboration during the development of the models used in this study.

## 7. References

Altinok, Y., Ersoy, Ş., Yalçıner, A. C., Alpar B., and Kuran, U., (2001), " Historical tsunamis in the Sea of Marmara, International Tsunami Symposium 2001, Seattle

Altinok, Y. and Ersoy, S., (2000), Tsunamis observed on and near Turkish Coasts, Kluwer Academic Publishers, Journal of Natural Hazards Vol. pp 185-199, 2000

Altinok, Y., Ersoy, Ş., Yalciner, A. C., Alpar B., Kuran U., (2001) "Historical tsunamis in the Sea of Marmara", International Tsunami Symposium ITS 2001, Session 4, Paper 4-2, Seattle, August 7-9, 2001, pp: 527-535

Hardy D. A. ve Renfrew A. C., (1990a), Thera and the Aegean World Archaeology', Proceedings of Third International Congress Santorini, Greece, September, 3-9, 1989, Vol. 1

Hardy D. A. ve Renfrew A. C., (1990b), Thera and the Aegean World Earth

Sciences', Proceedings of Third International Congress Santorini, Greece, September, 3-9, 1989, Vol. 2

Doumas C., (1980a), Thera and the Aegean World', ISBN 0 956 06133 3 4 Vol.1

Doumas C., (1980b), Thera and the Aegean World', ISBN 0 956 06133 3 9 Vol.2

Hardy D. A. ve Renfrew A. C., (1990c), Thera and the Aegean World Chronology', Proceedings of Third International Congress Santorini, Greece, September, 3-9, 1989, V 3

KOERI (2004), "Earthquake database of Bosphorus University, Kandilli Observatory and Earthquake Research Institute (KOERI), The link of Earthquake Database http://www.koeri.boun.edu.tr/jeofizik/defaulteng.htm

Kuran, U. and Yalciner, A. C.,(1993), "Crack Propagations Earthquakes and Tsunamis in the Vicinity of Anatolia", "Tsunamis in the World", Advances in Natural and Technological Hazards Research by Kluwer Academic Publisher, (1993), Ed. Stefano Tinti,pp:159-175.

Kurkin A.A., Kozelkov A.c., Zaitsev A.I., Zahibo N., and Yalciner A. Tsunami risk for the Caribbean Sea Coast. Izvestiya, Russian Academy of Engineering Sciences, 2003, vol. 4, 126 – 149 (in Russian).

Lander J.F, Whiteside L.S., Lockridge P.A., (2002), "A Brief History of Tsunamis in the Caribbean Sea", The International Journal of the Tsunami Society, Science of Tsunami Hazards, Volume 20, No:2, Published Electronically, 2002, ISSN 8755-6839

Minoura, K., Imamura, Kuran, U., Nakamura, T., Papadopoulos, G., Takahashi, T., Yalçıner, A. C., (2000), "Discovery of Minoan Tsunami Deposits" Geology, v. 28, no. 1, p.p: 59-62, January 2000

Okada, Y. (1985), Surface deformation due to shear and tensile faults in a half-space, *Bull. Seism. Soc. America*, *75*, 1135-1154.

Yalciner A. C., Kuran, U., Akyarli, A. and Imamura F., (1995), "An Investigation on the Generation and Propagation of Tsunamis in the Aegean sea by Mathematical Modeling", Chapter in the Book, "Tsunami: Progress in Prediction, Disaster Prevention and Warning", in the book series of Advances in Natural and Technological Hazards Research by Kluwer Academic Publishers, (1995), Ed. Yashuito Tsuchiya and Nobuo Shuto, pp 55-71

Yalciner A. C. Alpar B., Altinok Y., Ozbay I., Imamura F., (2002), "Tsunamis in the Sea of Marmara: Historical Documents for the Past, Models for Future" Marine Geology, 2002, 190, pp:445-463

Yalciner A.C, Pelinovsky E.N., Talipova T.G, Kurkin A.A.,. Kozelkov A.C., Zaitsev A.I., (2002), "Tsunamis in the Black Sea", Abstract published and paper presented in International Workshop on Local Tsunami Warning and Mitigation, September 14-17, Petropavlovsky, Kamcahtka, Russia.

Yalciner, A. C. ; Imamura, F. ; Synolakis, E. C, (2002), "Simulation of tsunami related to caldera collapse and a case study of Thera volcano in Aegean sea" Abstract Published and paper presented in EGS XXVII General Assembly, Nice, France, April 2002 Session NH8. Tsunamis,

Zahibo N., Pelinovsky E., Yalciner A. C., Kurkin A., Kozelkov A., and Zaitsev A., (2003), 2003), (2003), "The 1867 Virgin Island tsunami: observations and modeling", Oceanologica Acta, 26, 2003, 609-621

# ADVANCES IN TSUNAMI INUNDATION MAPPING IN CALIFORNIA

JOSÉ BORRERO and COSTAS SYNOLAKIS
*University of Southern California*

Since the International Tsunami Symposium held in Seattle, Washington in August 2001, significant progress has been made in tsunami inundation mapping for the state of California. While the mapping effort continues to focus on creating a "worst case scenario" for use by emergency management in evacuation planning and guidance, the modeling technologies being developed will be instrumental in creating a future generation of tsunami hazard assessments that include probabilities of occurrence. Significant progress over the last two years has been achieved with the completion of a uniform, 3 arc-second combined bathymetry and topography data set that extends from San Diego County to north of San Francisco Bay. This data set also includes a 1 arc-second resolution grid over the San Francisco Bay region. Over the last two years, the inundation mapping program has continued to focus on near-shore tsunamis sources. These include submarine mass movements, offshore thrust faulting and complex ruptures on near-shore strike slip faults. Sources recently modeled include the Palos Verdes debris avalanche, the Catalina Island restraining bend, the Lasuen Knoll uplift structure and the San Mateo Thrust Fault. The latter three sources in particular highlight the possibility for a local tsunami warning system - something previously assumed to be of little use due to the proximity of the tsunami sources to the coast. Numerical modeling shows that first wave arrival in to the Ports of Los Angeles and Long Beach is delayed by up to 20 minutes after tsunami generation due to the shallow offshore bathymetry of the San Pedro Shelf. Thus far the inundation mapping program has completed inundation maps for Los Angeles, Santa Barbara, San Diego, Ventura, Orange and San Luis Obispo counties. The remainder of the state is to be mapped in the coming year. Other projects being undertaken in this effort include examining sources within San Francisco Bay to investigate the tsunami hazard for marine oil transfer terminals located inside the Bay.

www.ingramcontent.com/pod-product-compliance
Ingram Content Group UK Ltd.
Pitfield, Milton Keynes, MK11 3LW, UK
UKHW020350150926
8624IPUK00004B/89

9 789812 565358